Also by Michael Shermer

Science Friction

In Darwin's Shadow

The Skeptic *Encyclopedia of Pseudoscience* (general editor)

The Borderlands of Science

Denying History

How We Believe

Why People Believe Weird Things

THE SCIENCE OF GOOD AND EVIL

The Science of
GOOD and EVIL

Why People Cheat, Gossip, Care, Share, and Follow the Golden Rule

MICHAEL SHERMER

An Owl Book

HENRY HOLT AND COMPANY • NEW YORK

Henry Holt and Company, LLC
Publishers since 1866
115 West 18th Street
New York, New York 10011

Henry Holt® is a registered trademark of
Henry Holt and Company, LLC.

Distributed in Canada by H. B. Fenn and Company Ltd.

For further information on the Skeptics Society and *Skeptic* magazine,
contact P.O. Box 338, Altadena, CA 91001. 626-794-3119;
fax: 626-794-1301. E-mail: skepticmag@aol.com.

Library of Congress Cataloging-in-Publication Data

Shermer, Michael.
The science of good and evil : why people cheat, gossip,
care, share, and follow the golden rule / Michael Shermer.
p. cm.
Includes bibliographical references and index.
ISBN-13: 978-0-8050-7769-8 (pbk.)
ISBN-10: 0-8050-7769-3 (pbk.)
1. Ethics, Evolutionary. I. Title.

BJ1311.S48 2004
171'.7—dc21 2003049343

Henry Holt books are available for special promotions
and premiums. For details contact: Director, Special Markets.

Published in hardcover by Times Books in 2004

First Owl Books Edition 2005

Designed by Victoria Hartman

Printed in the United States of America
3 5 7 9 10 8 6 4

To my beloved parents:
My late mother, Lois Godbold,
and my stepfather, Richard Godbold
My late father, Richard Shermer,
and my stepmother, Betty Shermer
Sine quibus non
In the very literal sense

Good and evil we know in the field of this world grow up together almost inseparably; and the knowledge of good is so involved and interwoven with the knowledge of evil, and in so many cunning resemblances hardly to be discerned, that those confused seeds which were imposed upon Psyche as an incessant labor to cull out, and sort asunder, were not more intermixed.

—John Milton, *Areopagitica*, 1644

CONTENTS

THE SCIENCE OF GOOD AND EVIL

PROLOGUE: ONE LONG ARGUMENT

> Our whole dignity consists in thought. Let us endeavor, then, to think well: this is the principle of ethics.
>
> —Blaise Pascal, *Pensées, II,* 1670

In 1959 professional astronomers were polled for their opinions on the then-undecided debate between two competing cosmological theories: "Did the universe begin with a Big Bang several thousand million years ago?" One-third answered yes. "Is matter continuously created in space?" Nearly half answered yes. Most telling, to the question "Is a Gallup poll of this kind helpful to scientific progress?" one hundred percent answered no.[1]

The reason for unanimity on the final question is that scientific debates are not settled by consensus opinion. Unfortunately, in complex human and social issues, separating fact from opinion is not so easy, and for no subject is this more apparent than morality and ethics. Thus, throughout this book I apply a principle I call *Darwin's Dictum,* which states: *All observation must be for or against some view if it is to be of any service.* The British naturalist and evolutionary theorist Charles Darwin, in a letter to his friend Henry Fawcett, dated September 18, 1861, responded to an accusation made against him in a meeting of the British Association for the Advancement of Science. Fawcett, who attended in Darwin's defense, reported that a critic claimed that *The Origin of Species* was too theoretical, and that he should have just "put his facts before us and let them rest." Darwin's response is now a classic: "About thirty years ago there was much talk that geologists ought only to observe and not theorize; and I well remember someone saying that at this rate a man might as well go into a gravel-pit and count the pebbles and describe the colours. How odd it is that anyone should not see that all observation must be for or against some view if it is to be of any service!"[2]

All observation must be for or against some view if it is to be of any service. My approach in this book is to apply Darwin's Dictum with a judicious use of data and theory in an attempt to understand both the why and the how of morality, acknowledging that since we are doing science, all claims to any validity for both data and theory are provisional. It is much like Caltech physicist and Nobel laureate Richard Feynman's observation: "I can live with doubt and uncertainty and not knowing. I think it is much more interesting to live not knowing than to have answers that might be wrong. If we will only allow that, as we progress, we remain unsure, we will leave opportunities for alternatives. We will not become enthusiastic for the fact, the knowledge, the absolute truth of the day, but remain always uncertain. . . . In order to make progress, one must leave the door to the unknown ajar." This applies, I think, to moral principles themselves as well as to any theory that purports to explain the how and the why of moral principles and behavior.

This balance between doubt and certainty, between open-mindedness and closed-mindedness, is what I call skepticism. "Modest doubt is call'd the beacon of the wise," William Shakespeare cleverly noted in *Troilus and Cressida* (act 2, scene 2). In many ways, the *search* for the origins of morality and the basis of right and wrong moral principles is as important as any discovery we might make; with morality, the journey counts as much as the destination. The two, more so, are inseparable. To be a fully functioning moral agent, one cannot passively accept moral principles handed down by fiat. Moral principles require moral reasoning. Unlike many other fields of human endeavor, where one may reasonably hope to arrive at a consistent set of principles, morality is so fraught with complexities and subtleties that few moral principles are without exception, and contradictions are as common as consistencies. While I have struggled mightily to be consistent in my thinking on the issues encountered in this book—many of which are the deepest of all questions of the human condition—I am reminded of Ralph Waldo Emerson's 1841 observation in "Self-Reliance" that "A foolish consistency is the hobgoblin of little minds." Walt Whitman, in his elegant *Song of Myself*, offered this out that I take as my own defense in those (hopefully few) places where constancy does not always carry the day:

> *Do I contradict myself?*
> *Very well, then, I contradict myself;*
> *(I am large—I contain multitudes.)*

Given this caveat, as brief intellectual autobiography in the form of an *apologia pro vita sua* (an apology for one's own life), the how and the why of morality form the basis of the third volume in a trilogy on the power of belief. The first volume, *Why People Believe Weird Things*,[3] dealt with a variety of subjects within the primary penumbra of my job as the editor in chief of *Skeptic* magazine and contributing editor and a monthly columnist for *Scientific American*. In exploring, explaining, and occasionally debunking pseudoscience and superstitions, the book lays the foundation for good science and, in showing how thinking goes wrong, indicates, by implication, how thinking goes right. Because the book deals with certain subjects uncomfortably close and tangentially interdigitated with religion—such as reincarnation, ghosts, near-death experiences, theories of immortality, psychic mediums who claim to talk to the dead, creationism, and alleged scientific proofs of God's existence—I was inevitably challenged to present my views on the nature and existence of God, the possibility of an afterlife, and the relationship of science and religion. At first I was flippant in my responses. "What's your position on life after death?" a questioner would inquire. "I'm for it," I would quip with a wink. But over the years it dawned on me, after hundreds of public lectures and thousands of letters, that for most people, life's ultimate questions are no joking matter.

A couple of years of serious research and reflection—tying together the experiences and thoughts of a lifetime spent thinking about, reading on, and actively committing myself to a variety of religious belief systems (from born-again Christian to born-again atheist to my current position of agnostic nontheist)—resulted in the second volume in the trilogy, *How We Believe: Science, Skepticism, and the Search for God.*[4] In this work I explicated the nature of belief systems, particularly with regard to the God question, and outlined the various positions one can take in attempting an answer. The general categories are obvious: *theism* is "belief in a deity, or deities" and "belief in one God as creator and supreme ruler of the universe." *Atheism* is "disbelief in, or denial of, the existence of a God." And *agnosticism* is "unknowing, unknown, unknowable." I present these common and historical usages from the *Oxford English Dictionary* because they represent how people have understood and used these terms, not how theologians and philosophers have finely nuanced them (in *How We Believe* I provide an extensive bibliographic essay just on these terms and how they can be parsed into dozens of finely graded positions). Since I shall be discussing morality primarily from the position of an agnostic and a nontheist, I should clarify what I mean by these terms.

The word *agnostic* (along with its derivative *agnosticism*) was coined by the British evolutionary biologist Thomas Henry Huxley in 1869 as: "one who holds that the existence of anything beyond and behind material phenomena is unknown and so far as can be judged unknowable, and especially that a First Cause and an unseen world are subjects of which we know nothing."[5] Based on this usage, and my own lifelong quest (ending in utter futility) to prove or disprove God's existence, I defined the God question as *the art of the insoluble,* and made what I think is an important distinction between a statement about the universe and a statement about one's personal beliefs. As a statement about the universe, agnostic seems to me to be the most rational position to take on the God question because, by the criteria of science and reason, God is an unknowable concept. We cannot prove or disprove God's existence through empirical evidence or rational analysis (although, in my opinion, atheists have slightly better arguments for the nonexistence of God than theists have for the deity's reality). Therefore, from a scientific perspective, theism and atheism are both indefensible positions as statements about the universe. One must think and act on a personal belief or a disbelief, however, so when forced to apply a label (which I generally try to avoid) I call myself a nontheist; if forced to bet on whether there is a God or not, I bet that there is not, and I live my life accordingly. Nontheism has the added advantage of making an end run around a common misunderstanding about agnosticism: that it is a position of waiting for more evidence about God's existence (or nonexistence) before making a decision, assuming that additional evidence will (or may) suddenly arise to prove or disprove God's existence. I could be wrong, of course, and proof may materialize, but until then I shall assume that there is no God and that the God question is an insoluble one. Is there a God? is not even the appropriate way to ask the question. A better question is this: Is it possible to know if there is a God or not? My answer is firmly negative. *Nontheism* also avoids the pejorative spin doctoring typically applied to *atheism,* associating it with communism, liberalism, postmodernism, and the general decay of morals and culture. Such associations are risible and insulting to atheists, but are common in modern culture.

In *How We Believe* I addressed many more issues than God's existence or nonexistence, including the nature and structure of belief systems; the psychology, anthropology, and sociology of religion; why people believe in God and why they think other people believe in God; the relationship of science and religion; the origins of myths and the storytelling impulse; the relationship of religion and morality; and how we can find meaning in a gloriously contingent universe. As with my experiences following the publication of *Why People Believe Weird Things, How We Believe*'s release in 1999 triggered a surfeit of correspondence challenging my claim that the primary function of religion can be found in the twofold purpose that religions serve in both traditional hunter-gatherer communities and modern state societies: (1) explanation and (2) social cohesion. In fulfilling the first purpose, myths explicate the origin and nature of the world and life, and have been, for the most part, displaced by science. We live in the Age of Science, and scientism is our mythology. Explanations of the origin and nature of the world and life are not final truths passed down through generations by mendicant monks preserving the knowledge and wisdom of the ancients; instead, they are always provisional and ever changing, and are best couched in empirical evidence, experimental testing, and logical reasoning.

Despite the triumph of science and the cultural diffusion of scientism, religion thrives as never before. In America in particular, but in many other countries as well, never have so many—and such a high percentage of the population—professed a belief in a deity. Although explanations for this remarkable trend are as varied and complex as the theorists proffering them, a general causal vector can be found in the second purpose of religion, that is, its social mode. Ever since the

Scientific Revolution and the Age of Enlightenment, religion has slowly but irrevocably gotten out of the business of explaining the world and instead has focused on what it has always done best—providing a foundation for social order and moral edification. Most people don't go to church to hear an explanation for the origin of the cosmos and life (and if they did, and they knew something about the findings of modern science, they would be dismayed to be told that the Genesis myth of a six-day creation less than ten thousand years ago was to be taken literally). Instead, most folks go to socialize with like-minded friends, neighbors, and colleagues to contemplate the meaning of their lives and of life and to glean moral messages from the homilies presented in stories, myths, and anecdotes of the knotty problems that daily life presents to us all. To date science—even scientism—has had little to do or say in this social mode, and it is here, especially, where we find no conflict between religion and science. When commentators argue that there is a conflict, they are thinking of religion in the explanatory mode where, for example, arguing that the earth is only 6,000 years old will indeed find you in direct contradiction with data from geology and astronomy indicating that it is in excess of 4.5 billion years old, or older. As long as religion does not make quasi-scientific claims about the factual nature of the world, then there is no conflict between science and religion.

This assumes, of course, a God who is not actively involved in some measurable way in the world and our lives. If, for example, prayer induced God to intervene in a person's recovery from disease or accidents, then prayed-for patients should recover from disease and accidents sooner than non-prayed-for patients. To date the prayer and healing studies have all proved either nonsignificant or significant but harboring deep methodological flaws.[6] Given this qualification, however, peace should reign in the valley of science and religion . . . and the sheep shall lie peacefully with the lion.

The How and the Why of Morality

The study of morality is, at its core, the study of why humans do what we do, particularly on the social level, since almost all moral issues revolve around how we interact with others. What do we mean by morality and ethics? I define *morality* as *right and wrong thoughts and behaviors in the context of the rules of a social group*. I define *ethics* as *the scientific study of and theories about moral thoughts and behaviors in the context of the rules of a social group*. In other words, morality involves issues of right and wrong thought and behavior, and ethics involves the study of right and wrong thought and behavior. The first half of this book deals primarily with morality, the second half with ethics. That is, part 1, "The Origins of Morality," focuses on the evolution of thoughts, behaviors, and emotions related to what we call the moral sentiments; part 2, "A Science of Provisional Ethics," focuses on the evolution of ethical systems—absolute, relative, and my own provisional ethical theory—and how we can reconcile them with our evolutionary and cultural heritage.

The Science of Good and Evil picks up where *How We Believe* left off. It defines religion as a social institution, one that evolved as an integral mechanism of human culture to create and promote myths, to encourage altruism and cooperation, to discourage selfishness and competitiveness, and to reveal the level of commitment to cooperate and reciprocate among members of a community. In the first half of this book I shall unpack that sentence in great detail, but for now what I mean is that religion evolved as a social structure to enforce the rules of human interactions before there were such institutions as the state or such concepts as laws and rights. Long before there were state-enforced constitutional rights for the protection of basic freedoms, humans devised various mechanisms of behavioral control to facilitate goodwill and to attenuate excessive greed, avarice, and other vices. The religious foundation of human virtues and vices, saints and sinners, in fact, is a codification of an informal psychology of moral and immoral behavior. Humans are a hierarchical social primate species, and as such we need rules and morals and a social structure to enforce them. In the social mode, religion is that social structure, and God—even a God that exists only in the heads of those who believe in Him—is the ultimate enforcer of the rules.

In a study I conducted with University of California, Berkeley, social scientist Frank Sulloway (the results of which are presented in *How We Believe*), we discovered that one of the most common reasons people give for believing in God is that without the existence of a deity there would be no ultimate basis for morality. The source of this belief might be that these three components—morality, God, and religion—have been intertwined for so long that there may be an evolutionary foundation that lies beneath the connection between these three cultural entities. Secular moral systems, such as those expressed in the French Revolution ("Liberté, Egalité, Fraternité" and the "Rights of Man") and the American Revolution (the right to "life, liberty, and the pursuit of happiness" and "all men are created equal") are centuries old. Religious moral systems such as those expressed in Judaism, Christianity, and Islam are millennia old. Evolutionary moral systems such as those expressed by indigenous peoples (the modern remnants of Paleolithic societies) are tens of millennia old. To find out why we are moral in an ultimate sense (and not just a proximate sense), we must return to those long-gone epochs when anatomically modern humans were living simultaneously with other hominid species, collected in tiny bands as hunter-gatherers eking out a living and struggling to survive in a physical environment filled with predators, parasites, diseases, accidents, and nature's quirks; and a social environment filled with hierarchies, conflicts, and competition for dominance, status, recognition, and mates.

When I ask why we are moral I am asking the question in the same manner that an evolutionary biologist might ask why we are hungry (to motivate us to eat) or why sex is fun (to motivate us to procreate). *Why* questions are different from *how* questions. How questions are concerned with proximate causes—the immediate or nearest cause or purpose of a structure or function. We are hungry because when our blood sugar is low our hypothalamus detects this drop and is stimulated to release chemicals that cause a sensation of hunger, motivating us to consume food. The fun of sex is similarly explained through physiological causes, such as the release during orgasm of oxytocin that enhances the bonding between a couple, an especially adaptive function because human infants are helpless for so long that they need the efforts of two parents (rather than one parent and one sperm or egg donor). But these are proximate explanations. Evolutionary biologists are also interested in ultimate causes—the final cause (in an Aristotelian sense) or end purpose (in a teleological sense) of a structure or behavior. We are hungry and horny because, ultimately, the survival of the species depends on food and sex, and those organisms for whom healthy foods tasted good and for which sex was exquisitely delightful left behind more offspring. Since differential reproductive success is the ultimate product of natural selection, and natural selection is the primary driving force behind evolution, we have reached an ultimate level of causal thinking in trying to answer these questions.

Of course, hunger and sex are relatively easy targets for evolutionary theorists. Psychological and social behavior—including and especially moral behavior—is another genera of problem altogether. But this does not exclude it from an evolutionary analysis. Although our species is arguably the most complex on earth (at least in terms of brain and behavior, especially as expressed in social systems), we are nonetheless animals, and as such we are not exempt from the forces of evolution. Ultimate why questions about social and moral behavior, while considerably more challenging, must nevertheless be subjected to an evolutionary analysis. There is a science dedicated specifically to this subject called evolutionary ethics, founded by Charles Darwin a century and a half ago and continuing as a vigorous field of study and debate today. Evolutionary ethics is a subdivision of a larger science called evolutionary psychology, which attempts a scientific study of all social and psychological human behavior. The fundamental premise of these sciences is that human behavior evolved over the course of hundreds of thousands of years during our stint as hominid hunter-gatherers, as well as over the course of millions of years as primates, and tens of millions of years as mammals. As such, evolutionary psychology is itself a branch of sociobiology and ethology, the sciences that study all animal behavior. Since we are, first and foremost, animals, the findings from all these fields are applicable to the study of human moral behavior, although humans are unique in the fact that the most advanced primates—chimpanzees, bonobos, gorillas, and orangutans—show only the most rudimentary forms of moral behavior. Finally, since *The Science of Good and Evil* is a work of science, it employs the evidence and findings from those sciences most allied with evolutionary ethics, evolutionary psychology, sociobiology, and ethology, including archaeology, history, anthropology, sociology, cognitive and social psychology, neurophysiology, behavior genetics, and evolutionary biology.

This scientific theory of morality will build a case for how humans evolved from social primates to moral primates, and how the foundation of moral principles can be built upon empirical evidence and logical reasoning. That is, this book tackles two deep and essential problems: (1) the origins of morality and (2) the foundations of ethics.

This is the why and the how of morality. Embedded within these are questions that have occupied the greatest minds in human history: Is it our nature to be moral or immoral? If we evolved by natural forces, then what was the natural purpose of morality? If we live in a determined universe, then how can we make free moral choices? Do good and evil exist, and if so, from whence do they come? Why do bad things happen to good people? If there is no outside source to validate moral principles, does anything go? Can we be good without God? How can we tell the difference between right and wrong?

In part 1, "The Origins of Morality," a theory on the origins of morality is presented in four chapters. This part addresses the why question of morality. Chapter 1, "Transcendent Morality," presents an answer to the challenge that without a transcendent source of validation (for most people, this is God), all ethical systems are reduced to moral relativism or moral nihilism. I demonstrate that evolutionary ethics can be ennobling and morality transcendent by virtue of the fact that the deepest moral thoughts, behaviors, and sentiments belong not just to individuals, or to individual cultures, but to the entire species. Chapter 2, "Why We Are Moral," presents my theory, based on a model of bio-cultural evolution, to explain the development of the moral sentiments and moral behavior. The chapter reviews the million years over which premoral sentiments evolved in our ancestors under primarily biogenetic control, the hundred thousand years over which the moral sentiments evolved in our species alone, the transition about 35,000 years ago when sociocultural factors became increasingly dominant in shaping our moral behavior, and the shift within the past 10,000 years when the moral sentiments were codified into formal ethical systems. Chapter 3, "Why We Are Immoral," addresses the darker side of humanity: war, violence, and the ignoble savage within, showing that we are both moral and immoral animals. Here I address the classic problem of evil: If God is all-powerful and all good, then why does evil exist? If God is neither all-powerful nor all good, then evil can logically exist; but this is not a deity most believers would profess belief in or make a commitment to. If there is no God, then how are we to deal with evil on the scale of the Holocaust? Do bad people ultimately get away with doing bad things if there is no final judgment? I suggest a way around this conundrum, as well as debunk the myth of the Noble Savage and peaceful native, showing how all humans share a common humanity. Finally, chapter 4, "Master of My Fate," considers how moral choices can be made in a determined universe. I suggest several solutions (since I do not believe that any single one is adequate) to the problem of free will—if God is all-powerful, or if nature is ultimately guided by the law of causality (where all effects have causes), then how can we be expected to make free moral choices, much less be accountable for making the wrong moral choices? Both philosophy and science provide viable solutions.

In part 2, chapter 5, "Can We Be Good Without God?," addresses head on one of the most common arguments made by believers that without an outside divine source of validation and objectification, moral principles cannot be universally held or consistently applied. Chapter 6, "How We Are Moral," reviews the various absolute and relative ethical systems that have been developed throughout human history, showing the strengths and weaknesses of each one, and concluding that because of the complexity of human society and culture, no single ethical system can be all-encompassing or thoroughly consistent; this chapter also presents a science of provisional ethics that is neither absolute nor relative, showing how moral principles can be applicable to most people in most circumstances most of the time. With provisional ethics there is no abdication of moral responsibility, but at the same time there is room for tolerance and diversity in recognizing that although we are all responsible for our moral actions, there is scope for forgiveness and redemption in recognizing the fallibility of humans and human social systems. Chapter 7, "How We Are Immoral," examines a number of principles that help us tell the difference between right and wrong, and then applies these principles to a number of ethical issues, including truth telling and lying, adultery, pornography, abortion, cloning and genetic engineering, and animal rights. Chapter 8, "Rise Above," considers the evidence that our species is on a long evolutionary trajectory that will lead to greater amity toward members of our own group, and a long historical path toward more liberties for more people in more places, whether they are members of our group or not. Out of this analysis arise two recommendations, one on personal tolerance and the other on political freedoms, based on extensive scientific data that demonstrate why and how humans can and should be more cooperative.

Finally, two appendices accompany the book. Appendix 1, "The Devil Under Form of Baboon," is a history of the evolution of evolutionary ethics and the background to the scientific study of moral behavior, starting with Darwin and working our way to the latest findings from evolutionary psychologists. Appendix II supplements chapter 2 in providing additional evidence of the evolutionary nature of our moral behavior in the form of human universals related to religion and morality.

That's it. The rest is details. But as the nineteenth-century French novelist Gustave Flaubert observed, "Le bon Dieu est dans le détail," a phrase astutely (and appropriately, considering his profession) reiterated by the twentieth-century architect Ludwig Mies van der Rohe, "God dwells in the details." Those details are the sum and substance of this book.

I

THE ORIGINS OF MORALITY

I fully subscribe to the judgment of those writers who maintain that of all the differences between man and the lower animals, the moral sense or conscience is by far the most important. . . . It is the most noble of all the attributes of man, leading him without a moment's hesitation to risk his life for that of a fellow-creature; or after due deliberation, impelled simply by the deep feeling of right or duty, to sacrifice it in some great cause. . . . The following proposition seems to me in a high degree probable—namely, that any animal whatever, endowed with wellmarked social instincts, the parental and filial affections being here included, would inevitably acquire a moral sense or conscience.

—Charles Darwin, *The Descent of Man*, 1871

1

TRANSCENDENT MORALITY: HOW EVOLUTION ENNOBLES ETHICS

One should accept the truth from whatever source it proceeds.

—Moses Maimonides, *Eight Chapters,* twelfth century

In one of the most starkly honest and existentially penetrating statements ever made by a scientist, Oxford University evolutionary biologist Richard Dawkins opined that "the universe we observe has precisely the properties we should expect if there is, at bottom, no design, no purpose, no evil and no good, nothing but blind, pitiless indifference."[1] Here we cut to the heart of what is, in my opinion, the single biggest obstacle to a complete acceptance of the theory of evolution, especially its application to human thought and behavior, particularly in the realm of morality and ethics: the equating of evolution with ethical nihilism and moral degeneration. If we are nothing more than the product of sightless natural forces operating within a mercilessly uncaring cosmos, from whence can we find absolute ethical standards or ultimate moral meaning? My answer to this question follows, although Charles Darwin said it more succinctly in his magnum opus, *The Origin of Species,* in which he presciently provided an answer to Dawkins's observation: "When I view all beings not as special creations, but as the lineal descendants of some few beings which lived long before the first bed of the Cambrian system was deposited, they seem to me to become ennobled."[2] What can possibly be ennobling about evolution, and how can we construct a transcendent morality out of evolutionary ethics? Here is how.

A Moral Dilemma

In his dialogue *The Euthyphro,* the Greek philosopher Plato presented what has come to be known as "Euthyphro's Dilemma," in which his favorite protagonist—the cantankerous political gadfly Socrates—asks a young man named Euthyphro the following question: "The point which I should first wish to understand is whether the pious or holy is beloved by the gods because it is holy, or holy because it is beloved of the gods?" The underlying assumption for Plato, as it has been ever since for most philosophers and theologians, is that moral principles are and must be linked to a God or gods in order to be considered absolute, eternal, and meaningful. Socrates is trying to show Euthyphro that a dilemma exists over whether God embraces moral principles naturally occurring and external to Him because they are sound ("holy") or that these moral principles are sound because He created them. It cannot be both.[3]

Regardless of which choice is made, under this paradigm, theologians and religious philosophers have made God an integral part of the moral process. The thirteenth-century Catholic scholar Thomas Aquinas laid the foundation for a natural law theory of moral development by arguing that God supports moral principles that occur naturally, instills them in us, which we then discover through rational analysis, prayer, and our God-given intuitive mental faculty for moral reasoning. William of Ockham and Samuel Pufendorf, by contrast, preferred the second choice in Euthyphro's Dilemma, arguing in what has come to be known as *Divine Command Theory* and *voluntarism* that God freely created moral principles through divine fiat. Hugo Grotius gave the nod to the first choice, claiming that God endorses certain already-existing moral principles. The assumption made by all these commentators, of course, is that at some level God is involved in the process of creating and/or sanctioning moral principles.[4]

By the eighteenth century, however, particularly within the intellectual and cultural movement known as the Enlightenment, a number of philosophers challenged the very premise of Euthyphro's Dilemma—most notably the atheist Scottish philosopher David Hume—by taking God out of the moral equation altogether. Thomas Jefferson, John Locke, and others, while not appropriately classified as atheists (rather, deism was a common belief among many Enlightenment intellectuals, in which God created the world and then stepped aside to allow matters to run their course), attempted to ground moral principles in natural law—a sort of deification of nature to make ethics transcendent of mere human convention. We see this in one of the founding statements of the United States of America: "We hold these Truths to be self-evident, that all men are created equal, that they are endowed by their Creator with certain unalienable Rights, that among these are Life, Liberty, and the Pursuit of Happiness." Even if the creator (small *c*) is nature, rights are still unalienable—society or people cannot take them away, simply because society or people did not create them.

An additional problem arises in Euthyphro's Dilemma, and that is if God is linked to moral principles—indeed, many argue that He must be in order to create a meaningful ethical system—then does that mean every moral statement of right and wrong ever made is infused with divine inspiration? What about moral principles espoused by Osama bin Laden, Adolf Hitler, Joseph Stalin, or Tomás de Torquemada? Did their questionable ethics derive from on high as well? Or, alternatively, did their "right" moral actions arise from a correct understanding of God's will and their "wrong" moral actions from a corruption in their understanding of the divine process? And what about the morality of nonbelievers, atheists, and agnostics? Did their moral sentiments arise simply from the surrounding religious culture, whereas believers' principles came from God? Problems of consistency in Euthyphro's Dilemma are legion.

There is a simple way around the dilemma: leave God out of the discussion altogether and adopt the *methodological naturalism* of science, in which all effects have natural causes subject to scientific analysis. The supposition is that the moral sense in humans and moral principles in human cultures are the result of laws of nature, forces of culture, and contingencies of history. I am not interested here in placing a value judgment on whether God exists or not, because it is not relevant to a scientific approach. Believers need not feel alienated, however, since if there is a God, it is acceptable to believe that He created and utilized the laws of nature, forces of culture, and contingencies of history to generate within humans a moral sense, and within human cultures moral principles. Thankfully for the future of our species—and perhaps all species—science can illuminate an answer, or at least many testable answers that can be confirmed or rejected based on the evidence.

Yet even in a strictly scientific explanation of morality we encounter an apparent dilemma. If morals have a natural instead of a supernatural origin, then there apparently can be no transcendent being or force to objectify them with absolute standards. If there are no absolute standards, then morality must be relative. Harvard evolutionary biologist Edward O. Wilson presented the dilemma this way: "Either ethical precepts, such as justice and human rights, are independent of human experience or else they are human inventions." On the one side, says Wilson, are the transcendentalists, "who think that moral guidelines exist outside the human mind." On the other side are the empiricists, "who think them contrivances of the mind." Wilson is an empiricist: "I believe in the independence of moral values, whether from God or not, and I believe that moral values come from human beings alone, whether or not God exists." For Wilson, there is no debate on the table with greater import. "The choice between transcendentalism and empiricism will be the coming century's version of the struggle for men's souls. Moral reasoning will either remain centered in idioms of theology and philosophy, where it is now, or shift toward science-based material analysis. Where it settles will depend on which world view is correct, or at least which is more widely perceived to be correct."[5] But must we accept either-or (either transcendentalism or empiricism)? Can't we have both-and (both transcendentalism and empiricism)? I would like to propose a both-and consilience of transcendentalism and empiricism.

Ennobling Evolutionary Ethics: A Moral Dilemma Resolved

In virtually every dialogue I have about religion and science, I am inevitably challenged to explain how any ethical system not rooted in divine inspiration can be anything but relative, and thus meaningless. Without transcendence, believers argue, moral acts and principles can have no firm foundation on which to stand.

My thesis is that morality exists outside the human mind in the sense of being not just a trait of individual humans, but a human trait; that is, a human universal. Think about it this way: evolution created moral sentiments and concomitant behaviors over hundreds of thousands of years, so that today even though we agree that humans created morality and ethics (and thus we are empiricists), it is not *us* who created the moral sentiments and behaviors, it was our Paleolithic ancestors who did so in those long-gone millennia. We simply inherit them, fine-tune and tweak them according to our cultural preferences, and apply them within our unique historical circumstances. In this sense, moral sentiments and behaviors exist beyond us, as products of an impersonal force called evolution. In the same way that evolution transcends culture, morality and ethics transcend culture, because the latter are direct products of the former. Given this presupposition it seems reasonable to be both a transcendentalist and an empiricist, or what I call a transcendent empiricist. Transcendent empiricism avoids supernaturalism as an explanation of morality, and yet grounds morality on something other than the pure relativism of culturally determined ethics. It has the added advantage of being a testable hypothesis in the same manner that any evolutionary trait might be subject to the scrutiny of empirical science.

An Exegesis of Why and How We Are Moral

If, perforce, I had to explain the why and the how of morality—that is, the origins of morality and how we can be good without God—in five minutes, the following exegesis encapsulates the theory presented in this book.[6]

1. Moral Naturalism. This is a secular and scientific approach to the study of morality. As such, whether there is a God or not is irrelevant to the theory because in science our approach is a naturalistic one—all effects have natural causes subject to scientific analysis. Since I am a nontheist, my assumption is that the moral sense in humans and moral principles in human cultures are the result of laws of nature, forces of culture, and the unique pathways of history; theists who embrace the findings of science may assume that God created and utilized the laws of nature and forces of culture to generate within humans a moral sense and within human cultures moral principles.

2. An Evolved Moral Sense. Moral sentiments in humans and moral principles in human groups evolved primarily through the force of natural selection operating on individuals and secondarily through the force of group selection operating on populations. The moral sense (the psychological feeling of doing "good" in the form of positive emotions such as righteousness and pride) evolved out of behaviors that were selected for because they were good either for the individual or for the group; an immoral sense (the psychological feeling of doing "bad" in the form of negative emotions such as guilt and shame) evolved out of behaviors that were selected for because they were bad either for the individual or for the group. While cultures may differ on what behaviors are defined as good or bad, the moral sense of feeling good or feeling bad about behavior X (whatever X may be) is an evolved human universal.

3. *An Evolved Moral Society*. Humans evolved as a social primate species with an ascending hierarchy of needs from self-survival of the individual (basic biological needs), to the extension of the individual through the family (the selfish gene), to a sense of bonding with the extended family (driven by kin selection of helping those most related to us), to the reciprocal altruism of the community (direct and obvious payback for good behaviors), to indirect altruism of society (doing good without direct payback), to species altruism and bioaltruism as awareness of our membership of the species and biosphere continues to develop. The most basic human needs and moral feelings are largely under biological control, whereas the more social and cultural human needs and moral feelings are largely under cultural control.

4. *The Nature of Moral Nature*. Humans are, by nature, moral and immoral, good and evil, altruistic and selfish, cooperative and competitive, peaceful and bellicose, virtuous and nonvirtuous. Such moral traits vary within individuals as well as within and between groups. Some people and populations are more or less moral and immoral than other people and populations, but all people have the potential for all moral traits. Most people most of the time in most circumstances are good and do the right thing for themselves and for others. But some people some of the time in some circumstances are bad and do the wrong thing for themselves and for others. The codification of moral principles out of the psychology of the moral traits evolved as a form of social control to ensure the survival of individuals within groups and the survival of human groups themselves. Religion was the first social institution to canonize moral principles, but morality need not be the exclusive domain of religion.

5. *Provisional Morality*. Moral principles, derived from the moral sense, are not absolute, where they apply to all people in all cultures

under all circumstances all of the time. Neither are moral principles relative, entirely determined by circumstance, culture, and history. Moral principles are provisionally true—that is, they apply to most people in most cultures in most circumstances most of the time.

6. *Provisional Right and Wrong*. In provisional morality we can discern the difference between right and wrong through three principles: the ask-first principle, the happiness principle, and the liberty principle. The ask-first principle states: to find out whether an action is right or wrong, ask first. The happiness principle states: *it is a higher moral principle to always seek happiness with someone else's happiness in mind, and never seek happiness when it leads to someone else's unhappiness.* The liberty principle states: *it is a higher moral principle to always seek liberty with someone else's liberty in mind, and never seek liberty when it leads to someone else's loss of liberty.* To implement social change, the moderation principle states: *when innocent people die, extremism in the defense of anything is no virtue, and moderation in the protection of everything is no vice.*

7. *Provisional Justice*. Although we are all subject to laws of nature and forces of culture and history that shape our thoughts and behaviors, we are free moral agents responsible for our actions because none of us can ever know the near-infinite causal net that determines our individual lives. Good things and bad things happen to both good and bad people. There is no absolute and ultimate judge to mete out rewards and punishments at some future date beyond human existence. But since moral principles are provisionally true for most people most of the time in most circumstances, there is individual culpability and social justice within human communities that produce feelings of righteousness and guilt, and administer rewards and punishments such that there is at least provisional justice in the here and now.

8. *Ennobling Evolutionary Ethics*. As an evolved mechanism of human psychology, the moral sense transcends individuals and groups and belongs to the species. Moral principles exist outside of us and are products of the impersonal forces of evolution, history, and culture.

Free Rider: Facing the Judge

In a review of *How We Believe* that appeared in the *Washington Post*, Michael Novak made this observation about my confidence in science and the freedom I found through a scientific worldview: "Science is a method for gaining important forms of knowledge; scientism is the reduction of all forms of knowing to scientific method. Shermer certainly comes perilously close to the latter. Still, he tries valiantly to maintain a sense of the sublime, the sacred, even the mystical, as in describing his exchange of eternal love with his soul mate over lit candles inside Chartres Cathedral, or standing 'beneath a canopy of galaxies, atop a pillar of reworked stone, or inside a transept of holy light,' when 'my unencumbered soul was free to love without constraint' and was 'emancipated from the bonds of restricting tradition, and unyoked from the rules written for another time in another place and for another people.'" Novak's denouement is as thought provoking as it is poetic: "The beauty of being Shermer is that he faces no Judge, undeceivable, transcendent of nature, and within him as well as beyond him; and stands in no long pilgrim community, struggling down the ages, falling, rising, and throwing cathedrals like Chartres up against the sky cathedrals. He is a free rider."[7]

Indeed, I am a free rider, but only in the freedom from one set of cultural traditions usually gathered under the umbrella of religion. But, like everyone else, I face judges that are in their own ways transcendent and powerful: family and friends, colleagues and peers, mentors and teachers, and society at large. My judges may be lowercased and occasionally deceivable, but they are transcendent of me as an individual, even if they are not transcendent of nature; as such, together, we all stand in a long pilgrim community struggling down the evolutionary and historical ages trying to live and love and learn to temper our temptations and do the right thing. I may be free from God, but the god of nature holds me to her temple of judgment no less than her other creations. I stand before my maker and judge not in some distant and future ethereal world, but in the reality of *this* world, a world inhabited not by spiritual and supernatural ephemera, but by real people whose lives are directly affected by my actions, and whose actions directly affect my life.

Throughout this book I attempt to peel back the inner layers covering our core of being to reveal a complexity of human motives—selfish and selfless, cooperative and competitive, virtue and vice, good and evil, moral and immoral. Along the way I attempt to show how these motives came into being as a product of both our evolutionary heritage and cultural history, and how we can construct a moral system that is neither dogmatically absolute nor irrationally relative—a provisional morality for an age of science that provides empirical evidence and a rational basis for belief. Such a system of morality suggests a more universal and tolerant ethic, an ethic that will ensure the well-being and survival of all members of the species, and of all species.

2

WHY WE ARE MORAL: THE EVOLUTIONARY ORIGINS OF MORALITY

Who are we? The answer to this question is not only one of the tasks but *the* task of science.

Erwin Schrödinger, *Science and Humanism*, 1951

In an episode of the classic television comedy series *The Honeymooners,* Jackie Gleason and Art Carney's characters—the boisterously bellicose Ralph Kramden and the benignly bumbling Ed Norton—engage in one of their archetypal over-the-top arguments, this time over a universal problem found in all primate species: food sharing and its anticipated consequent reciprocity. The conflict arises when the two families decide they are going to save on rent by sharing an apartment, which then involves dining together. Alice has just served dinner.

RALPH: When she put two potatoes on the table, one big one and one small one, you immediately took the big one without asking what I wanted.

NORTON: What would you have done?

RALPH: I would have taken the small one, of course.

NORTON: You would?

RALPH: Yes, I would!

NORTON: So, what are you complaining about? You got the little one!

In its essence this comedic routine symbolizes an enormous source of tension in human relations that led to the evolution of what was almost certainly the first moral principle—the Golden Rule. In 1690 the English political philosopher John Locke, in his classic work *Essay Concerning Human Understanding,* inquired about whether this "unshaken rule of morality, and foundation of all social virtue, 'that one should do as he would be done unto,' be proposed to one who never heard of it before, but yet is of capacity to understand its meaning, might he not without any absurdity ask a reason why?" Certainly, but as I shall argue, no such individual would find the Golden Rule surprising in any way because at its base lies the foundation of most human interactions and exchanges and it can be found in countless texts throughout recorded history and from around the world—a testimony to its universality. Table 1 presents just a few of its manifestations in chronological order (B.C.E. and C.E. are Before Common Era and Common Era).

TABLE 1

The Historical and Universal Expression of the Golden Rule

Lev. 19:18, c. 1000 B.C.E.: "Thou shalt not avenge, nor bear any grudge against the children of thy people, but thou shalt love thy neighbor as thyself."

Confucius, *The Doctrine of the Mean,* 13, c. 500 B.C.E.: "What you do not want others to do to you, do not do to others."

Isocrates, c. 375 B.C.E.: "Do not do to others what would anger you if done to you by others."

Tob. 4:15, c. 180 B.C.E.: "What thou thyself hatest, do to no man."

Diogenes Laertius, *Lives of the Philosophers,* c. 150 B.C.E.: "The question was once put to Aristotle how we ought to behave to our friends; and his answer was, 'As we should wish them to behave to us.'"

The Mahabharata, c. 150 B.C.E.: "This is the sum of all true righteousness: deal with others as thou wouldst thyself be dealt by. Do nothing to thy neighbor which thou wouldst not have him do to thee hereafter."

Matt. 7:12, c. first century C.E.: "All things whatsoever ye would that men should do to you, do ye even so to them: for this is the law and the prophets."

Luke 6:31, c. first century C.E.: "As ye would that men should do to you, do ye also to them likewise."

Epictetus, *Encheiridion,* c. 100: "What thou avoidest suffering thyself seek not to impose on others."

The Didache, or Teaching of the Twelve Apostles, c. 135: "All things whatsoever that thou wouldst not wish to be done to thee, do thou also not to another."

John Wycliffe, translation of Luke 7:31, 1389: "As ye will that men do to you, and do ye to them in like manner."

David Fergusson, *Scottish Proverbs,* 1641: "Do as ye wald be done to."

Thomas Hobbes, *Leviathan,* 1651: "Whatsoever you require that others should do to you, that do ye to them."

The Book of Common Prayer (catechism), 1662: "My duty towards my neighbor is to love him as myself, and to do to all men as I would they should do unto me."

Henry More, *Enchiridion ethicum,* 4, 1667: "The evil which you do not wish done to you, you ought to refrain from doing to another, so far as may be done without injury to some third person."

Baruch Spinoza, *Ethica,* 4, 1677: "Desire nothing for yourself which you do not desire for others."

John Wise, *A Vindication of the Government of New England Churches,* 1717: "If a man any ways doubt whether what he is going to do to another man be agreeable to the law of nature, then let him suppose himself to be in that other man's room."

John Stuart Mill, *Utilitarianism,* 1863: "To do, as one would be done by, and to love one's neighbor as one's self, constitute the ideal perfection of utilitarian morality."

Henry Sidgwick, *The Methods of Ethics,* 1874: "Reason shows me that if my happiness is desirable and a good, the equal happiness of any other person must be equally desirable."

Peter Kropotkin, *La Morale Anarchiste,* 1891: "Do unto others as you would have others do unto you in like case."

Hillel Ha-Babli, in the thirty-first book of *The Sabbath* in 30 B.C.E., raised the Golden Rule to the ultimate moral principle: "Whatsoever thou wouldst that men should not do to thee, do not do that to them. This is the whole Law. The rest is only explanation." That explanation, however, forms the basis of the evolutionary origins of morality, beginning with the evolution of the premoral sentiments.

The Evolution of Premorality

Our moral sentiments—the moral emotions contained within our mental armory—evolved out of premoral feelings of our hominid, primate, and mammalian ancestors, the remnants of which can be found in modern apes, monkeys, and other big-brained mammals. I consider these sentiments to be premoral because morality involves *right and wrong thoughts and behaviors in the context of a social group.* To date, it does not appear that nonhuman animals can consciously assess the rightness or wrongness of a thought, behavior, or choice in themselves or fellow members of their species. Thus, I hold that morality is the exclusive domain of *Homo sapiens.*

Of course, we are animals and so it is not surprising to discover premoral sentiments in other animals. And, equally unsurprising, the more like us a species is the more moral-like are their premoral sentiments, such as those observed in the great apes (e.g., chimpanzees, bonobos, and gorillas), monkeys (e.g., rhesus, baboons, and macaques), and other big-brained mammals (e.g., whales and dolphins). As we travel across the phylogenetic (evolutionary) scale away from humans we see that mammals exhibit more premoral behaviors than nonmammals (e.g., reptiles). In fact, when we examine the evolutionary sequence in detail we see that there is no distinct place to draw the line between moral and nonmoral, a problem created by the restrictive nature of binary logic. In binary logic one sees the world as black or white, up or down, in or out. In fuzzy logic one sees the world in shades of gray, between up and down, in and out. Instead of dividing the world into binary digits of 0 or 1, we can nuance it with shades in between (.1, .2, .3 . . . .9). Instead of simply saying that humans are moral and animals are not moral (1 and 0), we could describe humans as .9 or .8 moral, the great apes as .7 or .6 moral, monkeys as .5 or .4 moral, whales and dolphins as .3 or .2 moral, and so forth. This is why a scientific analysis of morality can be more fruitful than a philosophical one. An evolutionary perspective grants other animals degrees of morality (or premorality) that allows us to discover how we developed our moral sentiments; it also grants them greater dignity and status than does a nonevolutionary perspective.

Let us begin with household pets such as dogs and cats, since they are arguably the animals of which we are most intimately knowledgeable. Surely few outside of extreme animal rights' activists would claim that dogs and cats are moral animals, yet anyone who has had one for a pet for very long recognizes that they quickly learn the difference between right and wrong, and that dogs, especially, feel some sense of shame or guilt when being scolded for bad behavior and openly express joy and pride when being praised for good behavior. That sense of shame and guilt, or joy and pride, is what I mean by moral sentiments. But it appears that dogs and cats are not self-aware and self-conscious of their and other dogs' and cats' good and bad behavior—at least to an extent that would allow them to assess a moral judgment upon another member of their species (even if they cannot articulate it due to lack of language and speech apparatus). It might be more appropriate to call these premoral sentiments. (Some would debate this point, noting that wolf parents appear to teach their offspring the difference between right and wrong behavior, at least in terms of survival.)[1]

The difference between dogs and cats in the expression of such premoral sentiments reveals the importance of another component in morality—the social group. Cats are notoriously individualistic and independent, not typically herd or hierarchical social animals, especially when compared to dogs, who in the wild travel in packs with pecking orders (particularly noted in wolves, from which all dogs evolved) and in domestic situations establish social bonds with members of the household. (It is well known among dog trainers that the first rule of obedience is to establish yourself as the alpha member of the home, thereby relegating your dog to a lower rung in the pecking order.)

Examples of premoral sentiments among animals abound. It has been well documented that vampire bats, for example, exhibit food-sharing behavior and the principle of reciprocity. They go out at night in hordes seeking large sleeping mammals from which they can suck blood. Not all are successful, yet all need to eat regularly because of their excessively high metabolism. On average, older experienced bats fail one night in ten, younger inexperienced bats fail one night in three. Their solution: successful individuals regurgitate blood and share it with their less fortunate comrades, fully expecting reciprocity the next time they come home sans bacon. Gerald Wilkinson, in his extensive study of cooperation in vampire bats, has even identified a "buddy system" among bats, in which two individuals share and reciprocate from night to night, depending on their successes or failures. He found that the degree of affiliation between two bats—that is, the number of times they were observed together—predicted how often they would share food. Since bats live for upwards of eighteen years among the same community, they know who the cooperators are and who the defectors are.[2] Of course, the bats are not aware they are being cooperative in any conscious goodwill sense. All animals, including human animals, are just trying to survive, and it turns out that cooperation is a good strategy. This is especially apparent in the primates.

Psychologist and primatologist Frans de Waal has documented hundreds of examples of premoral sentiments among apes and monkeys. At the Yerkes Field Station where he and his colleagues observe their primate charges from on high, de Waal reveals the high level of conflict among these social species, as well as how those conflicts are resolved: "These records show that once the dust has settled after a fight, combatants are often approached by uninvolved bystanders. Typically the bystanders hug and touch them, pat them on the back, or groom them for a while. These contacts are aimed at precisely those individuals expected to be most upset by the preceding event." This is most apparent in the chimpanzees, our closest evolutionary relatives: "When upset, chimpanzees pout, whimper, yelp, beg with outstretched hand, or impatiently shake both hands so that the other will hurry and provide the calming contact so urgently needed. If all else fails, chimpanzees resort to their ultimate weapon, the temper tantrum. They lose control, roll around screaming pathetically, hitting their own head or beating the ground with their fists, regularly checking the effects on the other."[3]

Aristotle described humans as political animals. De Waal has discovered that chimpanzees are political animals as well (figure 1). When he shifted his research from macaques to chimps, he writes, "I was totally unprepared for the finesse with which these apes con each other. I saw them wipe undesirable expressions off their face, hide compromising body parts behind their hands, and act totally blind and deaf when another tested their nerves with a noisy intimidation display."[4] Politics—human and chimp—depends on reciprocity, and this too has been cleverly studied by de Waal in experiments whereby in order to obtain a heavy bucket of food two chimps must work together to pull it up with ropes. Although both are needed to haul the bucket up to the cage (where they are separated by a barrier), only one gets the food. De Waal discovered that if the recipient of the food did not share it with his companion, the companion refused to cooperate in the rope pulling in subsequent trials. In the long run, however, "cooperative behavior between two primates leads to greater food sharing after the task in which the cooperation occurred."[5] Quid pro quo.

Figure 1. Food Sharing in Chimpanzees

As documented by psychologist Frans de Waal at the Yerkes Field Station, four chimpanzee adults and one infant (seen in the food pile) participate in a vital social exchange process of food sharing that encourages cooperation and reciprocal altruism. In this group, the exchange process is controlled by the possessor female in the upper right corner. The female in the lower left reaches for food; whether it will be granted or not depends on her relationship to the possessor female. (Photograph by Frans de Waal)

Other scientists have documented premoral sentiments in other higher mammals. Dolphins have been seen to push sick or wounded members of a pod to the surface so that they may catch their breath. Whale fishermen know—and capitalize on the fact—that whales will put themselves in harm's way by coming to the defense of a wounded member of their group, circling them and striking the water with their flukes (thereby alerting the hunters to their whereabouts). One theory for why whales beach themselves in what often appears to be mass suicide is that, in fact, one member is disoriented and gets beached, and the others follow trying to help. In 1976 a pod of thirty false killer whales heaved themselves up on a Florida beach where they remained for three days; after one of their members died, the rest returned to the sea (able to do so because of a fortunate timing of the tides).[6] Cynthia Moss recorded the responses of a community of elephants to one of their members being shot by a poacher. As the struck elephant's knees buckled and she began to go down, her elephant comrades struggled to keep her upright. "They worked their tusks under her back and under her head. At one point they succeeded in lifting her into a sitting position but her body flopped back down. Her family tried everything to rouse her, kicking and tusking her, and Tallulah even went off and collected a trunkful of grass and tried to stuff it into her mouth." After she died, her friends and family members covered the corpse in dirt and branches.[7]

Hundreds of such examples exist in the scientific literature, and thousands more in popular literature.[8] The following characteristics appear to be shared by humans and other mammals, including and especially the apes, monkeys, dolphins, and whales: *attachment and bonding, cooperation and mutual aid, sympathy and empathy, direct and indirect reciprocity, altruism and reciprocal altruism, conflict resolution and peacemaking, deception and deception detection, community concern and caring about what others think about you, and awareness of and response to the social rules of the group*. Species differ in the degree to which they express these sentiments, and with our exceptionally large brains (especially the well-developed and highly convoluted cortex) we clearly express most of them in greater degrees than other species. Nevertheless, the fact that such premoral sentiments exist in our nearest evolutionary cousins may be a strong indication of their evolutionary origins. Still, something profound happened in the last 100,000 years that made us—and no other species—moral animals to a degree unprecedented in nature.

The Evolution of Morality

For the first 90,000 years of our existence as a species—that is, as anatomically modern humans distinct from other hominid species such as the Neanderthals—humans lived in small bands of tens to hundreds of individuals. In the last 10,000 years—only 10 percent of *Homo sapiens'* existence—some bands evolved into tribes of thousands of individuals, some tribes developed into chiefdoms of tens of thousands of individuals, some chiefdoms coalesced into states of hundreds of thousands of individuals, and a handful of states conjoined together into empires of millions of individuals. Within a rough order of magnitude comparison, the evolution of our species proceeded as depicted in table 2. (Recent fossil finds in Ethiopia by paleoanthropologist Tim White and his team indicate that anatomically modern humans may date to 160,000 years. The DNA evidence also supports this claim, although such figures come equipped with some error, meaning that the rough estimates presented here are legitimate.)[9]

TABLE 2

The Social Evolution of Humans Over the Past 100,000 Years

100,000–10,000 years	Bands	10s–100s of individuals
10,000–5,000 years	Tribes	100s–1,000s of individuals
5,000–3,000 years	Chiefdoms	1,000s–10,000s of individuals
3,000–1,000 years	States	10,000s–100,000s of individuals
1,000–present	Empires	100,000s–1,000,000s of individuals

Somewhere along the way, moral sentiments evolved out of the premoral sentiments of our primate and hominid ancestors and moral codes were created. Generally speaking, moral sentiments as expressed in thoughts and behaviors evolved during those first 90,000 years when we lived in bands. In the last 10,000 years, these moral thoughts and behaviors were codified into moral rules and principles by religions that arose as a direct function of the shift from tribes to chiefdoms to states.[10] How and why did this happen?

The human story begins roughly six to seven million years ago when a hominid primate branched off from the common ancestor shared by our primate contemporary, the chimpanzees. Two to three million years ago hominids in Olduvai Gorge in east Africa began chipping stones into tools and altering their environment. Around one million years ago *Homo erectus* added controlled fire to hominid technology, and between half a million and 100,000 years ago other hominids, such as *Homo neanderthalensis* and *Homo heidelbergensis*, lived in caves, had relatively elaborate tool kits, and developed throwing spears with finely crafted spear points. The evidence is now overwhelming that many hominid species lived simultaneously, and at present we can only speculate what speciation pressures these changing technologies put on natural selection.[11]

Sometime around 35,000 years ago one hominid species improved its tool kit dramatically, making it noticeably (in the fossil record) more complex and varied. Suddenly clothing covered their bodies, art adorned their caves, bones and wood formed the structure of their living abodes, and, perhaps most significantly, language produced sophisticated symbolic communication. They even buried their dead in prepared grave sites with burial ceremonies (figure 2). By 13,000 years ago our species had spread to nearly every region of the globe and all people everywhere lived in a condition of hunting, fishing, and gathering (HFG). Most were nomadic, staying in one place for no longer than a few weeks at a time. Small bands grew into larger tribes, and with this shift, possessions became valuable, rules of conduct grew more complex, and population numbers climbed steadily upward. When population pressures in numerous places around the globe grew too intense for the HFG lifestyle to support, an agricultural revolution sprouted.[12] New food-production technologies allowed populations to increase dramatically. With those increased populations came new social technologies for governance and conflict resolution. The creation of tribes from bands began around 13,000 years ago. Archaeological evidence indicates that the coalition of tribes into chiefdoms occurred in the Fertile Crescent of western Asia around 7,500 years ago. Within 2,000 years—roughly 5,500 years ago—chiefdoms began to merge into states in the same general area.

The concomitant leap in food production and population that accompanied the shift to chiefdoms and states allowed for a division of labor to develop in both economic and social spheres. Full-time artisans, craftsmen, and scribes worked within a social structure organized and run by full-time politicians, bureaucrats, and, to pay for it all, tax collectors. Organized religion came of age—along with these other social institutions—to fill many roles, not the least of which was the justification of power for the ruling elite. The "divine right of kings" is not the invention of early-modern European monarchs. In fact, every chiefdom and state society known to archaeologists from around the world, including those in the Middle East, Near East, Far East, North

Figure 2. The Evolutionary Origins of Morality 100,000 Years Ago

The moral sentiments evolved roughly 100,000 years ago among bands of hunter-gatherers who learned to cooperate with each other and compete with other hominid groups. Depicted here is the burial of a child at Qafzeh Cave in Israel, the oldest-known decorated grave site in the world, where paleoanthropologists discovered the skeleton of a child grasping the skull of a fallow deer. (Courtesy of W. W. Norton)

and South America, and the Polynesian Pacific islands, justified political power through divine sanction, in which the chief, pharaoh, king, queen, monarch, emperor, sovereign, or ruler of whatever title claimed a relationship to God or the gods, who allegedly anointed them with the power to act on behalf of the divinity. In a type of reciprocal exchange program, the masses would pay for this divine connection through taxes, loyalty, and service to the chiefdom or state (for example, through military inscription). As states evolved into bona fide civilizations and the centuries witnessed small cults evolve into world religions, behavior commitments evolved into standardized rituals, accompanied by the appropriate architectural displays of both political and religious power.[13] Organized religion as we know it was born.

This historical development supports what is known as the rational choice theory of religion.[14] Fundamental to this theory is that the beliefs, rituals, customs, emotions, commitments, and sacrifices associated with religion are best understood as a form of exchange relations between humans and God or gods. In this model, humans are assumed to be rational, making mental calculations to maximize resources and rewards. Where resources and rewards are available through secular avenues, religion is not needed. Where resources and rewards are scarce (for example, rain for crops) or nonexistent (for example, immortality) through secular sources, then religion becomes the accepted venue for the exchange of goods and services. Sociologists Rodney Stark and William Bainbridge present examples from various religions that reveal that the greater the number of gods in the religion the lower the price of exchange (the less that has to be given for the product or service received). They have also found that the more dependable and responsive the gods are in delivering goods and services the higher the price people are willing to pay. Finally, they found that religious rituals and the confidence of fellow believers serve to reinforce the confidence people have that their gods will deliver the goods and services, that such confidences are enhanced through miracles and mystical experiences associated with religions, and that religious organizations require commitment from believers in order to sustain their power as the primary source for such exchanges.

With the exception of the American experiment of separating church and state, politics and religion have always been tightly interdigitated. The reason has to do with the even more important role that religion has played in the development of morality. That is, in addition to the sanction of political power, religion has also served as an institution of social order and behavior control. The reinforcement of positive moral sentiments (and the punishment of negative ones) became a central role of religious leaders and organizations; they codified subjective social and moral norms and, with the invention of writing, literally canonized them in sacred scrolls and texts. From Moses proclaiming that God dictated to him the Ten Commandments, to Joseph Smith claiming that the angel Moroni delivered to him the golden plates, to L. Ron Hubbard's pulp science fiction repackaged as sacred religious texts said to be inspired by advanced alien intelligences, most religions tend to decree divine inspiration in order to enhance social and political power.

Such elaborate behavior controls yoked to religious rituals are not needed in bands and tribes, whose numbers are small enough that other less formal methods are more effective. In his three decades of research in New Guinea, for example, Jared Diamond says he has "never heard any invocation of a god or spirit to justify how people should behave toward others." Social obligations, he explains, depend on human relationships. "Because a band or tribe contains only a few dozen or a few hundred individuals respectively, everyone in the band or tribe knows everyone else and their relationships. One owes different obligations to different blood relatives, to relatives by marriage, to members of one's own clan, and to fellow villagers belonging to a different clan." Conflicts are directly resolved within these small bands because everyone is related to one another or knows one another. Members of the band are distinctly different from nonmembers on all levels. "Should you happen to meet an unfamiliar person in the forest, of course you try to kill him or else to run away; our modern custom of just saying hello and starting a friendly chat would be suicidal," Diamond reflected.[15] Populations in the many thousands of people, however, made such informal behavioral control mechanisms ineffectual. This led to the wedding of God and mammon.

This historical trajectory makes good sense in an evolutionary model. In bands and tribes the declaration of love for one's neighbors means something rather different than it does in chiefdoms, states, and empires. In the Paleolithic social environment in which our moral sentiments evolved, one's neighbors were family, extended family, and community members who were well known to all. To help others was to help oneself. In chiefdoms, states, and empires the biblical admonition "Love thy neighbor" meant only one's immediate in-group. Out-groups were not included. In a group selection model of the evolution of religion and morality, those groups who were particularly adept at amity within the group and enmity between groups were likely to be more successful than those who haphazardly embraced total strangers. This evolutionary interpretation also explains the seemingly paradoxical nature of Old Testament morality, where on one page high moral principles of peace, justice, and respect for people and property are promulgated, and on the next page raping, killing, and pillaging people

Figure 3. A Group of New Guinea Hunter-Gatherers Prepares for Battle

Before state societies, "love thy neighbor" meant one's immediate family, extended family, and community of fellow in-group members. "Out-groupers" were to be dealt with cautiously. Trust was tentative. Conflict was frequent. Death by war was common. (Courtesy of Film Study Center)

who are not one's "neighbors" are endorsed. In terms of evolutionary group selection, religious violence, genocide, and war are adaptive because they serve to unite in-group members against enemy out-groups.

Consider what Moses thought God meant by "neighbor" (in Lev. 19:18): "Thou shalt not avenge, nor bear any grudge against the children of thy people, but thou shalt love thy neighbour as thyself." Here "the children of thy people" are the neighbors one is not to kill (King James Version). In other translations, neighbors are "the sons of your own people" (Revised Standard Version) and "your countrymen" (Tanakh), in other words, thy fellow in-group members. By contrast, where Deut. 5:17 admonishes readers "Thou shalt not kill," fifteen chapters later, in Deut. 20:10–18, the Israelites are commanded to lay siege to an enemy city, steal their cattle, enslave all citizens who surrender, and kill the men and rape the women who do not surrender:

> When you draw near to a city to fight against it, offer terms of peace to it. And if its answer to you is peace and it opens to you,

> then all the people who are found in it shall do forced labor for you and shall serve you. But if it makes no peace with you, but makes war against you, then you shall besiege it; and when the LORD your God gives it into your hand you shall put all its males to the sword, but the women and the little ones, the cattle, and everything else in the city, all its spoil, you shall take as booty for yourselves; and you shall enjoy the spoil of your enemies, which the LORD your God has given you. Thus you shall do to all the cities which are very far from you, which are not cities of the nations here.
>
> But in the cities of these peoples that the LORD your God gives you for an inheritance you shall save alive nothing that breathes, but you shall utterly destroy them, the Hittites and the Amorites, the Canaanites and the Perizzites, the Hivites and the Jebusites, as the LORD your God has commanded; that they may not teach you to do according to all their abominable practices which they have done in the service of their Gods, and so to sin against the LORD your God.

As believers in sham gods and worshippers of false idols, heathens are to be extirpated by those who follow the One True God. The first of the Ten Commandments, in fact, states that one is to put no other gods before the One True God.

The rabbinical commentaries in the Talmud also support this in-group/out-group evolutionary model. The law (Mishnah) clarifies intentional and unintentional murder of an Israelite by an Israelite as such: "If he intended killing an animal but slew a man, or a heathen and he killed an Israelite . . . he is not liable" (Sanhedrin 79a). Further discussion of this law provides additional examples: "This excludes [from liability] the case of one who threw a stone into the midst of a company of Israelites and heathens. How is this? Shall we say that the company consisted of nine heathens and one Israelite? Then his non-liability can be inferred from the fact that the majority were heathens" (Gemara). So, thou shalt not kill a fellow Israelite, unless one is trying to kill a heathen, in which case the sacrifice was worth it. By contrast, the venerable rabbinical scholar Maimonides says that "if a resident alien slays an Israelite inadvertently, he must be put to death in spite of his inadvertence." In other words, good intentions apply only to members of our in-group. The book of Judges (5:9) is even more extreme in its in-group inclusiveness and out-group exclusiveness: "A Noahide [non-Jew] who kills a person, even if he kills an embryo in the mother's womb, is put to death. So too, if he kills one suffering from a fatal disease . . . he is put to death. In none of these cases is an Israelite put to death." The notion of "God's chosen people" resonates with evolutionary in-groupness.[16]

Because our deep moral sentiments evolved as part of our behavioral repertoire of responses for survival in a complex social environment (and not simply as infinitely plastic socially conditioned moral codes relative to one's culture), we carry the seeds of such in-group inclusiveness today. Israeli psychologist Georges Tamarin tested this hypothesis in 1966 on 1,066 schoolchildren ages eight to fourteen, by presenting them with the story of the battle of Jericho in Josh. 6. Joshua told his people to rejoice because God granted them access to Jericho: "Then they utterly destroyed all in the city, both men and women, young and old, oxen, sheep, and asses, with the edge of the sword. . . . And they burned the city with fire, and all within it; only the silver and gold, and the vessels of bronze and or iron, they put into the treasury of the house of the LORD." Tamarin wanted to test how biblical stories influenced children, particularly those stories that touted the superiority of monotheism, focused on the notion of the "chosen people," and made acts of genocide heroic. After presenting the Joshua story, the children were asked: "Do you think Joshua and the Israelites acted rightly or not?" Tamarin offered them three answers from which to choose: A, approval; B, partial approval or disapproval; and C, total disapproval. The results were disturbing: 66 percent of the children completely approved of the Israelites' murderous actions, while only 8 percent chose B and 26 percent chose C. One youngster wrote: "In my opinion Joshua and the Sons of Israel acted well, and here are the reasons: God promised them this land, and gave them permission to conquer. If they would not have acted in this manner or killed anyone, then there would be the danger that the Sons of Israel would have assimilated among the 'Goyim.'" Change the in-group to an out-group, on the other hand, and approval ratings for genocide drop precipitously. Substituting "General Lin" for Joshua and a "Chinese Kingdom 3,000 years ago" for Israel, Tamarin found that only 7 percent of his Israeli subjects (a control group of 168 different children) approved of the genocide (with 18 percent in the middle and 75 percent disapproving totally).[17]

The evolution of in-group morality, of course, is not restricted to any one religion, nation, or people.[18] It is a universal human trait common throughout history, from the earliest chiefdoms and states to modern nations and empires. As we shall see in the next chapter, the Yąnomamö people of the Amazon consider themselves to be the ultimate chosen people—in their language their name represents humanity, with all other peoples as something less than human. As the novelist and social commentator Aldous Huxley noted: "The propagandist's purpose is to make one set of people forget that the other set is human. By robbing them of their personality, he puts them outside the pale of moral oligation."[19] As we shall also see in the next chapter, the Nazis did this in spades. As many science fiction authors have suggested, perhaps an extraterrestrial threat to the entire planet Earth is what is needed to unite all humans into one giant in-group.

Morality and the Magic Number 150

Humans are, by nature, pattern-seeking, storytelling animals that evolved in both a physical and a social environment. As we have seen, morality is inextricably bound to religion, the first social structure to codify moral behaviors into ethical systems. Given the amount of time spent in the environment of our evolutionary history, we need to look more closely at the nature of human relationships in these small bands and tribes. In order to survive, these small hunter-gatherer bands would have had to employ considerable skills in cooperation and communication. Anthropologist Robert Bettinger demonstrates how, compared with individuals, "groups may often be more efficient" not only "in finding and taking prey, particularly large prey," but also in coordinating the activities of individuals, who might otherwise unduly interfere with each other. Foraging groups "that pool and share resources have the effect of 'smoothing' the variation in daily capture rates between individuals."[20] That is, as the group grows larger, "lucky" individuals share their take with "unlucky" individuals, and everyone benefits. (Think again of those vampire bats who share regurgitated blood.) Cooperation would have been as powerful, if not more powerful, a drive in human evolution as competition. And communication is an essential tool of cooperation, so it makes sense that Paleolithic hunter-gatherers, as well as their modern counterparts, would have employed language to solve the problems of survival in both the physical and the social environments.

How large were these communities? Most modern hunter-gatherer bands and tribes range in size from 50 to 400 residents, with a medium range of 100 to 200 people. Anthropologist Napoleon Chagnon, in his extensive studies of the Yąnomamö people in the Amazonian rain forest, found the typical group to be roughly 100 people in size, with 40 to 80 living together in the rugged mountain regions, and 300 to 400 members living together in the largest lowland villages. He has also noted that when groups get excessively large for the carrying capacity of their local environment and level of technology, they fission into smaller groups.[21] Such bifurcations are the result of exceeding the carrying capacity of both the physical and social environments. Psychologist Robin Dunbar suggests that these limits on social group size are related to the carrying capacity of human memory and thus have a deep evolutionary basis.[22] It turns out that 150 is roughly the number of living descendants (children, grandchildren, great-grandchildren) a Paleolithic couple would produce in four generations at the birthrate of hunter-gatherer peoples. In other words, this is how many people they knew in their immediate and extended family, which is corroborated by archaeological evidence from the Near East showing that agricultural communities typically numbered about 150 people. Even modern farming communities, like the Hutterites, average about 150 people.

When groups get large they split into smaller groups. Why? The answer is moral discipline and behavior control. According to the Hutterites, shunning as a primary means of social control does not work well in large groups. Sociologists know that once groups exceed 200 people, a hierarchical structure is needed to enforce the rules of cooperation and to deal with offenders, who in the smaller group could be dealt with through informal personal contracts and social pressure. Still larger groups need chiefs and a police force, and rule enforcement involves more violence or the threat of violence. Even in the modern world with a population exceeding six billion individuals, most of whom are crowded into dense cities, people find themselves divided into small groups. Studies on optimal group size (in terms of finding a balance between autonomy and control) by the military during the Second World War found that the average-size company in the British Army was 130 men, and in the U. S. Army it was 223 men. The 150 average also fits the size of most small businesses, departments in large corporations, and efficiently run factories. A Church of England study, conducted in an attempt to balance the financial resources provided by a large group with the social intimacy of a small group, concluded that the ideal size for congregations was 200 or less. The average number of people in any given person's address book also turns out to be about 150 people.

It would appear that 150 is the number of people each of us knows fairly well. Dunbar claims that this figure fits a ratio of primate group size to their neocortex ratio (the volume of the neocortex—the most recently evolved region of the cerebral cortex—to the rest of the brain). Extremely social primates (like us) need big brains to handle living in big groups, because there is a minimum amount of brain power needed to keep track of the complex relationships required to live in relatively peaceful cooperation. Dunbar concludes that these groupings "are a consequence of the fact that the human brain cannot sustain more than a certain number of relationships of a given strength at any one time [figure 4]. The figure of 150 seems to represent the maximum number of individuals with whom we can have a genuine social relationship, the kind of relationship that goes with knowing who they are and how they relate to us. Putting it another way, it's the number of people you would not feel embarrassed about joining uninvited for a drink if you happened to bump into them in a bar."[23]

Gossip and the Enmityville Horror

Morality evolved in these tiny bands of 100 to 200 people as a form of conflict resolution, social control, and group cohesion. In this social mode of religion and morality, amity is promoted over enmity. Without a system to reinforce cooperation and altruism and to punish excessive competitiveness and selfishness, Amityville becomes Enmityville. One means of accomplishing this social control is through what is known as reciprocal altruism, or "I'll scratch your back if you'll scratch mine." But, as Lincoln noted, men are not angels. There are cheaters. Individuals defect from informal agreements and social contracts. Reciprocal altruism, in the long run, only works when you know who will cooperate and who will defect. In small groups, cooperation is regulated through a complex feedback loop of communication between members of the community. (This also helps to explain why people in big cities can get away with being rude, inconsiderate, and uncooperative—they

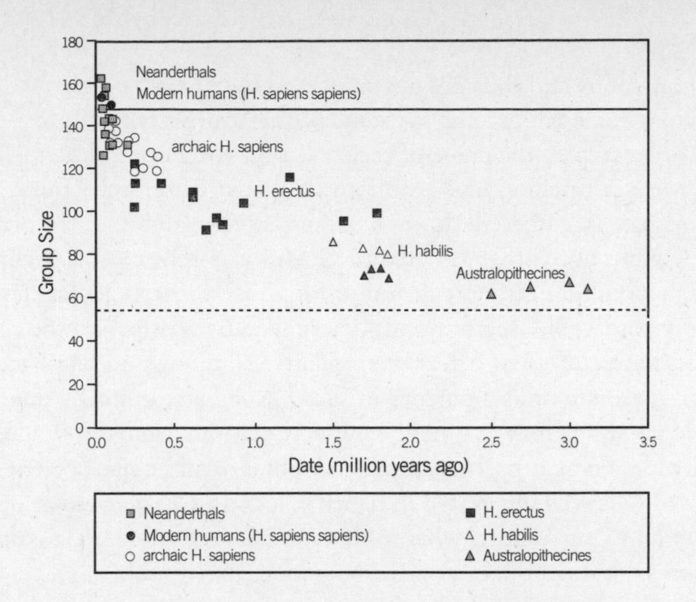

Figure 4. The Relationship Between Group Size, Brain Size, and Evolutionary History

Humans evolved as hierarchical social primate species living in tiny bands of hunter-gatherers. These groups grew larger over the past three million years, as hominid brains grew larger (*top*). Psychologist Robin Dunbar has identified a powerful relationship between group size in primates as a function of the ratio of their neocortex (defined as neocortex volume divided by volume of the entire brain), which is the seat of higher learning and memory (*bottom*). Dunbar theorizes that in order to live in large, complex social groups, primates need more memory and mental power to keep track of various relationships and social hierarchies. (From Robin Dunbar, "Brain on Two Legs," in *Tree of Origin*)

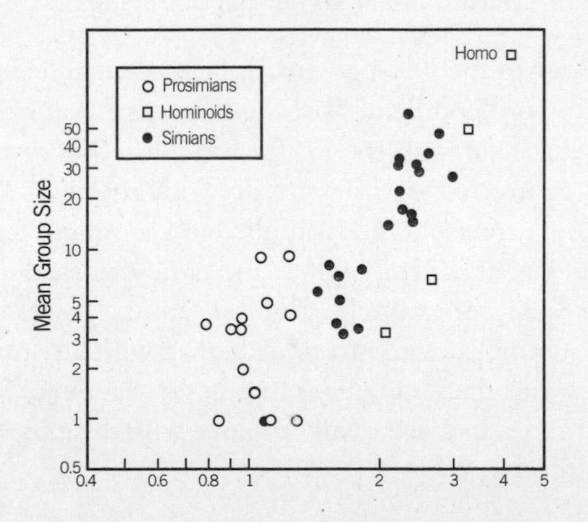

are anonymous and thus are not subject to the normal checks and balances that come with seeing the same people every day.)

In order to play the game of reciprocation you need to know whose back needs scratching and who you will trust to scratch yours. This information is gathered through telling stories about other people, more commonly known as gossip. Gossip is a tool of social control through communicating cultural norms, as anthropologist Jerome Barkow observed: "Reputation is determined by gossip, and the casual conversations of others affect one's relative standing and one's acceptability as a mate or as a partner in social exchange. In Euro-American society, gossiping may at times be publicly disvalued and disowned, but it remains a favorite pastime, as it no doubt is in all human societies."[24] This theory is well illustrated in figure 5, a 2003 *Crock* cartoon by Bill Rechin and Don Wilder, where the main character doesn't gossip; he transmits "pertinent data via a verbal mode." Well spoken.

Figure 5. Gossip as the Transmission of Data Via a Verbal Mode
(Courtesy of Bill Rechin and Don Wilder)

The etymology of the word *gossip*, in fact, is enlightening. The root stem is "godsib," or "god" and "sib," and means "akin or related." Its early use, as traced through the *Oxford English Dictionary*, included "one who has contracted spiritual affinity with another," "a godfather or godmother," "a sponsor," and "applied to a woman's female friends invited to be present at a birth" (where they would gossip). (In one of its earliest uses in 1386, for example, Chaucer wrote: "A womman may in no lasse synne assemblen with hire godsib, than with hire owene flesshly brother.") The word then mutated into talk surrounding those who are akin or related to us, and eventually to "one who delights in idle talk," as we employ it today.

Not surprisingly, we are especially interested in gossiping about the activities of others that most affect our *inclusive fitness,* that is, our reproductive success, the reproductive success of our relatives, and the reciprocation of those around us. Normal gossip is about relatives, close friends, and those in our immediate sphere of influence in the community, plus members of the community or society that are high ranking or have high social status. It is here where we find our favorite subjects of gossip—sex, generosity, cheating, aggression, violence, social status and standings, births and deaths, political and religious commitments, physical and psychological health, and the various nuances of human relations, particularly friendships and alliances. Gossip is the stuff of which not only soap operas but also grand operas are made.

My colleague Kari Konkola, a scholar researching the psychology of religious reinforcement and punishment of moral and immoral behavior, upon reading my initial theoretical foray into the connection between religion and morality (in *How We Believe*), made these important observations:

> The Protestants of early modern England knew very well the habit to gossip and regarded it as a personality trait that absolutely had to be eliminated. Indeed, the commandment "thou shalt not give false witness" was believed to be specifically a prohibition of gossiping. (In early modern religious terminology gossiping was called "backbiting.") The early modern interest in the roots of sin produced quite a bit of "research" on what today would be called the psychology of backbiting. The desire to talk about others was believed to be produced by envy, because the destructive competitiveness of envious people made them eager to spread rumors that would damage the reputation of those whom they envied. The competitiveness of proud people mostly manifested itself in efforts to surpass their peers, which made them less destructive to people around them than the envious. The desire to get ahead of others, however, made the proud an eager audience for malicious gossip, because they loved to hear disparaging news about their competitors.
>
> In early modern England, gossiping thus was a grave sin—a breach of one of the Ten Commandments. The strict religious prohibition against these behaviors is likely to have been quite effective, because a trained observer could easily notice them. Indeed, one commonly recommended method to detect hypocrisy was to observe people's favorite subjects of discussion: if a person liked to talk and hear about the flaws of others, this signified envy, pride

> and hypocrisy. On the other hand, an eagerness to dwell on one's own faults and to disparage one's achievements was a sign of humility and true religiosity. Normative evidence from early modern England leaves no doubt that at least some religions have been very emphatic about—and possibly also quite effective in—rooting the habit to gossip out of human nature.[25]

What is the relationship between gossip, morality, and ethics? Moral sentiments and behaviors were initially codified into ethical systems by religion. That is, long before there were such institutions as states and governments, or such concepts as laws and rights, religion emerged as the social structure to enforce the rules of human interactions. The history of the modern nation-state with constitutional rights and protection of basic human freedoms can be measured in mere centuries, whereas the history of organized religion can be measured in millennia, and the history of the evolution of moral sentiments can be measured in tens of millennia. When bands and tribes gave way to chiefdoms and states, religion developed as the principal social institution to facilitate cooperation and goodwill. It did so by encouraging altruism and selflessness, discouraging excessive greed and selfishness, promoting cooperation over competition, and revealing the level of commitment to the group through social events and religious rituals. If I see you every week in church, mosque, or synagogue, consistently participating in our religion's activities and following the prescribed rituals and customs, it is a positive indication that you can be trusted and you are a reliable member of our group that I can count on. As an organization with codified moral rules, with a hierarchical structure so well suited for hierarchical social primates like humans, and with a higher power to enforce the rules and punish their transgressors, religion responded to a need. In this social and moral mode I define religion as *a social institution that evolved as an integral mechanism of human culture to encourage altruism and reciprocal altruism, to discourage selfishness and greed, and to reveal the level of commitment to cooperate and reciprocate among members of the community.*

The divinity evolved along a parallel track. As bands and tribes coalesced into chiefdoms and states, animistic spirits gave way to anthropomorphic and polytheistic gods, and in the eastern Mediterranean the anthropomorphic gods of the pastoral people there lost out to the monotheistic God of Abraham. In addition to serving as an explanation for the creation of our universe, our world, and ourselves, God became the ultimate enforcer of the rules, the final arbiter of moral dilemmas, and the pinnacle object of commitment. God, religion, and morality were inseparable. People believe in God because we are pattern-seeking, storytelling, mythmaking, religious, moral animals.[26]

The Bio-Cultural Evolutionary Pyramid: A Model for the Origins of Morality

The Bio-Cultural Evolutionary (BCE) Pyramid in figure 6 depicts how morality evolved in small bands and tribes as individuals cooperated and competed with one another to meet their needs. Individuals belonged to families, families to extended families, extended families to communities, and, in the last several millennia (in parallel with the rise of chiefdoms and states), communities to societies. This natural progression is now in its latest evolutionary stages: perceiving societies as part of the species, and the species as part of the biosphere.

The BCE Pyramid is a hybrid of Abraham Maslow's hierarchy of needs and Peter Singer's expanding circle of ethical sentiments. It depicts the 1.5 million years over which moral sentiments evolved among social primates primarily under biogenetic control. Around 35,000 years ago a transition took place, and sociocultural factors increasingly assumed control in shaping our moral behavior and ethical systems. Keep in mind that this is a continuous process. There was no point at which an Upper Paleolithic Moses descended from a glacier-covered mountain and proclaimed to his fellow Cro-Magnons, "I've just invented culture. We no longer have to obey our genes like those stupid Neanderthals. From now on we obey THE LAW!"

The "bio-cultural transitional boundary" marks the shift from mostly biological control to mostly cultural control. It divides time and the dominant source of influence. In the early phases of our evolution, the individual, family, extended family, and Paleolithic communities were primarily molded by natural selection. In the later phases, Neolithic communities and modern societies were and are primarily shaped by cultural selection. Starting at the bottom of the BCE Pyramid, the individual's need for survival and genetic propagation (through food, drink, safety, and sex) is met through the family, extended family, and the community. The nuclear family is the foundation. Despite recent

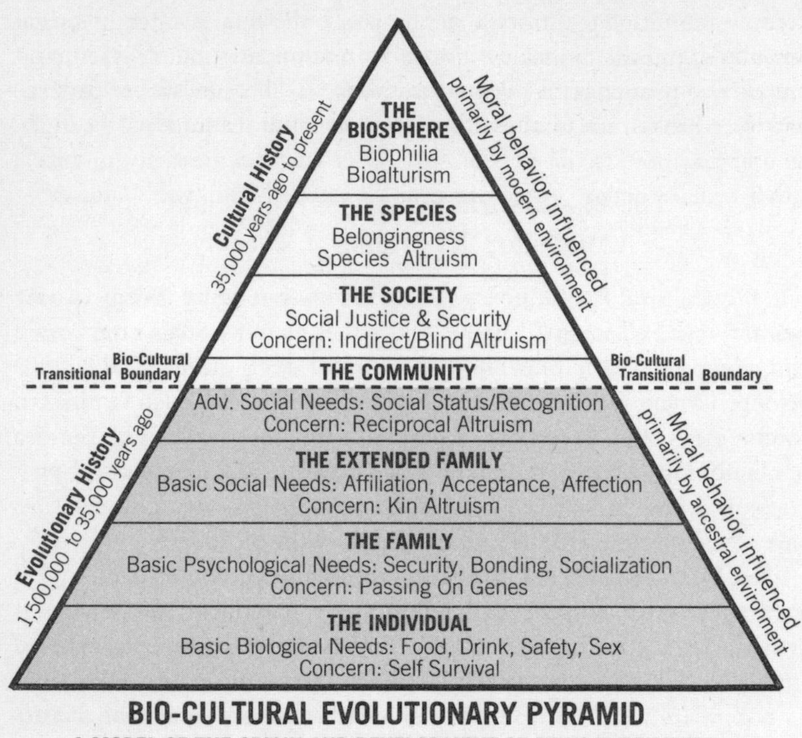

Figure 6. The Bio-Cultural Evolutionary Pyramid

The Bio-Cultural Evolutionary Pyramid models the origin and evolution of moral sentiments and ethical systems. The pyramid depicts the 1.5 million years over which moral sentiments evolved biogenetically among social primates and the transition about 35,000 years ago when sociocultural factors increasingly assumed control in shaping *Homo sapiens'* moral behavior and ethical systems. The "transitional boundary" shows the time range during which social groups grew larger and cultural selection began to take precedence over natural selection. In early *Homo sapiens* (the bottom of the pyramid), the individual's need for survival and genetic propagation is met through the family, extended family, and the community. Over time, basic psychological and social needs evolved that aided and reinforced cooperation, altruism, and, subsequently, genetic propagation through children. This *inclusive fitness* applies to anyone who is genetically related to us. In larger communities and societies, where there is no genetic relationship, *reciprocal altruism* and *indirect/blind altruism* supplement *kin altruism*. The natural progression leads to *species altruism* and *bioaltruism*. (Rendered by Pat Linse) claims that the traditional family is going the way of the Neanderthals, it remains the most common social unit around the world. Even within extremes of cultural deprivation—slavery, prison, communes—the structure of two-parents-with-children emerges: (1) African slave families that were broken up retained their structure for generations through the oral tradition; (2) in women's prisons pseudofamilies self-organize, with a sexually active couple acting as "husband" and "wife" and others playing "brothers" and "sisters"; (3) even when communal collective parenting is the norm (for example, kibbutzim), many mothers switch to the two-parent arrangement and the raising of their own offspring.[27] Our evolutionary history is too strong to overcome this foundational social structure. Conservatives need not bemoan the decline of families. They will be around as long as our species survives.

Moving up the BCE Pyramid, basic psychological and social needs such as security, bonding, socialization, affiliation, acceptance, and affection evolved to aid and reinforce cooperation and altruism, all of which facilitate genetic propagation through children. *Kin altruism* works indirectly—siblings and half siblings, grandchildren and great-grandchildren, cousins and half cousins, nieces and nephews, all carry portions of our genes.[28] This *inclusive fitness* applies to anyone who is genetically related to us. In larger communities and societies, where there is no genetic relationship, *reciprocal altruism* and *indirect/blind altruism* (if you scratch my back now, I may scratch yours much later) supplements kin altruism. Inclusive fitness gives way to what we might call *exclusive fitness.* The natural progression of exclusive fitness may be the adoption of *species altruism* and *bioaltruism* (we will prevent extinction and destruction now for a long-term payoff), which Ed Wilson argues in *Biophilia,* may even have a genetic basis.[29] But, Wilson confesses, this should probably still be grounded in self-interest arguments—my children will be better off in a future with abundant biodiversity and a healthy biosphere—since inclusive fitness is more powerful than exclusive fitness.

The width of the BCE Pyramid at any point indicates the strength of ethical sentiment and the degree to which it is under evolutionary control. The height of the BCE Pyramid at any point indicates the degree to which that ethical sentiment extends beyond our own genome (ourselves). But the pyramid also shows that these two sets of sentiments are inversely related. The more a sentiment reaches beyond ourselves,

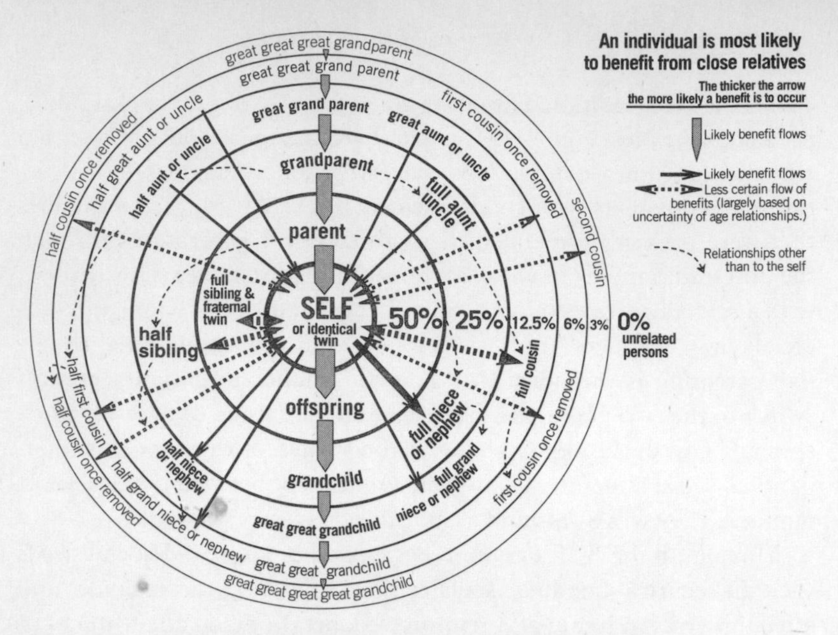

Figure 7. The Expanding Circle of Inclusiveness

Blood is thicker than water. According to evolutionary psychology theory, the percentage of genes (on average) shared by various degrees of kinship should predict the amount of benefits received from a given individual. The right side of the diagram shows relatives resulting from monogamy; the left side, polygyny. The assumption behind the theory is that organisms (including humans) act to enhance their inclusive fitness, that is, to increase the frequency and distribution of their genes in future generations. (Rendered by Pat Linse, adapted from Richard D. Alexander, *Darwinism and Human Affairs*)

the further it goes in the direction of helping someone genetically less related, and the less support it receives from underlying evolutionary mechanisms. This relationship is grounded in the aphorism that blood is thicker than water, as visualized in figure 7.

Group Selection and the Evolution of Morality

Since I first began work a decade ago on a theory to explain the origins of morality, new research has emerged that leads me to think that there might be an additional force at work in the evolution of morality, and that is group selection. Natural selection is believed by most evolutionary biologists to operate strictly on the organism level: the individual organism is the primary target of selection because it is the only thing that nature "sees." Genotypes, or genes, are simply code for phenotypes, or bodies. Nature cannot see genes, but it can see bodies running around and can select for or against those individuals based on their characteristics. In group selection, a group of individuals is the target of selection as a group of individuals competes against another group of individuals. When one succeeds and the other fails, the successful group passes along the genes of the individuals more than the unsuccessful group.

Among evolutionary theorists this is a volatile subject because group selection has, for the past thirty years, been next to creationism as the doctrine strict Darwinians most love to hate. There is some irony in this because the first person to propose group selection was none other than Charles Darwin. In *The Descent of Man* Darwin began by making a case against applying his own theory of natural selection at the individual level, noting that in trying to explain the origins of morality, "It is extremely doubtful whether the offspring of the more sympathetic and benevolent parents, or of those who were the most faithful to their comrades, would be reared in greater number than the children of selfish and treacherous parents of the same tribe. He who was ready to sacrifice his life, as many a savage has been, rather than betray his comrades, would often leave no offspring to inherit his noble nature. . . ." Darwin concluded that "it seems scarcely possible (bearing in mind that we are not here speaking of one tribe being victorious over another) that the number of men gifted with such virtues, or that the standard of their excellence, would be increased through natural selection, that is, by survival of the fittest."[30] *Within* a group, Darwin argued, natural selection would not foster cooperation and virtue. From whence did it come? It came from competition *between* groups, Darwin concluded:

> It must not be forgotten that although a high standard of morality gives but a slight or no advantage to each individual man and his children over the other men of the same tribe, yet that an increase in the number of well-endowed men and advancement in the standard of morality will certainly give an immense advantage to one tribe over another. There can be no doubt that a tribe including many members who, from possessing in a high degree the spirit of patriotism, fidelity, obedience, courage and sympathy, were always ready

> to aid one another, and to sacrifice themselves for the common good, would be victorious over most other tribes; and this would be natural selection. At all times throughout the world tribes have supplanted other tribes; and as morality is one important element in their success, the standard of morality and the number of well-endowed men will thus everywhere tend to rise and increase.[31]

Subsequent to Darwin, group selection was occasionally invoked until the mid-1960s, when the highly influential evolutionary biologist George C. Williams published his widely read book *Adaptation and Natural Selection,* in which he demonstrated that natural selection at the individual level was all that was needed to explain nature's diversity, including human social and moral behavior. Since the 1960s, group selection was vilified as the pap of bleeding-heart liberals who couldn't deal with the reality of "nature red in tooth and claw." Evolutionary theorist Michael Ghiselin described the "economy of nature" as competitive exclusively at the individual level: "The impulses that lead one animal to sacrifice himself for another turn out to have their ultimate rationale in gaining advantage over a third. . . . Given a full chance to act in his own interest, nothing but expediency will restrain him from brutalizing, from maiming, from murdering—his brother, his mate, his parent, or his child. Scratch an 'altruist,' and watch a 'hypocrite' bleed."[32]

In the late 1990s, however, group selection made something of a comeback. In their 1998 book *Unto Others,* anthropologist David Sloan Wilson and his colleague, philosopher Elliott Sober, demonstrated through a sophisticated mathematical model and a series of logical arguments that group selection is viable. They began by defining a group as "a set of individuals that influence each other's fitness with respect to a certain trait but not the fitness of those outside the group." They then explained how "individual selection favors traits that maximize relative fitness within single groups," and that "group selection favors traits that maximize the relative fitness of groups." In this model, "altruism is maladaptive with respect to individual selection but adaptive with respect to group selection." Therefore, they conclude, "altruism can evolve if the process of group selection is sufficiently strong."[33] For example, they cite William Hamilton's analysis of how consciousness might have provided a group selective advantage for certain human populations with regard to the ethical enforcement of rules: "The more consciences are lacking in a group as a whole, the more energy the group will need to divert to enforcing otherwise tacit rules or else face dissolution. Thus considering one step (individual vs. group) in a hierarchical population structure, having a conscience is an 'altruistic' character."[34] The reason individual selection will not work becomes apparent in this thought experiment proposed by David Sloan Wilson:

> Imagine a population that consists of solid citizens and shirkers. The solid citizens produce a public good that is available to everybody, including themselves. For purposes of the example, let's say that the public good can be produced at no cost to the solid citizens. Not only do they share the bounty, but they lose nothing by creating it. Even so, the solid citizens will not be favored by natural selection in this example because the solid citizens and shirkers do not differ in their survival or reproduction. Natural selection requires differences in fitness so raising or lowering the fitness of everyone in the population has no effect. If, as seems likely, the public good is costly to produce, the solid citizens will go extinct, even if they share the benefits, because their private cost reduces their fitness relative to the shirkers. Behaviors that are "for the good of the group" are at best neutral (if the public good is cost-free) and at worst maladaptive (if there is any cost associated with producing the public good).

Wilson calls this "the fundamental problem of social life" and provides the following group selection solution:

> Imagine not one but many populations that vary in their proportions of solid citizens and shirkers. Even if shirkers fare better than solid citizens within each population, populations with an excess of solid citizens fare better than populations with an excess of shirkers. In short, there is a process of natural selection at the group level that favors solid citizens, just as there is a process of natural selection at the individual level (within each group) favoring shirkers. Group-level adaptations will evolve whenever group-level selection is stronger than individual-level selection.[35]

Part of the problem in this debate is the all-too-human tendency to dichotomize. Instead of viewing this as a forced choice between individual selection and group selection, we can readily adopt a hierarchical theory of evolution, where we recognize and acknowledge both forces at work. A second challenge is in how we define certain terms, such as *altruism* and *cooperation,* and the temptation to force these categories into either-or choices, where people are (generally/situationally/purely) altruistic or selfish, cooperative or competitive. As descriptive terms, *altruistic* and *cooperative* are not reified things; they are behaviors. And like all behaviors, there is a broad range of expression, from a little to a lot. Here again a fuzzy logic analysis helps clarify this complex human phenomenon. Depending on the circumstances, some people in some situations are .2 altruistic and .8 nonaltruistic (or selfish), or .6 cooperative and .4 noncooperative (or competitive). Fuzzy fractions apply to individual and group selection as well. Individuals and groups can be altruistic and nonaltruistic, cooperative and noncooperative in degrees of expression changing over varying circumstances. It all depends on the situation. In this context, it might be useful to settle the group selection debate, at least provisionally, by acknowledging that group selection might work in a limited set of circumstances for some species. Group selection is not in opposition to individual selection; it is complementary, giving whole populations a selective advantage over other whole populations.

In applying group selection to the origins of religion, Wilson argues that "Religions exist primarily for people to achieve together what they cannot achieve alone. The mechanisms that enable religious groups to function as adaptive units include the very beliefs and practices that make religion appear enigmatic to so many people who stand outside of them." Going inside religion, Wilson argues, allows us to see what their practical function is—the group itself becomes a living organism, subject to the forces of natural selection. "Through countless generations of variation and selection, they acquire properties that enable them to survive and reproduce in their environments."[36] He notes as support, for example, anthropological studies of meat sharing practiced by all modern hunter-gatherer communities around the world. It turns out that these small bands and tribes—which can cautiously be used as a model for our own Paleolithic ancestors—are remarkably egalitarian. Using portable scales to precisely measure how much meat each family within the group received after a successful hunt, anthropologists discovered that the immediate families of successful hunters got no more meat than the rest of the families in the group, even when this was averaged over several weeks of regular hunting excursions. "Hunter-gatherers are egalitarian, not because they lack selfish impulses but because selfish impulses are effectively controlled by other members of the group," Wilson explains. "In human hunter-gatherer groups, an individual who attempts to dominate others is likely to encounter the combined resistance of the rest of the group. In most cases even the strongest individual is no match for the collective, so self-serving acts are effectively curtailed."[37]

How gossip, morality, and group selection link together can be seen in studies of a number of hunter-gatherer societies. Anthropologist Chris Boehm, for example, has demonstrated the use of gossip to ridicule, shun, and even ostracize individuals whose competitive drives and selfish motives interfere with the cooperative needs and altruistic tendencies of the group. In other words, people are competitive and selfish, and individual selection has created these important and powerful drives. But people are also cooperative and altruistic, and these drives are created and reinforced by the group in which the individual lives. Thus, an emotional sense of "right" and "wrong" action is ingrained into individuals back to the earliest days in human evolutionary history, through genetic transmission of such traits supplemented by culture transmission through modeling and learning. A moral "sense," then, need not be culturally codified in some formal fashion (for example, in writing), or even be a conscious effort on the part of individuals. "Good" and "bad" behaviors are rewarded and punished by the group, individuals succeed or fail as members of the group depending on their moral or immoral behaviors, and groups with more moral individuals than immoral individuals gain a collective advantage over other groups who are less successful at fostering a healthy balance between cooperation and competition, altruism and selfishness.

An anthropological example of how this process works can be seen in the Malaysian rain forest tribe called the Chewong. Like other hunter-gatherer groups, the Chewong (who also employ limited agriculture) are egalitarian, a way of life that is governed by a system of superstitions called *punen*. In the words of anthropologist Signe Howell, who conducted an extensive ethnography of the Chewong, *punen* is "a calamity or misfortune, owing to not having satisfied an urgent desire." In the Chewong world, for example, strong desires are connected with food, and powerful norms about food sharing are associated with the mythical being Yinlugen bud, who supposedly brought the Chewong out of a more primitive state by insisting that eating alone was improper human behavior. Myth, gods, religion, and morality are all integrated in the Chewong culture through the concept of *punen* and are linked to a very practical matter of individual and group survival: eating and sharing food. Thus, the Chewong avoid provoking *punen* at all costs. When game is caught away from the village, it is promptly returned, publicly displayed, and equitably distributed among all households and even among all individuals within each home. To reinforce the sanction against *punen,* someone from the hunter's family touches the catch and then proceeds to touch everyone present, repeating the word *punen.* In this system, religious superstitions and gods oversee the exchange process, generating within the individuals an overall sense of right and wrong action as related to the success or failure of the group.[38]

The Psychology of Morality: Emotions, Tipping, and the Prisoner's Dilemma

At the foundation of the Bio-Cultural Evolutionary model is an *evolved moral sense.* By a moral sense, I mean a moral feeling or emotion generated by actions. For example, positive emotions such as righteousness and pride are experienced as the psychological feeling of doing "good." These moral emotions likely evolved out of behaviors that were reinforced as being good either for the individual or for the group. Negative emotions such as guilt and shame are experienced as the psychological feeling of doing "bad." These moral emotions probably evolved out of behaviors that were reinforced as being bad either for the individual or for the group. This is the psychology of morality—the feeling of being moral or immoral. These moral emotions represent something deeper than specific feelings about specific behaviors. While cultures may differ on what behaviors are defined as good or bad, the general moral emotion of feeling good or feeling bad about behavior X (whatever X may be) is an evolved emotion that is universal to all humans.

Consider some of the more basic emotions that represent something deeper than specific feelings. When we need to eat we do not compute caloric input/output ratios; we simply feel hungry. That feeling is an evolved hunger sentiment that triggers eating behavior. When we need to procreate to pass on our genes into the next generation we do not calculate the genetic potential of our sexual partner, we just feel horny and seek out a partner we find attractive. The sexual urge—the undeniably powerful feeling of wanting to have sex—is an evolved sexual sentiment that triggers sexual behavior. In other words, we are hungry and horny because, ultimately, the survival of the species depends on food and sex, and those organisms for whom healthy foods tasted good and for which sex was exquisitely delightful left behind more offspring. To be blunt, we are the descendants of hungry and horny hominids.

Theists often ask, "If there is no God why should we be moral?" In this evolutionary theory of morality, asking "Why should we be moral?" is like asking "Why should we be hungry?" or "Why should we be horny?" For that matter, we could ask, "Why should we be jealous?" or "Why should we fall in love?" The answer is that it is as much a part of human nature to be moral as it is to be hungry, horny, jealous, and in love.

Again, to punch home the important distinction between the how and the why in the search for the origins of morality, specific behaviors in a culture may be considered right or wrong and these vary over cultures and history. But the *sense* of being right or wrong in the emotions of righteousness and pride, guilt and shame, is a human universal that had an evolutionary origin. There is variation within human populations on this evolved trait, just as there is variation in any personality trait, where some people feel more or less guilt or more or less pride than others. This variation, like the variation in personality traits, is accounted for by roughly half genetics and half environment.

Consider tipping at a restaurant in a city where you have never been and will never return. Since I travel a fair amount for my work I am faced with this moral question nearly every trip: why should I bother to tip a restaurant server I will never see again? There is no anticipated reciprocity since the tip comes after the service. I often dine alone so there is no one to impress with my generosity. Since I do not believe in God my answer cannot be "because God will know." (Even if you are a believer this seems like a rather shallow reason in any case.) Ethical egoism theory states that I will leave a tip because it makes me feel good. That is, tipping is not an altruistic act at all, but a purely selfish one. But what does it mean to feel good about an act, regardless of whether we consider it selfish or altruistic? In my theory this sense of feeling good about doing something good for someone else is an evolved moral sense that has a perfectly reasonable evolutionary explanation. Humans practice both deception and self-deception. Research shows that we are better at deception than at deception detection, but deceivers get caught often enough that it is risky to attempt to deceive others. Research also shows that the normal cues we give off when we are attempting to deceive others (particularly nonverbal cues like taking a deep breath, looking away from the person you are talking to, hesitating before answering, and so forth) are less likely to be expressed if you actually believe the deception yourself.[39] Liars are not liars if they believe the lie. This is the power of self-deception.

It is not enough to fake doing the right thing in order to fool our fellow group members, because although we are fairly good deceivers, we are also fairly good deception detectors. We cannot fool all of the people all of the time, and we do learn to assess (through gossip, in part) who is trustworthy and who is not trustworthy, so it is better to actually *be* a moral person because that way you actually believe it yourself and thus there is no need for deception. What I am saying is that the best way to convince others that you are a moral person is not to fake being a moral person but to actually *be a moral person*. Don't just go through the motions of being moral (although this is a good start), *actually be moral*. Don't just pretend to do the right thing, *do the right thing*. It is my contention that this is how moral sentiments evolved in our Paleolithic ancestors living in small communities.

Modern game theory grants us a deeper understanding of the tension between competitiveness and cooperation, the most common example of which is the "Prisoner's Dilemma," first developed in the 1950s by the Rand Corporation to model global nuclear strategy. The scenario is that each of two prisoners arrested for a crime is independently made an offer (and both know the other has been presented with the same deal). To make the game more real, pretend you are one of those charged with the crime. Here are the four possible options:

1. If you and your partner cooperate with each other and take the Fifth, then you each get one year in jail.
2. If you confess and admit that your partner was in on the crime with you, then you get off and your partner gets three years in the slammer.
3. If your partner confesses and you don't, then you receive a three-year stay in the pokey while he walks.
4. If you both confess, then you each get a two-year stay in the gray-bar hotel.

What should you do? If you defect on your partner and confess, then you will get either zero or two years in the pen, depending on what he does. If you cooperate and stay quiet, you get either one or three years, again depending on his response. In this scenario the logical choice is to defect. Of course, your partner is likely going to make the same calculation as you, which means he too will defect, guaranteeing that you will receive a two-year stint in prison. Knowing that he is probably computing the same strategy as you, then surely he will realize that you should both cooperate. Of course, perhaps he'll figure that this will be your conclusion as well, so he'll defect in hopes that you will cooperate, getting him off the hook and sending you to wallow in the general population for three big ones. Herein lies the dilemma.

When the game is noniterated—that is, just one round is played—defection is the norm. When the game is iterated over a fixed number of trials, defection is also the norm because awareness of how many rounds there will be means that both players know that the other one will defect on the last round, which pushes another defection strategy to the second to the last round, and so forth back to the start of the game. But when the game is iterated over an unknown number of trials, and both players keep track of what the other has been doing throughout the history of the game—that is, it more closely resembles real life—cooperation prevails.

Still more realistic simulations include the "Many Person Dilemma," in which one player interacts with multiple other players. Of particular interest to the study of ethical theory is the "Tit-for-Tat" strategy in which you start off cooperating and then do whatever the other player does. Although Tit-for-Tat resembles the Golden Rule, it more closely models the Old Testament "Eye for an Eye" morality than it does the idealistic New Testament strategy of turning the other cheek. Jesus, Gandhi, and Martin Luther King heroically employed an unalloyed New Testament ethic and all paid with their lives (receiving what is called the "sucker's payoff" in game theory). Turning the other cheek only works if the opposition is inherently benevolent or has chosen a purely cooperative game strategy. In most cases, defections creep in often enough to make a purely Gandhian morality ineffective, even dangerous. On the optimistic side, the more experience the players have with one another—that is, the better they know each other—the more cooperation dominates as the preferred strategy. This is why the scenario of informal and noncodified rules of conduct developed within small communities so accurately describes how morality evolved—knowing all the other players in the game leads to the evolution of cooperation.[40]

The Prisoner's Dilemma model has been applied to everything from cold war strategies to marital conflict. It turns out that in both computer simulations and real-world experiments, not only is being a cooperator better than being a defector, but being a real cooperator is better than being a fake cooperator because being genuine about cooperating more readily convinces others of the genuineness of the action. And action backed by good intentions pays off in bigger dividends for all individuals in the group (and for the group as a whole in relation to other groups). So, tipping a stranger makes you feel good because it really is a good thing to do, for the recipient, for yourself, and for the group.

Human Moral Universals: The Transcendency of Morality and Universality of Moral Sentiments

In his great philosophical work *An Enquiry Concerning the Principles of Morals,* David Hume speculated on the universal nature of human morality: "The notion of morals implies some sentiment common to all mankind, which recommends the same object to general approbation, and makes every man, or most men, agree in the same opinion or decision concerning it."[41] Is there a moral sentiment common to all humans? Are there moral universals?

In anthropology, human universals "comprise those features of culture, society, language, behavior, and psyche for which there are no known exceptions to their existence in all ethnographically or historically recorded human societies."[42] Common and well-known universals include tools, myths and legends, sex roles, social groups, aggression, gestures, grammar, and emotions. There are general universals, such as social status and emotional expressions, and specific universals within broader universals, such as kinship statuses and facial expressions like the smile, frown, or eyebrow flash. Since cultures vary dramatically, the supposition made about "universals" is that there is an evolutionary and biological basis behind them (or, at the very least, that they are not strictly culturally determined). As such, we can presume that there is a genetic predisposition for these traits to be expressed within their respective cultures, and that these cultures, despite their considerable diversity, nurture these genetically predisposed natures in a consistent fashion.

For an analysis of universals we turn to anthropology, because it is this science that documents the diversity of ways that humans live and illuminates so clearly which traits are universal and which traits are not. "Universals exist, they are numerous, and they engage matters unquestionably of anthropological concern," explains anthropologist Donald E. Brown, who has arguably done more work on human universals than anyone in his field. "Universals can be explained, and their ramified effects on human affairs can be traced. But universals comprise a heterogenous set—cultural, social, linguistic, individual, unrestricted, implicational, etc.—a set that may defy any single overarching explanation. If, however, a single source for universals had to be sought, human nature would be the place to look."[43] Culture matters, of course, but not in some token fashion tossed off by sociobiologists and evolutionary psychologists to display proper political correctness, rather as wholly integrated and fully interdigitated with nature such that you cannot speak of one without the other. This *interactionism* is the only reasonable position to take on the so-called (and artificial) nature-nurture debate, despite the limitations most anthropologists have set on understanding this interaction. As Brown notes in castigating his fellow anthropologists: "these interactions can only be properly explored if there are ways to distinguish nature from culture, and I submit that there is little if anything in the way of an established and valid method in anthropology for doing this. Typically, anthropologists simply do not concern themselves with this problem, because they assume (in accordance with other propositions) that what humans do, unless it is 'obviously' natural, is essentially cultural."[44]

In his comprehensive study of human universals, Brown compiled a list of 373. From these I count 202 (54 percent) moral and religious universals, which I list in appendix 2 along with parenthetical notes explaining what I think the relationship is between the universal trait and morality or religion. (I include religious universals because in my theory on the origins of morality, religion and morality were inseparable in their coevolution.) Although some traits are more obviously related to morality and religion than others, in general it is strikingly clear just how much of what we do has some bearing on our state of being as social organisms in interaction with others of our kind. We are moral animals, and these moral universals belong to the species and are thus transcendent of the individual members of that species.

The expression of human moral behavior is a product of internal psychological traits related to morality, and external social states related to moral behavioral control. Going through the list of moral and religious universals in appendix 2, the sheer number is indicative of their undeniable role in human biological and cultural evolution. Some moral psychological traits (or sentiments) include: *affection expressed and felt* (necessary for altruism and cooperation to be reinforced); *attachment* (necessary for bonding, friendship, pro-social behavior); *coyness display* (courtship, moral manipulation); *crying* (sometimes expression of grief, moral pain); *emotions* (necessary for moral sense); *empathy* (necessary for moral sense); *envy* (moral trait); *fears* (generate much religious and moral behavior); *generosity admired* (reward for cooperative and altruistic behavior); *incest between mother and son unthinkable or tabooed* (obvious evolutionary moral trait); *judging others* (foundation of moral approval/disapproval); *mourning* (expression of grief, part of symbolic moral reasoning); *pride* (a moral sense); *self-control* (moral assessment and judgment); *sexual jealousy* (foundation of major moral relations and tensions); *sexual modesty* (foundation of major moral relations and tensions); and *shame* (moral sense), to name just a few.

Going through the list again, this time picking out moral behavioral control mechanisms universal to human cultures, we are again awed by their numbers and importance. Some moral behavioral control mechanisms include: *age statuses* (vital element in social hierarchy, dominance, respect for elder's wisdom); *coalitions* (foundation of social and group morality); *collective identities* (basis of xenophobia, group selection); *conflict, consultation to deal with* (resolution of moral problems); *conflict, means of dealing with* (resolution of moral problems); *conflict, mediation of* (foundation of much of moral behavior); *corporate (perpetual) statuses* (moral ranking of groups); *customary greetings* (part of conflict prevention and resolution); *dominance/submission* (foundation of hierarchical social primate species); *etiquette* (enhances social relations); *family (or household)* (the most basic social and moral unit); *food sharing* (form of cooperation and altruism); *gift giving* (reward for cooperative and altruistic behavior); *government* (social morality); *group living* (social morality); *groups that are not based on family* (necessary for higher moral reasoning and blind altruism); *inheritance rules* (reduces conflict within families and communities); *institutions (organized coactivities)* (religion); *kin groups* (foundation of kin selection/altruism and basic social group); *law (rights and obligations)* (foundation of social harmony); *law (rules of membership)* (foundation of social harmony); *males engage in more coalitional violence* (gender differences in moral behavior); *marriage* (moral rules of foundational relationship); *murder proscribed* (moral judgment necessary in communities); *oligarchy (de facto)* (group social control); *property* (foundation of moral reasoning and judgment); *reciprocal exchanges (of labor, goods, or services)* (reciprocal altruism); *redress of wrongs* (moral conflict resolution); *sanctions* (social moral control); *sanctions for crimes against the collectivity* (social moral control); *sanctions include removal from the social unit* (social moral control); *taboos* (moral and social control); *tabooed foods* (element in moral and social control); *tabooed utterances* (communication of moral and social control); and *violence, some forms of proscribed* (moral and social control), to name just a few.

Finally, we began this chapter with the Golden Rule, a moral guideline found in all cultures that represents the very foundation of universal morality. Of the list of human moral universals, here are the traits that contribute to a behavioral expression of the Golden Rule: *cooperation* (part of altruism); *cooperative labor* (part of kin, reciprocal, and blind altruism); *customary greetings* (part of conflict prevention and resolution); *empathy* (necessary for moral sense); *fairness* (equity); *food sharing* (form of cooperation and altruism); *generosity admired* (reward for cooperative and altruistic behavior); *gestures* (signs of recognition of others, conciliatory behavior); *gift giving* (reward for cooperative and altruistic behavior); *good and bad distinguished* (necessary for moral judgment); *hospitality* (enhances social relations); *inheritance rules* (reduces conflict within families and communities); *insulting* (communication of moral approval/disapproval); *intention* (part of moral reasoning and judgment); *interpolation* (part of moral reasoning and judgment); *interpreting behavior* (necessary for moral judgment); *judging others* (foundation of moral approval/disapproval); *making comparisons* (necessary for moral judgments); *moral sentiments* (the foundation of all morality); *moral sentiments, limited effective range of* (parameters of moral foundation); *planning for future* (foundation for moral judgment); *pride* (a moral sense); *promise* (moral relations); *reciprocal exchanges* (of labor, goods, or services) (reciprocal altruism); *reciprocity, negative* (revenge, retaliation) (reduces reciprocal altruism); *reciprocity, positive* (enhances reciprocal altruism); *redress of wrongs* (moral conflict resolution); *shame* (moral sense); *turn taking* (conflict prevention).

The Nature of Human Moral Nature

Humans are, by nature, moral and immoral, good and evil, altruistic and selfish, cooperative and competitive, peaceful and bellicose, virtuous and nonvirtuous. Such moral traits vary between individuals and within and between groups. Some people and populations are more or less moral and immoral than other people and populations, but all people have the potential for all the moral traits. (Whether the evolution of such moral and immoral traits was inevitable is an important and interesting question, but it is an ancillary one to my analysis here.) Most people most of the time in most circumstances are good and do the right thing for themselves and for others. But some people some of the time in some circumstances are bad and do the wrong thing for themselves and for others.

The codification of moral principles out of the psychology of moral traits evolved as a form of social control to ensure the survival of individuals within groups and the survival of human groups themselves. Religion was the first social institution to canonize moral principles, but morality need not be the exclusive domain of religion. Religions succeeded in identifying the human universal moral and immoral thoughts and behaviors most appropriate for accentuating amity and attenuating enmity. But we can improve on the ethical systems developed thousands of years ago by people of agricultural societies whose moral codes are surely open to change. As we transition from kin and reciprocal altruism to species altruism and bioaltruism, and as religion continues to give ground to science, we need a new ethic for an Age of Science, a new morality that not only incorporates the findings of science, but applies scientific thinking and the methods of science to tackling moral problems and resolving moral dilemmas. We have done well thus far, but we can do better.

3

WHY WE ARE IMMORAL: WAR, VIOLENCE, AND THE IGNOBLE SAVAGE WITHIN

If you wish for peace, understand war.

—B. H. Liddell Hart, *Strategy*, 1967

In Rob Reiner's 1992 film *A Few Good Men,* Jack Nicholson's character—battle-hardened marine Colonel Nathan R. Jessup—is being cross-examined by Tom Cruise's naive rookie navy lawyer, Lieutenant Daniel Kaffee. Kaffee is defending two marines accused of killing a fellow soldier named Santiago at Guantanamo base in Cuba. He thinks Jessup ordered a "code red," an off-the-books command to rough up a lazy marine trainee in need of discipline, and that matters got tragically out of hand. Kaffee wants answers to specific questions about the incident. Jessup wants to lecture him on the meaning of freedom and the need to defend it, explaining: "Son, we live in a world that has walls. And those walls have to be guarded by men with guns. I have a greater responsibility than you can possibly fathom."

Jessup complains that he does not have the luxury to ignore the fact that Santiago's death ultimately saved lives by generating greater discipline, and that Kaffee does not really want the truth because "deep down, in places you don't talk about at parties, you want me on that wall. You need me on that wall." Disgusted that he should even have to bother with such elucidations, Jessup concludes: "I have neither the time nor the inclination to explain myself to a man who rises and sleeps under the blanket of the very freedom I provide, then questions the manner in which I provide it. I'd prefer you just said thank you and went on your way."[1]

The simple observation that we live in a world with walls—and have for the past 6,000 years of recorded history—implies that those walls are needed. The constitutions of states cannot completely alter the constitution of humanity. Is that constitution an evil one, and thus good walls will always be needed to make good neighbors? Or are we constitutionally good, but corrupted by evil circumstances and environments?

The Problem of Evil

A philosophical conundrum that has plagued theologians and moral philosophers is known as the "problem of evil." The Greek philosopher Epicurus, in his *Aphorisms,* stated it as early as 300 B.C.:

> The gods can either take away evil from the world and will not, or, being willing to do so cannot; or they neither can nor will, or lastly, they are both able and willing. If they have the will to remove evil and cannot, then they are not omnipotent. If they can, but will not, then they are not benevolent. If they are neither able nor willing, then they are neither omnipotent nor benevolent. Lastly, if they are both able and willing to annihilate evil, how does it exist?[2]

Here is a simpler way to state the problem. The following three conditions are incompatible:

God is Omnipotent
God is Omnibenevolent
Evil Exists

If God is all-powerful, *can* He not prevent evil from existing? If God is all good, *should* He not prevent evil from existing? If evil exists, then either God is not all powerful or not all good. The eighteenth-century English poet Alexander Pope poetically phrased a solution to the problem a different way in *An Essay on Man*:

> *All Nature is but Art, unknown to thee;*
> *All Chance, Direction, which thou canst not see;*
> *All Discord, Harmony, not understood;*
> *All partial Evil, universal Good:*

And, spite of Pride, in erring Reason's spite,
One truth is clear, "Whatever IS is RIGHT."[3]

To explain away the problem of evil, believers often invoke the final clause in a modified version—read "God's will" for "Whatever is." We heard it in the wake of the December 1997 shootings of eight teenagers in a Kentucky high school, when the local minister declared that it was "God's will" that his own son was spared a bullet. On the flip side of evil, Kenny McCaughey, father of the famed McCaughey septuplets, thanked God for the good health of the babies and their mother, and said it was "God's will" that they have seven children: "I'm just confident the Lord's going to handle this. He's brought them this far and I think he's going to carry it through." The McCaugheys were offered a standard "selective abortion" option to reduce the number of babies and thus guarantee the health of the remaining fetuses and mother, but they declined, explaining that as God-fearing Christians they were strong abortion opponents. "That's just the way we all feel about it. It's going to be. If for some reason he decides to change it, that's his will."[4]

There is an obvious inconsistency here. God did not will the conception and birth of the McCaugheys' seven children; modern medical technology did. If whatever is, is right, then Bobbi McCaughey's infertility was also God's will, along with His active intervention in generating the appropriate number of viable eggs in the in-vitro procedure. Why is it morally acceptable to alter God's will of infertility through the intervention of modern medicine and technology, but not to opt for selective abortion through the same modern medical techniques? If Mrs. McCaughey had died because of complications of childbirth, would it still be God's will? Would we not have been bombarded in the media by fertility technology Luddites for the hubris of modern medicine in trying to change what is "natural"? "Scientists should not play God," we are told.[5]

As for the Kentucky high school killings, why was it God's will to spare the minister's son, but not the three girls who died? Was it God's will that their lives be snuffed out as young teenagers? At the Columbine High School massacre, one of the victims, a young woman named Cassie Bernall, was gunned down after proclaiming her belief in God. A book about her life and proclamation, *She Said Yes,* rode the *New York Times* best-seller list for weeks.[6] Are we to presume that this was her "reward" for having the courage to openly display her Christian faith? That hardly seems fair. And, on the other side of the story, what about all those nonbelieving high schoolers who lived? That hardly seems divinely just to believers. These events would seem to make God's will both good and evil, or God's power limited, or both.

In the wake of the terrorist attacks on September 11, 2001, we saw similar declarations of a divine force that intervened to spare those who survived. Most of the survivors spoke of being saved by divine grace. In fact, a more likely explanation is that the law of large numbers kicked in. In a situation in which the lives of thousands of people are endangered, any who survive do so by an extraordinary turn of events, any one of which seems so improbable as to be a miracle. Yet, as in the lottery, even when the odds of winning are astronomically stacked against any one person taking home the loot, someone is going to win. Put your life in place of that money and the stakes are elevated that much higher, and it isn't surprising that divine intervention will be credited. But, as much as we may be willing to believe in miracles, from whence does our willingness to believe in evil arise?

The Myth of Pure Evil

The myth of pure evil is the belief that evil exists separately from individuals, or that evil exists within people as something like what we traditionally think of as an evil "force," driving them to perform evil acts. If pure evil exists, however, then how can we hold people morally culpable for their actions? Evil is intimately linked to the problem of free will and determinism: if we do not have complete free will in our actions, how can we be held morally accountable? Further (and even more distressing), if evil does exist, then will we always be plagued by violence, war, genocide, crime, rape, and other evils?

One solution to the problem of evil is a semantic one. *Evil* as a descriptive adjective merely modifies something else, as in evil thoughts or evil deeds. But *evil* as a noun implies an existence all its own, as in an "evil force" or even an "evil person," or "the force in nature that governs and gives rise to wickedness and sin," or "the wicked or immoral part of someone or something," and so on.[7] In this latter sense I claim that *there is no such thing as evil*. There is no supernatural force operating outside the realm of the known laws of nature and human behavior that we can call evil. Calling something or someone "evil" gets us nowhere. It leads to no greater understanding. In a scientific sense it is a term ultimately indefinable. That is, there is no way to establish quantifiable criteria by which we may distinguish between something or someone that is "evil" or "not evil," or shades of evil in between. The tendency to use the term at all comes from our Western Platonic tendency to think in terms of essences, or nonchanging "things" or "types" that are what they are by their very nature. This is one reason people have such a difficult time accepting the theory of evolution—we tend to view a species as an essence, a type, a fixed and unchanging entity. But, in fact, they do change, however slowly, and their essences are only temporary. Analogously, evil is not a fixed entity or essence. It is not a thing. *Evil* is a descriptive term for a range of environmental events and human behaviors that we describe and interpret as bad, wrong, awful, undesirable, or whatever appropriately descriptive adjective or synonym for evil is chosen. To call something "evil" does not lead us to a deeper understanding of the cause of evil behavior.

In this chapter, I want to focus not on evil as a metaphysical concept that exists outside the natural realm, but on evil as a physical concept that exists entirely within the natural realm as behavior. This is a shift from the supernatural to the natural. (On a larger scale, I go so far as to claim that there is no such thing as the paranormal and supernatural. There is only the normal and the natural, and mysteries we have yet to explain through them.) If there were no humans there would be no evil. Earthquakes that kill people are not, in and of themselves, evil. A shift between two tectonic plates that causes the earth to make a sudden and dramatic movement cannot possibly be considered evil outside the effects such an earthquake might have on the humans living near the fault line. It is the *effects* of the earthquake on our fellow humans that we judge to be evil. Evil as a physical concept requires human evaluation of a behavior and its effects on humans. As such, bacterial diseases cannot be inherently evil. By causing humans to sneeze, cough, vomit, and have diarrhea, bacteria are highly successful organisms, spreading themselves far and wide. As their human hosts, we may label the effects of a disease as evil, but the disease itself has no moral existence. Good and evil are human constructs. Which is not to say that a person is not morally responsible for his or her choices and their effects.

Oskar Schindler or Amon Goeth?

One morning in 1995 I had breakfast with Thomas Keneally, author of *Schindler's List.* Out of curiosity I asked him what he thought was the difference between Oskar Schindler, rescuer of Jews and hero of his story, and Amon Goeth, the antihero Nazi commandant of the Plaszow concentration camp. His answer was revealing. Not much, he said. Had there been no war, Schindler and Goeth might have been drinking buddies and business partners, morally questionable at times, perhaps, but relatively harmless and ineffectual as historical personages. What a difference a war makes.

This question, on a larger scale, is what spurred the debate in 1996 over the publication of Daniel Goldhagen's book *Hitler's Willing Executioners.* Goldhagen's thesis is that ordinary Germans participated in the mass murder of Jews, that anti-Semitism was pervasive and nearly exclusively German, and that we cannot blame a handful of extremists in the Nazi party for the Holocaust—all Germans share the blame. "My explanation . . . is that the perpetrators, 'ordinary Germans,' . . . having consulted their own convictions and morality and having judged the mass annihilation of Jews to be right, did not want to say 'no.'"[8] The problem with this thesis is that it does not explain the exceptions—the Germans who helped Jews and the non-Germans who participated in the Holocaust. Max Frankel, in assessing the Goldhagen thesis, recalls how his Jewish mother escaped Nazi Germany thanks to the help of no less than a Gestapo police chief who, after giving her the name of the Gestapo contact who would aid in her escape, told her, in reference to her goal of reaching America, "If you get there, will you tell them we're not all bad?"[9]

It's true, not all Germans—not even all Nazis—were bad. Likewise, and extrapolating to the larger issue on the table, humans are neither all bad nor all good. Most of us most of the time are good in most situations. Under extreme circumstances, however, the flexibility of our behavior may push us in the other direction. The Holocaust was not the product of "ordinary" Germans, but of Nazi Germans in extraordinary circumstances and conditions. A narrow focus on the proximate causes of the Holocaust misses the ultimate, evolutionary lesson that this tragic event in human history can teach us about the malleable condition of human nature.

Humans evolved to be moral animals, but by no means always moral. There are times when we are amoral, and even immoral. We have the potential for all three, and like any human trait, the degree of expression of the quality varies between individuals. Some people, for whatever reason, are more moral than others. A number of historical contingencies (and who knows what else in his genes and environment) drove Oskar Schindler to follow a completely different path from Amon Goeth, even though he could just as easily have gone the other way. From there the cascading consequences of their decisions took them down their alternately chosen tracks; the road not taken makes all the difference.

This is not mere just-so storytelling. Yale University social psychologist Stanley Milgram observed this range of moral flexibility in his famous "shock" experiments in the 1960s, conducted in the wake of the Adolf Eichmann trial in an attempt to make a scientific study of the banality of evil. How was it possible for educated, intelligent, and cultured human beings to commit mass murder? What environmental conditions would override our evolutionary propensity toward moral behavior and the repulsion most of us would (or at least should) feel in causing or witnessing the pain of another human? Milgram presented his subjects with a "learning" experiment that was purportedly to test the possible effect of punishment on memory. The subject would read a list of paired words to the "learner," then present the first word of each pair again, upon which the learner had to recall the second word. If the learner was wrong the teacher—the real subject in this experiment—was to deliver an electric shock. No one was really shocked of course (the learners were shills working for Milgram who purposely gave them wrong answers), but the subjects believed the shocks were real. Sitting in front of the subject was a panel of toggle switches that read: Slight Shock, Moderate Shock, Strong Shock, Very Strong Shock, Intense Shock, Extreme Intensity Shock, DANGER: Severe Shock, XXXX (this final category was labeled at 450 volts). The results were, well, shocking: 65 percent administered the strongest shock possible—XXXX—and 100 percent administered at least a Strong Shock of 135 volts (figure 8).

Milgram varied the conditions in order to control for other variables that might influence the outcome of the experiment. The physical proximity of the victim, for example, had an effect on how far the teachers would go in shocking the learners—closer, less shock; farther

Figure 8. A Shocking Experiment on Obedience to Authority

In a quest to understand how highly educated and richly cultured Germans could be turned into Nazi mass murderers, psychologist Stanley Milgram undertook the study of obedience to authority. Here subjects, playing the role of "teachers," were told by a scientist/authority figure to shock "learners" who made mistakes on a test. Sitting in front of the subject was a panel of toggle switches that read: Slight Shock, Moderate Shock, Strong Shock, Very Strong Shock, Intense Shock, Extreme Intensity Shock, DANGER: Severe Shock, XXXX (450 volts). Milgram discovered that 65 percent administered the strongest shock possible—XXXX—and 100 percent administered at least a Strong Shock of 135 volts. (From Stanley Milgram, *Obedience,* 1965. Courtesy of Penn State Media Sales)

away, more shock. Group pressure was also a factor—when two other "teachers" (Milgram's confederates) encouraged the subject to continue shocking the learner, they did so with impunity; but when the confederates refused to shock their own learners, the subject tended to refuse as well. In other words, Milgram discovered that moral behavior is extremely malleable. He expressed his own and others' amazement at what these experiments revealed about our moral natures: "What is surprising is how far ordinary individuals will go in complying with the experimenter's instructions. Despite the fact that many subjects experience stress, despite the fact that many protest to the experimenter, a substantial proportion continue to the last shock on the generator." Why? Because, Milgram continued, "It is psychologically easy to ignore responsibility when one is only an intermediate link in a chain of evil action but is far from the final consequences of the action." Even Eichmann, he noted, was repulsed when he first witnessed a camp killing, but to implement the Holocaust he had only to push papers across a desk. The SS guards who actually did the deed justified their behavior by claiming they were only following orders. "I am forever astonished when lecturing on the obedience experiments in colleges across the country," Milgram concluded. "I faced young men who were aghast at the behavior of experimental subjects and proclaimed they would never behave in such a way, but who in a matter of months, were brought into the military and performed without compunction actions that made shocking the victim seem pallid. In this respect, they are no better and no worse than human beings of any other era who lend themselves to the purposes of authority and become instruments in its destructive processes."[10]

Depending on the circumstances, perhaps any of us could become Nazis. Who is to say otherwise? Raised in a free, democratic society like America, how do any of us know how we might react in a totalitarian regime like Nazi Germany, in a time of war and under pressure to obey one's superior (what Milgram termed "obedience to authority")?

Stanford University social psychologist Philip Zimbardo tested this hypothesis in the "Stanford County Jail," a mock prison set up in the basement of his psychology building where randomly chosen students were assigned to be "prisoners" or "guards." To make it as real as possible, Zimbardo gave the guards sunglasses, a whistle, a club, and cell keys. Prisoners were arrested, sprayed for lice, forced to stand naked during orientation, given bland uniforms, and stuck in dreary six-by-nine-foot cells. What unfolded in a matter of days was disturbing. These psychologically normal undergraduate American students were transformed into the role of either violent, authoritative guards or demoralized, impassive prisoners. The experiment was to last for two weeks. Zimbardo was forced to call it off after six days, not just because he feared that the lives of his students would be permanently transformed by this *Lord of the Flies* experience, but because of what he discovered about the reality of the human moral condition: "I called off the experiment not because of the horror I saw out there in the prison yard, but because of the horror of realizing that I could have easily traded places with the most brutal guard or become the weakest prisoner full of hatred at being so powerless that I could not eat, sleep, or go to the toilet without permission of authorities."[11]

How realistic are these experiments conducted in American universities in the 1960s? The real-world example of Nazi Germany, in some sense, serves as a historical experiment. At the Nuremberg Trials following the war, psychologist G. M. Gilbert was assigned to study the men imprisoned for committing these "crimes against humanity." He discovered that not only were the perpetrators well-cultured and highly educated, they tested out two to three standard deviations above average on a standardized intelligence test used at the time, the Wechsler Adult Intelligence Scale. Where the average IQ is 100, Reichs-Commissioner Seyss-Inquart tested at 141, Reichsmarschall Hermann Göring at 138, Reich Chancellor Franz Von Papen at 134, Poland Governor-General Dr. Hans Frank at 130, Foreign Minister Joachim von Ribbentrop at 129, Hitler's architect and Reichsminister of Armaments Albert Speer at 128. The prison psychiatrist Douglas Kelley, after his evaluations, offered this observation about the moral character of the leading Nazis:

> As far as the leaders go, the Hitlers and the Görings, the Goebbels and all the rest of them were not special types. Their personality patterns indicate that, while they are not socially desirable individuals, their like could very easily be found in America. Neurotic individuals like Adolf Hitler, suffering from hysterical disorders and obsessive complaints, can be found in any psychiatric clinic. And there are countless hundreds of similar ones, thwarted, discouraged, determined to do great deeds, roaming the streets of any American city at this very moment. No, the Nazi leaders were not spectacular types, not personalities such as appear only once in a century. They simply had three quite unremarkable characteristics in common—and the opportunity to seize power. These three characteristics were: overweening ambition, low ethical standards, and a strongly developed nationalism which justified anything done in the name of Germandom.[12]

Hitler aside, arguably the most morally corrupt Nazi for the duration of the twelve-year Reich was SS chief Heinrich Himmler, as responsible as anyone for the horrors of the concentration and extermination camps and for the organization and implementation of the Final Solution. Yet even in the case of Himmler we can see how he might have gone down a different moral path, under dissimilar conditions. The renowned Holocaust historian and Himmler biographer Richard Breitman concluded that "The mass murders, the brutality, the sadism—

those were not what was unique about the Nazis. The brutal murder of whole populations, including children, has been with us since the beginning of recorded history and most probably before that." In fact, Breitman believes, if we cannot explain Himmler then we cannot explain most of human history. "We can put ourselves in the shoes of the perpetrators, as well as the shoes of the victims, because we all have in ourselves the potential for extreme good and extreme evil—at least, what we call good and evil. The real horror of Himmler is not that he was unusual or unique but that he was in many ways quite ordinary, and that he could have lived out his life as a chicken farmer, a good neighbor with perhaps somewhat antiquated ideas about people."[13]

From an evolutionary perspective this makes sense. Individuals in our ancestral environment needed to be both cooperative and competitive, depending on the context and desired outcomes. Cooperation may lead to more successful hunts, food sharing, and group protection from predators and enemies. Competition may lead to more resources for oneself and family, and protection from other competitive individuals who are less inclined to cooperate, especially those from other groups. Social psychologists have adequately demonstrated that moral behavior is tractable and that there is a range of potential for the expression of moral or immoral behavior.[14] We evolved to be moral, but have the capacity to be immoral some of the time in some circumstances with some people. Which direction any one of us takes in any given situation will depend on a complex array of variables.

An asymmetry in our moral observations about what people are really like comes from the fact that we have a tendency to focus on extreme acts of immorality and ignore the fact that most of the time, most people are gracious, considerate, and benevolent. For every act of violence or deception that appears on the nightly news, there are 10,000 acts of kindness that go largely unnoticed. In fact, violence and deception make the news precisely because they are so unusual in our daily experience. In the U.S. population of 280 million people, acts of cruelty happen daily—the law of large numbers says, in fact, that million-to-one odds happen 280 times a day. But how many of us have seen even one murder, carjacking, or kidnapping? Far more of us have probably been victims of some sort of scam or another, but this is because truth telling is the basis of almost all human interactions. The con artist can only be successful because most of us are honest most of the time. And once burned, twice shy. We learn from our mistakes. Shading and nuance, as we have seen in fuzzy logic, are at the foundation of the study of human behavior.

Explaining Evil

In the 1961 Israeli trial of Adolf Eichmann, one of the chief orchestrators of the Nazi "Final Solution to the Jewish Question," Hannah Arendt, covering the trial for the *New Yorker,* penned a phrase that has become infamous in the lexicon of social commentary: "the banality of evil." Expecting to see the raw viciousness of evil in the face of Eichmann—seated in a bulletproof glass box like a caged predatory beast—Arendt instead gazed upon a sad and pathetic-looking man who recounted in cold language and with dry statistics the collection, transportation, selection, and extermination of millions of human beings. Most surprising of all, Eichmann appeared to be a relatively normal human being—not a monster, not mentally deranged, not so different from many paper-pushing bureaucrats who go about their daily tasks like automata.

Indeed, a glance at Eichmann's working life shows a person who could share a smoke and a brandy with colleagues after a hard day at the office. Consider Eichmann's description at the end of the infamous Wannsee Conference held on January 20, 1942, to plan for the "Endlösung der Judenfrage":

> I remember that at the end of this Wannsee conference, Heydrich, Müller, and my humble self settled down comfortably by the fireplace, and that then for the first time I saw Heydrich smoke a cigar or a cigarette, and I was thinking: today Heydrich is smoking, something I have not seen before. And he drinks cognac—since I had not seen Heydrich take any alcoholic drink in years. After this Wannsee Conference we were sitting together peacefully, and not in order to talk shop, but in order to relax after the long hours of strain.[15]

What "*evil*" describes here is the banality of Heydrich and Eichmann's bureaucratic duties, which included the processing not just of paper but also of people. This is what I call the "evil of banality."[16]

Since 1945, in fact, the ultimate test of any moral theory is Hitler and the Holocaust. If ever there were an embodiment of evil behavior, it is surely Adolf Hitler, and if ever there were an act that should be labeled evil, it is surely the Holocaust. The images in figure 9 of Adolf Hitler alongside his SS henchman Heinrich Himmler, and the photograph of burning bodies in an open pit outside Crematoria 5 at Auschwitz-Birkenau, surely represent nothing if not an evil act.

Yet even here I would go so far as to say that calling Adolf Hitler evil moves us no closer to an understanding of the causes of what he did. What he did may be worse than almost anything anyone ever did to anyone else in history, but it is all still within the realm of human possibilities. The Holocaust may be the supreme act of inhumanity (indeed, the Nuremberg Trials established the legal precedence of convicting individuals for their acts of inhumanity), but we must always keep in mind that these inhuman acts were committed by humans, inhuman acts within our behavioral repertoire. Explaining the Holocaust, in fact, is intimately linked with explaining Hitler, both of which have become something of a scholarly and publishing industry. "The shapes we project onto the inky Rorschach of Hitler's psyche are often cultural self-portraits in the negative. What we talk about when we talk about Hitler is also who we are and who we are not," author Ron Rosenbaum writes.[17]

The explanations for Hitler, and by inference for the Holocaust (as in Milton Himmelfarb's catchy idiom, "No Hitler, No Holocaust"), have ranged from the ridiculous (Hitler's grandfather was Jewish) and the absurd (the "one ball" theory that Hitler had only one testicle) to the metaphysical (Hitler was evil). Some insist the explanation has been found (John Lukacs places the crystallization of Hitler's anti-Semitic personality as early as 1919), that it can be but has not yet been found (Yehuda Bauer: "Hitler is explicable in principle, but that does not mean that he has been explained"), that it cannot be found (Emil Fackenheim: "The closer one gets to explicability the more one realizes nothing can make Hitler explicable"), or that it should not be found (Claude Lanzmann: "There is even a book written . . . about Hitler's childhood, an attempt at explanation which is for me obscenity as such"). The Hitler of the Holocaust ranges wildly between intentional and functional evil. Lucy Dawidowicz's Hitler is a sole conductor who orchestrated the Holocaust with evil intent, deciding on his war against the Jews as early as November 1918 while still in the military hospital recovering from a gas attack. By contrast, Christopher Browning's

Figure 9. Hitler, Himmler, and the Holocaust as the Embodiment of Pure Evil

(*top*): Heinrich Himmler and Adolf Hitler, architects of the Final Solution. (Photograph by Estelle Bechhoefer. Courtesy of U.S. Holocaust Memorial Museum)

(*bottom*): A secret photograph of the burning of bodies in an open pit after gassing. When the crematoria were not working, or there were too many bodies for the capacity of the crematoria, the Nazis resorted to burning bodies in large communal ditches. (Courtesy of Yad Vashem, Jerusalem, Israel)

Hitler stumbles his way hesitatingly into the Holocaust, with "a sense that in the end he was scared of what he was doing. Now I interpret that as he didn't think it was wrong, but he was aware that he was now doing something that had never been done before."[18]

The problem scholars and historians have had in explaining Hitler and the Holocaust is the same one that plagues explanations of evil—that is the myth of pure evil. As Rosenbaum opines: "The search for Hitler has apprehended not one coherent, consensus image of Hitler but rather many different Hitlers, competing Hitlers, conflicting embodiments of competing visions, Hitlers who might not recognize each other well enough to say 'Heil' if they came face to face in Hell."[19] If Hitler can escape explanation in this sense, can the Holocaust? What about evil itself? We agree on the basic facts about the Holocaust, but interpretations about why it happened and what it means quickly become entangled in contradictory premises about human history and human nature. For Claude Lanzmann, the Holocaust "is a product of the whole story of the Western world since the very beginning."[20] But what does this tell us? If everything is the cause, then nothing is the cause.

These are not explanations as such. They are more like opinion editorials. A scientific approach to explaining evil can be found in social psychologist Roy Baumeister's thoroughly researched treatise on evil. Baumeister demonstrates that although for most people killing one human being is repulsive, killing millions can become routine: "The essential shock of banality is the disproportion between the person and the crime. The mind reels with the enormity of what this person has done, and so the mind expects to reel with the force of the perpetrator's presence and personality. When it does not, it is surprised."[21] The explanation for the surprise can be found by contrasting the victim's perspective with that of the perpetrator. For example, Maximillian Grabner, head of the Political Department at Auschwitz and associate of the camp commandant Rudolph Höss, explained the crime of the Holocaust from the perpetrator's perspective: "I only took part in this crime because there was nothing I could do to change anything. The blame for this crime lay with National Socialism. I myself was never a National Socialist. Nevertheless, I still had to join the party. . . . I only took part in the murder . . . out of consideration for my family. I was never an anti-Semite and would still claim today that every person has the right to live."[22] This is the evil of banality in its purest state.

In taking a broader perspective on the perpetrator's evil, Baumeister targets specific bad acts for scientific examination, including wife beatings, gang violence, drive-by shootings, rape, and other examples of what he calls the "breakdown of self-control." In an interesting twist on how most of us think about evil and violence, Baumeister suggests that "you do not have to give people reasons to be violent, because they already have plenty of reasons. All you have to do is take away their reasons to restrain themselves."[23] Most of us restrain ourselves most of the time, but there are circumstances when any of us has the potential to express extreme anger and violence. Are we all, then, evil—or at least potentially evil? No, not quite. We all have the *potential* to behave in ways that others might consider to be bad, cruel, mean, or violent. Baumeister makes this point in exploring seven individual myths about evil:

1. Evil is always intentional (from the victims' perspective; perpetrators always have a justification).
2. Evil is motivated by pleasure.
3. The victim of evil is innocent and good.
4. Evil is conducted by people completely different from us, wholly other.
5. Evil is the original sin, built into our natures.
6. Evil is the opposite not only of good but of order, peace, and stability.
7. Evil people are selfish egotists driven to improve their self-esteem by evil acts.[24]

It is important to note that in no way does debunking the myth of pure evil ignore the fact that human behaviors range broadly, or that in some cases a person may have serious mental problems or fall well away from the mean of normal human behavior toward genuinely wicked or sadistic actions. But while some may call Adolf Hitler a madman or a psychopath, I doubt that anyone would allow him to plea "not guilty by reason of insanity" for his crimes against humanity.

A deeper problem caused by the myth of pure evil, says Baumeister, is that it "conceals the reciprocal causality of violence." That is, as in divorces and most other human interactions involving conflict and resolution, there are two sides to almost every story. It turns out, for example, that research on perpetrators shows that they have, in their minds anyway, perfectly legitimate reasons for the violence. Ironically, Baumeister concludes, the myth of evil itself may lead to greater violence: "The myth encourages people to believe that they are good and will remain good no matter what, even if they perpetrate severe harm on their opponents. Thus, the myth of pure evil confers a kind of moral immunity on people who believe in it. . . . Belief in the myth is itself one recipe for evil, because it allows people to justify violent and oppressive actions. It allows evil to masquerade as good."[25]

September 11, 2001, comes to mind here. United States President George W. Bush described what happened that day as an act of pure evil. Yet millions of people around the world celebrate that day as a triumphant victory over what they perceive to be an evil American culture. What we are witnessing here is not a conceptual difference in understanding the true nature of evil. Nor is it simply a matter of who is in the right. It is, at least on one important level, a difference of perspective. To achieve true understanding and enlightenment it might help to understand what the other side was thinking. In a less emotionally charged example, if you lived in seventeenth-century Europe and you really believed that torturing religious heretics and burning women as witches would save their souls and restore peace to your community, then from that perspective the Spanish Inquisition and the European witch craze were supreme acts of morality. Similarly—and it seems almost blasphemous to suggest it—if you lived in the twenty-first century Muslim Middle East and you truly believed that killing American citizens was God's will to save your people and restore peace to your community, then from that perspective what could serve as a more visually striking statement than bringing down the twin symbols of your enemy? If there is a moral module in the brain (and I suspect there is something that at least corresponds to the concept of such a module in the brain, even if it is splayed out over a large portion of the cortex or consists of lots of smaller modules interconnected), then I have little doubt that Osama bin Laden and Muhammad Ata's moral modules were fully lit up on September 11.

This is not to argue that morality is reduced to one's perspective, or that events like the Holocaust do not represent an act of evil (in its adjectival form). But if we are to understand why the Holocaust happened, we must scientifically investigate the reasons behind such acts.

Fuzzy Logic, Fuzzy Evil

If we are not going to talk about evil as a metaphysical entity, then how shall we talk about it? One answer is to study evil as a scientist would, beginning with proper descriptive terms and employing fuzzy logic.[26] In fuzzy logic, shades of gray rule the universe, despite our heroic efforts over at least the past two and a half thousand years to dichotomize the world into Platonic categories. Aristotle said *A is A*, and that binary logic dictates a single overarching law: *A or not-A*. Either something is A or it is not A. It cannot be both. The sky is blue or it is not blue. It cannot be both blue and not blue. But what color is the sky at sunrise and sunset? What color is it overhead versus on the horizon? What color is it at 150,000 feet? And how is blue defined? What shade of blue was chosen as the defining essence of "blueness"? In point of fact, in the real world something can be A and not A. The sky can be blue and not blue. Thinking like a scientist, in statistical terms, we can assign a probability to the blueness of the sky. The sky is fuzzy blue. Directly overhead we might call it .9 blue. At a forty-five-degree angle from directly overhead we might assign it a fuzziness of .7 blue. On the horizon it might be a .5 blue.[27]

Fuzzy logic also allows for subtle nuances and shades of gray in the real-life complexities of ethical dilemmas.[28] The social psychologist Carol Tavris provides an apt example of fuzzy thinking in the moral realm in a 1998 *Los Angeles Times* op-ed on the Clinton morality debacle, with a title question that answers itself—"All Bad or All Good? Neither." Tavris explains that the scientific evidence points to the fact that "people's reasoning about moral dilemmas, like their moral behavior itself, is specific to given situations."[29] The false choice of either all bad or all good does not depict the subtleties and nuances of human behavior.

Consider Lawrence Kohlberg's famous stage theory of moral reasoning, for example, which suggests that, as people grow up, they pass through largely fixed rungs on a moral ladder:

1. Obedience and Punishment
2. Individualism, Instrumentalism, and Exchange
3. Good boy/girl
4. Law and Order

5. Social Contract
6. Principled Conscience[30]

Morality, says Kohlberg, develops within an individual, beginning with parental fear of punishment, moving to selfish hedonism, changing to conformity and loyalty to peers, developing into social law and order, rising to social contract reasoning, and finally reaching the highest rung of Gandhiesque moral principles. Research subsequent to Kohlberg's shows, however, that these stages are not as fixed and universal as they once appeared. One anthropologist, for example, found that these stages do not always apply to people in non-Western cultures. Moral development varies widely across the globe.[31] Two extensive studies on Kohlberg's theory by psychologists of religion found that specific cultural and religious values held by an individual influence where he or she may fall in the stage sequence.[32] In other words, the stages are not fixed developmental sequences so much as they are values that are context dependent. One study discovered that people make a distinction between moral and religious values in the various stages (since not all values in religion are related to morality), and another study reported that when the individual's religion emphasized principle-based moral decisions, they were more inclined to reach the "highest" rung of the moral ladder.[33] One psychologist even found a slight negative correlation between Kohlberg's stages and both intrinsic and extrinsic religiosity, meaning that whether one is motivated to be moral by intrinsic principles or by extrinsic rewards of one's religion depends on the context and circumstances of the moral issue; the stages of moral development were irrelevant.[34] And social psychologist Carol Gilligan noted that women's moral development differs from that of men. Women tend to emphasize care and responsibility in moral choices, whereas Kohlberg emphasized male "justice orientation" in his research.[35]

Humans can be morally principled in one circumstance, hedonistic in another, fear punishment in one context, exert our loyalty to friends in a different context. Referencing Clinton but generalizing to us all, Carol Tavris concludes that "the assumption that a moral failing in one domain reveals something profoundly important about a person's entire character, or predicts his or her behavior in other situations, is wrong."[36] One bad act does not an immoral person make. Perhaps this is what Jesus meant when he defied the Old Testament law that required the death penalty for adultery, by challenging a woman's accusers: "He that is without sin among you, let him first cast a stone at her." This may also represent the fundamental difference between Old Testament and New Testament morality: inflexible moral principles versus contextual moral guidelines—a stricter, draconian God versus a kinder, gentler God. As we shall see in the second half of this book, a fuzzy provisional moral system is another step in the contextualization of moral rules and behavior, where moral principles are ranked in terms of their fuzzy values, which can change under changing circumstances, yet still retain their core meaning. A case study on how fuzzy logic can be applied to the study of the origins of our moral and immoral nature can be found in the Yąnomamö peoples of Amazonia, variously described by their ethnographers as either erotic or fierce.

The Erotic-Fierce People: A Case Study in Fuzzy Logic

On July 7, 1947, a spacecraft crash-landed near Roswell, New Mexico. Aboard was an alien anthropologist sent here from an advanced civilization to study the newly emerging intelligence that calls itself "wise man." Whether or not that self-assessment is accurate was of great interest to the Galactic Federation, for this species had just achieved mastery of atomic fission, and thus could be a potential threat to the galactic peace. The Galactic Anthropological Association wanted to know if this formerly primitive people was basically "erotic" and thus there would be no need to worry about them, or if they were inherently "fierce," in which case further monitoring and missionary reeducation might be required.

From the subject's (our) perspective the fundamental flaw in the inquiry is that humans are not so easily pigeonholed into such clear-cut categories as "fierce" or "erotic." We are both (and a lot more), the nature and intensity of our behavior being dependent upon a host of biological, social, and historical variables. If an alien anthropologist had crashed in Europe in 1943, our intrepid observer would surely have called us the "fierce" people. But if, say, the landing cite was Woodstock, New York, or San Francisco, California, in 1968, ET would likely have labeled us the "erotic" people. Local and historical context matters, and any description based on an isolated context is grossly oversimplified and hopelessly incomplete. This is why we need anthropologists, the scientific observers of our planet's rich diversity of people and cultures.

There is a maxim anthropologists often cite about the geopolitics of diplomacy and warfare among indigenous peoples: *The enemy of my enemy is my friend.* In reality, of course, the maxim applies to virtually all groups, from bands to tribes to chiefdoms to states—recall the temporary friendship between the United States and the USSR from 1941 to 1945 that promptly dissolved into the cold war upon Germany's defeat. I thought of this maxim when I interviewed journalist Patrick Tierney about his book *Darkness in El Dorado: How Scientists and Journalists Devastated the Amazon.* Tierney had recently been pummeled by a panel of experts in front of a thousand scientists gathered at the annual meeting of the American Anthropological Association. At the heart of *Darkness in El Dorado* is the question of whether humans are by nature erotic, fierce, or both. Among the many scientists whom Tierney attacks is the anthropologist Napoleon Chagnon, whose study of the Yąnomamö people of Amazonia is arguably the most famous ethnography since Margaret Mead's Samoan classic, *Coming of Age in Samoa.* Since Chagnon has a reputation as an intellectual pugilist who had accumulated a score of enemies over the decades, I fully expected that the scientists would rally around Tierney in a provisional alliance. With a couple of minor exceptions, however, there was almost universal condemnation of the book. A British science writer who witnessed the verbal thumping said: "If I had taken such a beating as Tierney I would have crawled out of the room and cut my throat."[37]

Humans are storytelling animals. Thus, following Darwin's Dictum that "all observation must be for or against some view if it is to be of any service," we can recognize that Tierney's story argues against the view that he believes has been put forth by certain anthropologists about the Yąnomamö and, by implication, about all humanity. Chagnon, he points out, subtitled his best-selling ethnographic monograph on the Yąnomamö *The Fierce People.* Tierney spares no ink in presenting a picture of Chagnon as a fierce anthropologist who sees in the Yąnomamö nothing more than a reflection of himself (figure 10). Chagnon's sociobiological theories of the most violent and aggressive males winning the most copulations and thus passing on their genes for "fierceness," says Tierney, is a Rorschachian window into Chagnon's own libidinous impulses.

Tierney's strongest case may be against the French anthropologist

Figure 10. Napoleon Chagnon: The Man Who Called the Yąnomamö "Fierce"

Anthropologist Napoleon Chagnon accompanies two Yąnomamö men on a 1995 field study. (Courtesy of Napoleon Chagnon)

Jacques Lizot, who calls the Yąnomamö "the erotic people."[38] Lizot, Tierney claims, engaged in homosexual activities for years with so many Yąnomamö young men, and so frequently, that he became known in Yąnomamöspeak as *bosinawarewa,* which translates politely as "ass handler" and not so politely as "anus devourer."[39] In response to these claims not only did Lizot not deny the basic charges (that also included exchanging goods for sex), but he admitted to *Time* magazine: "I am a homosexual, but my house is not a brothel. I gave gifts because it is part of the Yąnomamö culture. I was single. Is it forbidden to have sexual relations with consenting adults?"[40] No, but Tierney disputes both the age of Lizot's partners and whether or not they consented, and suggests that even if it were both legal and moral this is hardly the standard of objectivity one might have hoped for in scientific research, and that it is Lizot who best deserves the descriptive adjective "erotic."

So which is it? Are the Yąnomamö fierce or erotic, or are these descriptive terms for their anthropological observers? Carping over minutiae in Chagnon's research methods and ethics has dogged him throughout his career, but it is secondary to a deeper, underlying issue in the anthropology wars. What Chagnon is really being accused of is biological determinism. To postmodernists and cultural determinists, calling the Yąnomamö "fierce" and explaining their fierceness through a Darwinian model of competition and sexual selection indicts all of humanity as innately evil and condemns us to a future of ineradicable violence, rape, and war. Are we really this bad? Are the Yąnomamö?[41]

The Yąnomamö skirmish is only the latest in a long line of battles that have erupted in the century-long anthropology wars. The reason such controversies draw so much public attention is that what's at stake is nothing less than the true nature of human nature and how that nature can most profitably be studied.

Anthropologist Derek Freeman's lifelong battle with the legacy of Margaret Mead, for example, was not really about whether Samoan girls are promiscuous or prudish. Mead's philosophy, inherited from her mentor, Franz Boas, that human nature is primarily shaped by the environment, was supported by her "discovery" that Samoan girls are promiscuous, whereas in other cultures promiscuity is taboo. Freeman argues that Mead was duped by a couple of Samoan hoaxers, and had she been more rigorous and quantitative in her research she would have discovered this fact before going to press with what became the all-time anthropological best-seller—*Coming of Age in Samoa*. According to Freeman, Mead's ideology trumped her science, and anthropology lost.[42] His 1983 book, *Margaret Mead and Samoa: The Making and Unmaking of an Anthropological Myth*, triggered a paroxysm within the anthropological community, as he recalled:

> The 1983 annual meeting of the American Anthropological Association in Chicago included a special session dedicated to my book, but strangely I was not invited to defend myself. Now I know why. One eyewitness described it as "a sort of grotesque feeding frenzy," while another told me "I felt I was in a room with people ready to lynch you." At the annual business meeting later that day, a motion denouncing my refutation as "unscientific" was moved, put to the vote, and passed. In the December, 1983 issue of the *American Anthropologist*, no fewer than five different critiques of

> my book were published, but I was denied the usual right of simultaneous reply. My rejoinder, when it did appear, some six months later, was limited to one tenth of the space that had been given to my critics.[43]

Anthropology is a sublime science because it deals with such profoundly deep questions as the nature of human nature. But to even ask such questions as "Are we by nature good or evil?" overlooks the complexity of human affairs. The failure of *Darkness in El Dorado* has less to do with getting the story straight and more to do with a fundamental misunderstanding of the plasticity and diversity of human behavior. Tellingly, the fourth edition of Chagnon's classic work *Yąnomamö* dropped the subtitle *The Fierce People*. Had Chagnon determined that the Yąnomamö were not "the fierce people" after all? No. He realized that too many people were unable to move past the moniker to grasp the complex and subtle variations contained in all human populations, and he became concerned that they "might get the impression that being 'fierce' is incompatible with having other sentiments or personal characteristics like compassion, fairness, valor, etc."[44] In fact, the Yąnomamö call themselves *waiteri* (fierce), and Chagnon's attribution of them as such was merely attempting "to represent valor, honor, and independence" that the Yąnomamö saw in themselves. As he notes in his opening chapter, the Yąnomamö "are simultaneously peacemakers and valiant warriors." Like all people, the Yąnomamö have a deep repertoire of responses for varying social interactions and differing contexts, even those that are potentially violent: "They have a series of graded forms of violence that ranges from chest-pounding and club-fighting duels to out-and-out shooting to kill. This gives them a good deal of flexibility in settling disputes without immediate resort to lethal violence."[45]

Chagnon has often been accused of using the Yąnomamö to support a sociobiological model of an aggressive human nature. Even here, returning to the primary sources in question shows that Chagnon's deductions from the data are not so crude, as when he notes that the Yąnomamö's northern neighbors, the Ye'Kwana Indians—in contrast to the Yąnomamö's initial reaction to him—"were very pleasant and charming, all of them anxious to help me and honor bound to show any visitor the numerous courtesies of their system of etiquette," and therefore that it "remains true that there are enormous differences between whole peoples."[46] Even on the final page of his chapter on Yąnomamö warfare, Chagnon inquires about "the likelihood that people, throughout history, have based their political relationships with other groups on predatory versus religious or altruistic strategies and the cost-benefit dimensions of what the response should be if they do one or the other." He concludes: "We have the evolved capacity to adopt either strategy."[47]

As an example of this moral plasticity, Chagnon summarized the data from his now-famous *Science* article revealing the positive correlation between levels of violence among Yąnomamö men and their corresponding number of wives and offspring. "Here are the 'Satanic Verses' that I committed in anthropology," Chagnon joked, as he reviewed his data:

> I didn't intend for this correlation to pop out, but when I discovered it, it did not surprise me. If you take men who are in the same age category and divide them by those who have killed other men (unokais) and those who have not killed other men (non-unokais), in every age category unokais had more offspring. In fact, unokais averaged 4.91 children versus 1.59 for non-unokais. The reason is clear in the data on the number of wives: unokais averaged 1.63 wives versus 0.63 for non-unokais. This was an unacceptable finding for those who hold the ideal view of the Noble savage. 'Here's Chagnon saying that war has something good in it.' I never said any such thing. I merely pointed out that in the Yąnomamö society, just like in our own and other societies, people who are successful and good warriors, who defend the folks back home, are showered with praise and rewards. In our own culture, for example, draft dodgers are considered a shame. Being a successful warrior has social rewards in all cultures. The Yąnomamö warriors do not get medals and media. They get more wives.[48]

Despite the mountains of data Chagnon has accumulated on Yąnomamö aggression, he is careful to note the many other behaviors and emotions expressed by the Yąnomamö: "When I called the Yąnomamö the 'fierce people,' I did not mean they were fierce all the time. Their family life is very tranquil. Even though they have high mortality rates due to violence and aggression and competition is very high, they are not sweating fiercely, eating fiercely, belching fiercely, etc. They do kiss their kids and are quite pleasant people."[49]

In contrast to Chagnon's depiction of the Yąnomamö as "fierce," many commentators (such as Tierney) hold up anthropologist Ken Good's book, *Into the Heart: One Man's Pursuit of Love and Knowledge Among the Yąnomami,* to argue the case that they are "erotic." *Into the Heart* is a page-turner because the very features of Yąnomamö culture that Chagnon's critics claim he overemphasizes are, in fact, present in spades in every chapter of Good's gripping tale. As Chagnon's graduate student, Good immersed himself in Yąnomamöland, but in time found himself falling in love with a beautiful young Yąnomamö girl named Yarima.[50] As the years passed and he was occasionally forced to leave Yąnomamöland (to renew his permit or attend conferences or work on his doctoral dissertation), he became emotionally distraught over leaving Yarima alone. Why? When Yarima came of age (defined in her culture as first menses), she and Good began living together and consummated their "marriage" (Yąnomamö do not have a marriage ceremony per se; instead a couple, usually the man, declares that they are married and the two begin living together). Good's problem was that he was all too aware of the very human nature of Yąnomamö men. "They will grab a woman while she is out gathering and rape her. They don't consider it a crime or a horrendously antisocial thing to do. It is simply what happens. It's standard behavior. In such a small, enclosed community this (together with affairs) is the only way unmarried men have of getting sex."[51]

Good's worries were justified and the universal emotion of jealousy was no less intense in this highly civilized, educated man than it was in any of the people he was studying to earn his Ph.D. In short, Good was on an emotional roller coaster from which he could not extricate himself.

> I felt the tension, and I tried to deal with it. I wanted to think that Yarima would be faithful to me. But I knew the limits of any woman's faithfulness here. Fidelity in Yąnomami land is not considered a standard of any sort, let alone a moral principle. Here it is every man for himself. Stealing, rape, even killing—these acts aren't measured by some moral standard. They aren't thought of in terms of proper or improper social behavior. Here everyone does what he can and everyone defends his own rights. A man gets up and screams and berates someone for stealing plantains from his section of the garden, then he'll go and do exactly the same thing. I protect myself, you protect yourself. You try something and I catch you, I'll stop you.[52]

Many antisocial behaviors, such as theft, are kept at a minimum through such social constraints as shunning or such personal constraints as fear of violent retaliation. But, as Good explains, sex is a different story because "The sex drive demands an outlet, especially with the young men. It cannot be stopped. Thus the personal and social constraints have less force; they're more readily disregarded." As a consequence women are often raped, an act they themselves must keep secret for fear of retaliation from their husbands against them. If the wife is young and childless, "the husband might find he cannot tolerate it; he might lose control utterly and embark on violent action. He badly wants to at least get his family started himself, rather than have someone else make her pregnant."[53]

Chagnon's ethnography of the Yą̈nomamö people is a case study in the application of fuzzy logic to human nature. In *Yąnomamö* Chagnon notes that the variation in violence observed by different scientists can be accounted for by a concatenation of intervening variables, such as geography, ecology, population size, resources, and especially the contingent history of each group, where "the lesson is that past events and history must be understood to comprehend the current observable patterns. As the Roman poet Lucretius mused, nothing yet from nothing ever came."[54]

Many other anthropologists who have studied the Yąnomamö corroborate Chagnon's data and interpretations. Even at their "fiercest," however, the Yąnomamö are not so different from many other peoples around the globe (recall Captain Bligh's numerous violent encounters with Polynesians and Captain Cook's murder at the hands of Hawaiian natives), even when studied by tender-minded, nonfierce scientists. Evolutionary biologist Jared Diamond, for example, told me that he found the role of warfare among the peoples of New Guinea that he has studied over the past thirty years quite similar to Chagnon's depiction of the role of warfare among the Yąnomamö.[55]

Finally, if the last five thousand years of recorded human history is any measure of a species' "fierceness," the Yąnomamö have got nothing on either Western or Eastern "civilization," whose record includes the murder of hundreds of millions of people. *Homo sapiens* in general, like the Yąnomamö in particular, are the erotic-fierce people, making love and war far too frequently for our own good as both overpopulation and war threaten our very existence.

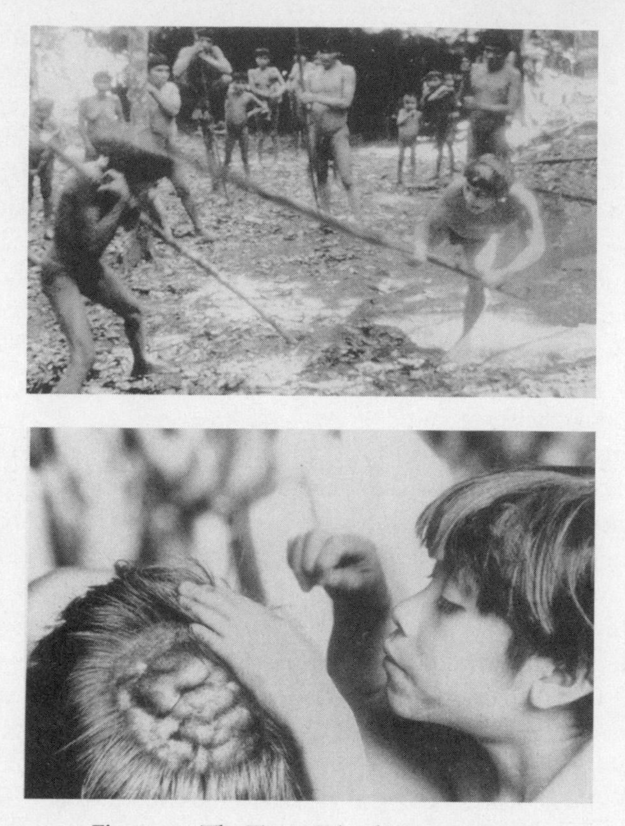

Figure 11. The Fierce Side of Human Nature

(*top*) Despite Patrick Tierney's claim that Chagnon has exaggerated the level of aggression and rape among the Yąnomamö, Kenneth Good documented both, here showing "two men duel over the infidelity of one of their wives." (Courtesy of Kenneth Good) (*bottom*) Many Yąnomamö men have deep scars on their heads from such battles. (Courtesy of Napoleon Chagnon)

The Myth of Pure Good: Noble Savages and Beautiful People

In 1670, the British poet John Dryden penned this expression of humans in a state of nature: "I am as free as Nature first made man / When wild in woods the noble savage ran." A century later, in 1755, the French philosopher Jean-Jacques Rousseau canonized the noble savage into Western culture by proclaiming, "nothing can be more gen-

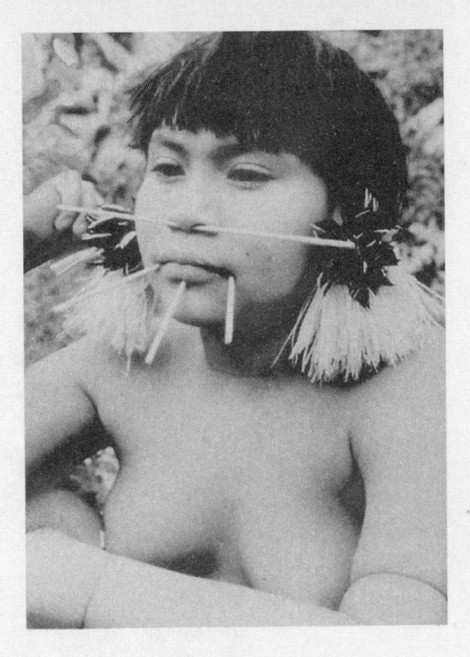

Figure 12. The Erotic Side of Human Nature

Into the Heart is a moving love story between anthropologist Kenneth Good and a young Yąnomamö woman named Yarima. They eventually married, had children, and returned to the United States. Yarima grew bored with American life and returned to the more stimulating environment of Amazonia. (Courtesy of Kenneth Good)

tle than him in his primitive state, when placed by nature at an equal distance from the stupidity of brutes and the pernicious good sense of civilized man." From the Disneyfication of Pocahontas to Kevin Costner's ecopacifist Native Americans in *Dances with Wolves,* and from postmodern accusations of corrupting modernity to modern anthropological theories that indigenous people's wars are just ritualized games, the noble savage remains one of the last epic creation myths of our time. Within this myth lies the antithesis of the myth of pure evil, and that is the myth of pure good. The latter is just as detrimental toward a deeper understanding of human moral nature as is the former. The evidence from all the human sciences overwhelmingly supports the view that humans are good and bad, cooperative and competitive, selfish and altruistic. The potential for the expression of both moral and immoral behavior is built into human nature. How, when, and where such behaviors are expressed depends on a host of variables. But the myth of the noble savage extends far beyond what Rousseau envisioned and is still embraced today by many scientists, academics, and social commentators in what I call the *Beautiful People Myth* (BPM). The BPM is the fable of pacifist and ecofriendly humans ruined only by

Figure 13. Splendor in Fierceness

Chagnon nuances the realities of Yąnomamö life, and of human life, in all its richness in his ethnography entitled simply *Yąnomamö*. Earlier editions included the subtitle *The Fierce People,* but readers missed the multilayered meaning of courage, valor, and compassion. However, warfare among the Yąnomamö is a reality, as it is for all of humanity. Here we see (*top*) Yąnomamö men dressed for ceremony and (*bottom*) dressed differently for war, without feathers and their faces painted black with masticated charcoal. (Courtesy of Napoleon Chagnon) the plight of modernity and the burden of Dead White European Males. I have characterized the myth in the following manner:

> Long, long ago, in a century far, far away, there lived beautiful people coexisting with nature in balanced ecoharmony, taking only what they needed, and giving back to Mother Earth what was left. Women and men lived in egalitarian accord and there were no wars and few conflicts. The people were happy, living long and prosperous lives. The men were handsome and muscular, well coordinated in their hunting expeditions as they successfully brought home the main meals for the family. The tanned, bare-breasted women carried a child in one arm and picked nuts and berries to supplement the hunt. Children frolicked in the nearby stream, dreaming of the day when they too would grow up to fulfill their destiny as beautiful people.
>
> But then came the evil empire—European White Males carrying the diseases of imperialism, industrialism, capitalism, scientism, and the other "isms" brought about by human greed, carelessness, and short-term thinking. The environment became exploited, the rivers soiled, the air polluted, and the beautiful people were driven from their land, forced to become slaves, or simply killed.
>
> This tragedy, however, can be reversed if we just go back to living off the land where people would grow just enough food for themselves and use only enough to survive. We would then all love one another, as well as our caretaker Mother Earth, just as they did long, long ago, in a century far, far away.

I have thoroughly deconstructed and debunked the Beautiful People Myth elsewhere, so I will not belabor the point here.[56] When it comes to how humans treat other humans and the environment, the Beautiful People have never existed except in myth. Humans are neither Beautiful People nor Ugly People, in the same way that we are neither moral nor immoral in some absolute categorical sense. Humans are only doing what any species does to survive; but we do it with a twist (and a vengeance)—instead of our environment shaping us through natural selection, we are shaping our environment through artificial selection. In a fascinating 1996 study, for example, University of Michigan ecologist Bobbi Low used the data from the Standard Cross-Cultural Sample to test the hypothesis that we can solve our ecological problems by returning to the mythological Beautiful People's attitudes of reverence for (rather than exploitation of) the natural world, and by opting fo[illegible] term group-oriented values (rather than short-term individual values).[57] Her analysis of 186 hunting-fishing-gathering (HFG) societies around the world showed that their use of the environment is driven by ecological constraints and not by attitudes, such as sacred prohibitions, and that their relatively low environmental impact is the result of low population density, inefficient technology, and the lack of profitable markets, not from conscious efforts at conservation. Low also showed that in 32 percent of HFG societies, not only were they not practicing conservation, environmental degradation was severe; again, it was limited only by the time and technology to finish the job of destruction and extinction.

Extending the analysis of the BPM to other areas of human culture, UCLA anthropologist Robert Edgerton surveyed the anthropological record and found clear evidence of drug addiction, abuse of women and children, bodily mutilation, economic exploitation of the group by political leaders, suicide, and mental illness in indigenous preindustrial peoples, groups not contaminated by Western values (allegedly the source of such "sick" behavior).[58]

Anthropologist Shepard Krech analyzed a number of Native American communities, such as the Hohokam of southern Arizona, and discovered that a large-scale irrigation program led to the salinization and exhaustion of the Gila and Salt River valleys, ultimately triggering the collapse of their society. Krech says that even the reverence for big game animals we have been led to believe was ingrained into the worldview of America's indigenous peoples is a myth. Many, if not most, Native Americans believed that common game animals such as elk, deer, caribou, beaver, and especially buffalo are replenished through divine physical reincarnation. Game populations bounced back after successful hunts not because Native Americans made it happen through ecological veneration, but because they believed the gods willed it.[59] Given the opportunity to overhunt big game animals, Native Americans were only too willing to do so.

One of the most poignant examples of this is the famous "Head-Smashed-In" buffalo kill site in southern Alberta, Canada. I had an opportunity to visit Head-Smashed-In (the name alone belies the Noble Indian myth). It is a most dramatic site. Standing on the edge of the cliff, one looks down upon a thirty-foot-thick deposit of buffalo bones that reflects five thousand years of Native American mass hunting. Looking back away from the cliff, one sees a vast and expansive V-shaped valley in which the hunters ambushed and drove their game for tens of miles. The terrain is on a slight decline toward the cliff, so these massive animals built up so much speed that upon reaching the cliff they were unable to stop themselves. They tumbled over, one after another, until there were so many carcasses that most were left unused. Buffalo populations were ultimately stable not because of a Native American conservation ethic, but because they simply did not have the numbers and technology to drive these big game animals into extinction. Other species were not so fortunate. The evidence is now overwhelming that woolly mammoths, giant mastodons, ground sloths, one-ton armadillo-like glyptodonts, bear-sized beavers, and beefy saber-toothed cats, not to mention American lions, cheetahs, camels, horses, and many other large mammals, all went extinct at the same time that Native Americans first populated the continent in the mass migration from Asia some 15,000 to 20,000 years ago. The best theory to date as to what happened to these mammals is that they were overhunted into extinction.[60]

The Ignoble Savage and the Nature of War

The evidence is overwhelming that violence, aggression, and warfare are part of the behavioral repertoire of most primate species. While most conflicts among monkeys end relatively peacefully, this is due primarily to the fact that they lack brute strength and deadly weapons. In their stead, screams, gestures, pushing, hitting, and biting result in struggles for and changes in social status and mate choices, but it is clear that the potential for deadly violence exists. Among the great apes it was long believed that "only man kills," but that is no longer the case. Murderous raids among chimpanzees have now been well documented, and they are not rare. While it would be inappropriate to compare the gang raids among chimpanzees to the wars of modern civilization (chimpanzee gangs mostly attack individuals or much smaller groups), the basic process of a gang of young and aggressive males fanning out into neighboring environments on a seek-and-destroy mission to gain resources and females is apparent in the species, genus, and family. As Jane Goodall famously observed: "If they [chimpanzees] had had firearms and had been taught to use them, I suspect they would have used them to kill."[61]

Even when anthropologists have admitted that there is evidence for prehistoric human warfare, they often portray it as rare, harmless, and little more than ritualized sport. Now even that noble image has taken a major hit from new data. For example, in his survey of and comparison between primitive and civilized societies, University of Illinois anthropologist Lawrence Keeley demonstrates that prehistoric war was, relative to population densities and fighting technologies, at least as frequent (as measured in years at war versus years at peace), as deadly (as measured by percentage of conflict deaths per population), and as ruthless (as measured by the killing and maiming of noncombatant women and children) as modern war.[62] At a bone bed site at Crow Creek in South Dakota, dated in the pre-Columbian fourteenth century, Keeley also found "the remains of nearly 500 men, women, and children. These victims had been scalped, mutilated, and left exposed for a few months to scavengers before being interred." Keeley also recounts the last moments of a young man who was shot in the back during a mass raid of a Neolithic village in Britain, during which he fell and crushed the infant he was carrying. In yet another example, seven thousand years ago in Talheim, Germany, a band of thirty-four adults and children were murdered by blows to the head and then tossed into an open pit, not so different from what the Nazis did many millennia later.[63] The Nazis had no monopoly on violence.

In *Constant Battles*, an exceptionally insightful study of this problem by Steven A. LeBlanc, the Harvard archaeologist quips, "anthropologists have searched for peaceful societies much like Diogenes looked for an honest man." That is, they are exceptionally rare. "In spite of the presumption that most societies were peaceful in the past, anthropologists have had a lot of trouble finding ethnographically known peaceful people. Despite all the effort that has been devoted to the search, the number of what can be considered classic cases of peaceful societies is quite small."[64] Consider the evidence from a 10,000-year-old Paleolithic site along the Nile River: "The graveyard held the remains of fifty-nine people, at least twenty-four of whom showed direct evidence of violent death, including stone points from arrows or spears within the body cavity, and many contained several points. There were six multiple burials, and almost all those individuals had points in them, indicating that the people in each mass grave were killed in a single event and then buried together."[65] LeBlanc presents evidence from a site in Utah that contains the remains of ninety-seven people killed violently: "six had stone spear heads in them . . . several breast bones shot through with arrows and many broken heads and arms. . . . Individuals of all ages and both sexes were killed, and individuals were shot with atlatl darts, stabbed, and bludgeoned, suggesting that fighting was at close quarters."[66]

LeBlanc's survey of our not-so-noble past reveals that even cannibalism, long thought to be a form of primitive urban legend and a myth to be debunked (noble savages would never eat each other, would they?), has now been supported by powerful physical evidence that includes broken and burned bones, bones with cut marks, bones broken open lengthwise to allow access to the marrow, and bones broken to fit inside cooking jars. Such evidence for prehistoric cannibalism has been uncovered in Mexico, Fiji, Spain, and other parts of Europe. The final (and gruesome) proof came with the discovery of the human muscle protein myoglobin in the fossilized human feces of a prehistoric Anasazi pueblo Indian.[67]

Savage yes. Noble no.

We are moral animals, yes, but we are also immoral animals, tragically but indubitably so. Figures 14 through 17 graphically depict this hard and factual reality of the human condition.

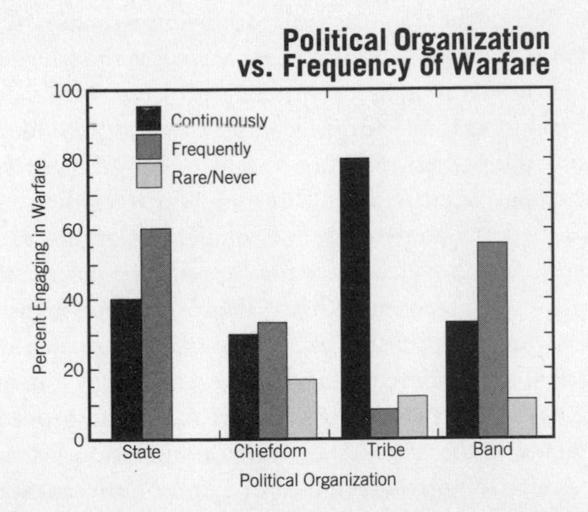

Figure 14. Political Organization and Frequency of Warfare

The level of warfare for ten organized states, six chiefdoms, twenty-five tribes, and nine bands. (Derived from Table 2.1 in Lawrence Keeley, *War Before Civilization*, 1996)

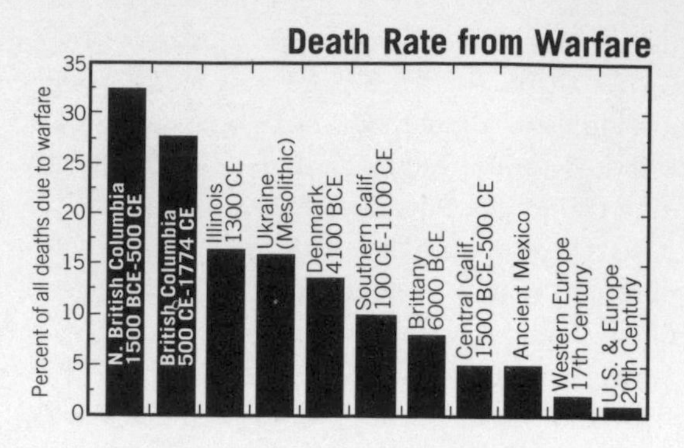

Figure 15. Death Rate from Warfare

Deaths due to warfare as a percent of all deaths recorded in representative societies. Ancient societies: Northern British Columbia, British Columbia, Southern California, Central California. Western Europe is for the seventeenth century; the United States and Europe are for the twentieth century. (Data from Table 6.2 in Lawrence Keeley, *War Before Civilization,* 1996)

Figure 14 presents Keeley's data showing that there is no obvious distinction between bands, tribes, chiefdoms, and states in terms of warfare frequency—humans typically and frequently solve social disputes with violence, prehistorically, historically, and today—and they do so regardless of the political structure. Figure 15 shows, counterintuitively, that if there are any historical trends it is that the death rate as a result of warfare is actually decreasing over time, with modern Western states representing the lowest death rate and premodern political organizations the highest. Figure 16 depicts the different sites within Native American Southwest cultures in which evidence of violent deaths, processed human remains, and mutilations of the dead occurred. These are not isolated events, since the sites studied include the remains of hundreds of murdered individuals. Figure 17 depicts the data from R. N. Holdaway's study of the extinction of New Zealand moa birds after the arrival of Polynesian Maoris, surely the perfect image of Beautiful People by anyone's standard in the West. When Europeans arrived in New Zealand in the 1800s, they found bones and eggshells of large extinct moa birds, an ostrichlike bird of a dozen different species ranging in size from three feet tall and 40 pounds to ten feet tall and 500 pounds. We now know, from preserved moa gizzards containing pollen and leaves of dozens of plant species and from archaeological digs of

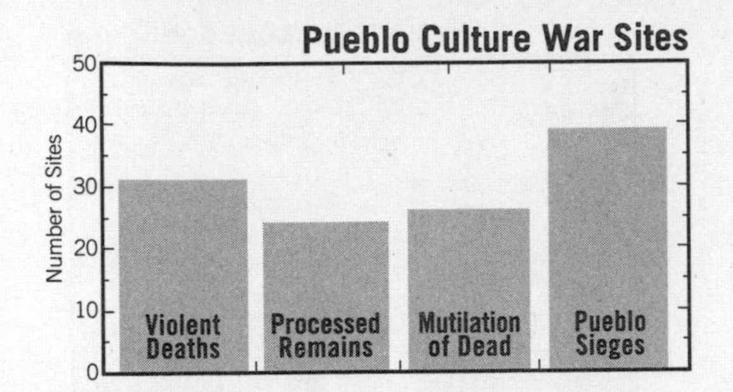

Figure 16. Pueblo Culture War Sites

The number of sites in the American Southwest showing excavated evidence of violent deaths, processed human remains, mutilations of the dead, and sieges. These sites include the remains of at least 128, 252, an unknown number, and 174 individuals, respectively. The "processed remains" represent individuals who were apparently butchered for consumption. (Data from Steven LeBlanc, *Prehistoric Warfare in the American Southwest,* 1999)

Polynesian trash heaps, that the Maoris committed a full-scale ecocide. Although some biologists have suggested a change in climate as the cause of the moa extinction, Jared Diamond makes the case that when the extinction occurred, New Zealand was enjoying the best climate it had had in a long time. Also, carbon-dated bird bones from Maori archaeological sites prove that all known moa species were still present in abundance when the Maoris arrived around 1000 C.E. But by 1200 C.E.—six centuries before the arrival of Europeans—they were all gone. The final piece of evidence of what happened to the moas came from archaeologists who uncovered Maori sites containing between 100,000 and 500,000 moa skeletons, ten times the number living at any one time. In other words, they had been slaughtering moas for many generations, until they were all gone in a mass extermination.[68]

There is nothing beautiful about the Beautiful People. Give them the plants, animals, and technologies—and the need through population pressures—to exploit their environment and they would do so; indeed, those that had that particular concatenation of elements did just that. In other words, centuries before and continents away from modern economies and technologies, and long before European White Males (dead or alive), humans consciously and systematically destroyed each other and their environments. The ignoble savage lies within.

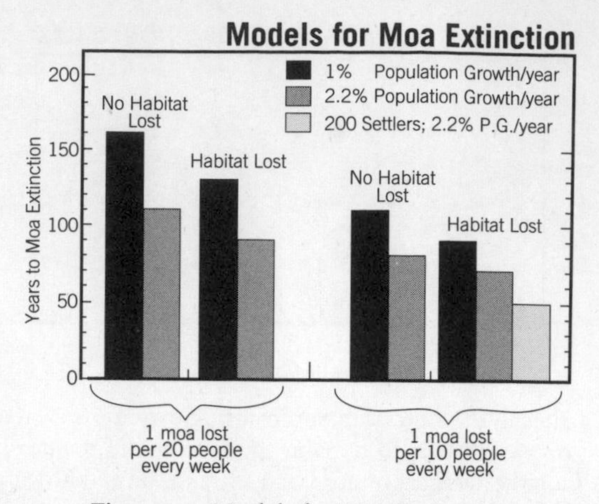

Figure 17. Models for Moa Extinction

The results represent modeling the time to extinction of the New Zealand moas after the arrival of the Maoris in the late thirteenth century. The model assumes 100 initial settlers, that no moa eggs were taken, and that no juvenile moas were killed. The model includes two moa consumption rates, two rates of human population growth, and whether moa habitat was destroyed. The light gray bar at the far right shows the results for the most likely of the tested conditions: 200 initial settlers, a 2.2 percent human population growth rate, and habitat destruction. (Data from R. N. Holdaway and C. Jacomb, "Rapid Extinction of the Moas" in *Science,* vol. 287, 2000)

In the Heart of Every Human

Now that we have dispensed with the myths of pure evil and pure good, with what are we left? What remains when we strip away the mythic fog that has for too long shrouded human nature is human behavior—the things people do. So, a final way to view the visage of humanity is to think about human behavior not as inherently good or evil, moral or immoral, but as actions that we like and actions that we do not like, as these actions may be provisionally defined and judged. That is, in most circumstances, for most people certain behaviors most of the time are considered moral or immoral, and we reward and punish those actions accordingly. Our political constitutions are formed according to our natural constitutions. As Preacher Casey tells Tom Joad in Steinbeck's *The Grapes of Wrath,* after he explains that he has given up holding revivals because of the obvious hypocrisy between the content of his preaching about fidelity and context of his own infidelities:

> I says, "Maybe it ain't a sin. Maybe it's just the way folks is." Well, I was layin' under a tree when I figured that out, and I went to sleep. And it come night, an' it was dark when I come to. They was a coyote squawkin' near by. Before I knowed it, I was sayin' out loud . . . "There ain't no sin and there ain't no virtue. There's just stuff people do. . . . And some of the things folks do is nice, and some ain't nice, but that's as far as any man got a right to say."[69]

This is not in any way to endorse a purely situational ethics or relative morality. The stuff some folks do really ain't nice in most circumstances, to most people, most of the time. But there is no such thing as pure sin or pure virtue, any more than there is pure evil or pure good. The purpose of this exercise in ethical debunking is to shift the focus from sin and virtue, evil and good as metaphysical Platonic essences to quantifiable human behaviors that can be scientifically studied, causally understood, and ultimately modified or dealt with according to the needs and dictates of society. Evil forces do not exist, but evil acts are an all-too-human expression. Walt Kelly's cartoon character Pogo put it simply: "We have met the enemy and he is us."[70]

Aleksandr Solzhenitsyn put it more elegantly in his analysis of the gulags of the Soviet Union, surely a den of evil if ever there was one: "If only there were evil people somewhere insidiously committing evil deeds, and it were necessary only to separate them from the rest of us and destroy them. But the line dividing good and evil cuts through the heart of every human being. And who is willing to destroy a piece of his own heart?"[71]

Surely none of us, as Aeschylus suggested in *Prometheus Bound*:

> *Prometheus, Prometheus,*
> *hanging upon Caucasus*
> *Look upon the visage*
> *Of yonder vulture:*
> *Is it not thy face,*
> *Prometheus?*[72]

But remember, it was Prometheus who brought us knowledge, and through knowledge comes power, including the power of cultural amity to override natural enmity. We may always live in a world with walls; in recent history, however, the stone and mortar walls enforced by men with guns are gradually being replaced by invisible boundaries enforced by social contracts; in the future even these invisible boundaries may be replaced by semipermeable lines of demarcation, kept open through negotiation and cooperative exchange. It is a visage worthy of humanity.

4

MASTER OF MY FATE: MAKING MORAL CHOICES IN A DETERMINED UNIVERSE

He that is good is free, though he be a slave; he that is evil is a slave, though he be a king.

—St. Augustine, *The City of God, IV,* A.D. 427

On March 30, 1981, as United States President Ronald Reagan emerged from the Washington Hilton Hotel, he was gunned down in his tracks by John W. Hinckley, Jr., an obsessive loner who, in the scuffle following the initial shots, also blew a hole in the head of Reagan's press secretary, James Brady, and wounded two others. Hinckley was immediately arrested. Remarkably, although Hinckley clearly fired the shots and never denied this fact, he pleaded not guilty. How can you not be guilty when your act is observed by dozens of eyewitnesses, filmed and seen by millions, and you admit you committed the crime?

Hinckley claimed that he was insane at the time of the assassination. His insanity? He was "crazy" about the movie star Jodie Foster, who he said obsessed him. His was the so-called insanity defense, known legally as NGRI, or "Not Guilty by Reason of Insanity." To determine whether Hinckley was insane—that he was not responsible for his actions and thus should be placed in a mental institution instead of jail—the court subjected him to an extensive psychological evaluation. Three government-appointed psychiatrists determined that he was sane at the time of the crime because of the considerable planning required to attempt a political assassination. But his defense-appointed psychiatrists diagnosed him with several severe mental disorders, including Schizophrenia Spectrum Disorder and Paradoxical Rage. Amazingly (although maybe we should not be amazed any longer), the jury agreed that Hinckley was not responsible for his actions because he was insane. Reflecting modern understandings of the "out of control" nature of some extreme human behaviors, the jury acquitted him. Rather than being imprisoned for shooting the president, Hinckley was sent to St. Elizabeth's Hospital in Washington, D.C., where he underwent psychological observation and treatment. In time, he even earned privileges to leave the facility for supervised visits to his parents' home, and eventually was granted unsupervised trips off the facility grounds.[1]

By sharp contrast, nearly two and a half centuries earlier, on January 5, 1757, French king Louis XV was charged by an unknown assailant, Robert-François Damiens, who broke through the king's protective guards, grabbed his shoulder with one hand, and stabbed him with a knife held in the other. Damiens was a one-time menial in the college of the Jesuits in Paris. During a religious dispute between Pope Clement XI and the parliament of Paris over whether sacraments should be granted to Jansenists and Convulsionnaires, Damiens got in his mind the idea that religious peace would be restored if the king were eliminated. For a crime resembling Hinckley's (albeit with a different motive), Damiens was convicted for attempted regicide and sentenced to receive a rather harsher punishment than Hinckley got:

> Pincered at the breasts, arms, thighs and calves, his right hand holding the knife with which he perpetrated the said act, he was to be burned on the hand with sulfur, to be doused at the pinion points with boiling oil, molten lead, and burning resin, and then to be dismembered by four horses, before his body was burned, reduced to ashes, and scattered to the winds. Then one of the executioners, a strong and robust man, grasped the metal pincers, each one foot long, and by twisting and turning them, tore out huge lumps of flesh, leaving gaping wounds which were doused from a red-hot spoon. Between his screams, Damiens repeatedly called out, "My God, take pity on me!" and "Jesus, help me!" The final operation lasted a very long time, because the horses were not used to it. Six horses were needed: but even they were not enough. The executioner asked whether they should cut him in pieces, but the Clerk ordered them to try again. After two or three more attempts, the executioners took out knives, and cut off his legs. . . . They said that he was dead. But when the body had been pulled apart, the lower jaw was still moving, as if to speak.[2]

As if that were not enough, Damiens's home was razed to the ground, his brothers and sisters were ordered to change their names, and the rest of his family, including his wife and daughter, were banished from the country.

These two dramatically different forms of punishment reflect changing social and cultural attitudes toward behavior and its causes over the past two centuries. Have we changed for the better? This is a moral question with broad and sweeping ramifications for psychology, sociology, social policy and legislation, and political and ethical theory.

In the previous chapter we discussed theodicy, or the problem of evil, where God's omnipotence, omnibenevolence, and the existence of evil were seen to be incompatible. The noted Oxford theologian and man of letters C. S. Lewis, in his moving posthumously published work *A Grief Observed,* reflected on this problem after the premature death from cancer of his beloved wife, Joy: "But is it credible that such extremities of torture should be necessary for us? Well, take your choice. The tortures occur. If they are unnecessary, then there is no God or a bad one. If there is a good God, then these tortures are necessary. For no even moderately good Being could possibly inflict or permit them if they weren't."[3]

A similar set of logical tenets arises in theology over the problem of free will, and in my opinion there is no satisfactory solution for either of them. Squaring free will with God's omniscience and omnipotence is problematic. How can He hold us responsible for making "choices" we could not have made freely if He is all-knowing and all-powerful? If we are volitional beings, then we can make free choices, which means that God is limited in knowledge, power, or both. And a limited God is not the God of Abraham, Jesus, and Muhammad. God Himself, as it were, offered this solution in Milton's *Paradise Lost,* in an explanation for how Adam and Eve could have freely chosen to disobey Him even though he already knew their disobedience was foreordained by his power:

> *They themselves decreed*
> *Their own revolt, not I. If I foreknew,*
> *Foreknowledge had no influence on their fault*
> *Which had no less proved certain unforeknown.*[4]

The French philosopher René Descartes suggested a similar way out in less poetic form: "We will be free from these embarrassments if we recollect that our mind is limited while the power of God, by which he not only knew from all eternity what is or can be, but also willed and preordained it, is infinite. It thus happens that we possess sufficient intelligence to know clearly and distinctly that this power is in God, but not enough to comprehend how he leaves the free actions of men indeterminate."[5] After tackling the problem of evil, C. S. Lewis turned his acumen to the problem of free will, expanding on Descartes's resolution of placing God outside of time:

> But suppose God is outside and above the Time-line. In that case, what we call "tomorrow" is visible to Him in just the same way as what we call "today." All the days are "Now" for Him. He doesn't remember you doing things yesterday; he simply sees you doing them, because, though you've lost yesterday, He has not. He doesn't foresee you doing things tomorrow; He simply sees you doing them: because, though tomorrow is not yet there for you, it is for Him. You never supposed that your actions at this moment were any less free because God knows what you are doing.[6]

Even without including God in the equation, a new paradox arises in its stead. If we live in a determined universe, how can we make free moral choices? If genetics, biology, culture, and history combine to create a suite of factors that determine our thoughts and behaviors, how can society hold us morally and legally responsible for our actions? Is it legally, philosophically, or scientifically tenable to argue that some or most of us are free most of the time to make moral decisions, while a few (like John Hinckley) are absolutely determined (in other words, they could not have acted otherwise) some of the time to make immoral decisions? Hinckley, it was decided, had lost his free will. He was so under the control of inner forces and outer circumstances that he was determined to commit this act of violence. Determined by what precisely? Presumably by some mix of his genes and his environment, of internal traits unique to him and external states to which he was subjected.

Since we are all subject to some blend of heredity and environment—internal traits and external states—then why couldn't any of us cop an insanity plea, or at least a determinism appeal, for any of our immoral actions? Indeed, lots of people do. Consider the various defenses employed to build a case against moral freedom and for criminal determinism: the Twinkie defense (high blood sugar caused Dan White to kill San Francisco's Mayor George Moscone and Supervisor Harvey Milk), the abuse excuse (the Menendez brothers murdered their parents because they were abused as children), black rage syndrome (Colin Ferguson shot six white people on a train because he snapped under the pressure of our racist society), pornography defense (watching other people have sex causes men to rape women), PMS defense (premenstrual syndrome caused a woman to assault a police officer), and television violence (watching other people being violent makes people more aggressive). In the case of John Hinckley, the American court held that he was a determined puppet, and while he didn't get off scot-free, his punishment was far less draconian than that of Robert-François Damiens, whom the French court held was responsible for his crime because he freely chose to commit it, even though his justification might just as easily be construed today as being insane. (Perhaps a modern attorney would argue that Damiens was the victim of "regirage"—uncontrollable anger over being subjected to the rule of a king.) Which view of human nature is correct? Are we free or are we determined? And if we are determined, how can we make free moral choices and be held accountable for them? This is what I call the paradox of moral determinism, a subset of the larger free will/determinism problem.

Appointment in Samara: Free Will and the Problem of Determinism

The English novelist William Somerset Maugham aptly expressed the paradox of free will and determinism in his thought-provoking parable "Appointment in Samara":

> *Death speaks:*
> There was a merchant in Baghdad who sent his servant to market to buy provisions and in a little while the servant came back, white and trembling, and said, "Master, just now when I was in the market place I was jostled by a woman in the crowd and when I turned I saw it was Death that jostled me. She looked at me and made a threatening gesture. Now, please lend me your horse and I will ride away from this city and avoid my fate. I will go to Samara and there Death will not find me."

> The merchant lent him his horse and he dug his spurs in its flanks and as fast as the horse could gallop, he went. Then the merchant went down to the market place and saw me standing in the crowd and approached me and said, "Why did you make a threatening gesture to my servant when you saw him this morning?"
>
> "That was not a threatening gesture," I replied. "It was only a start of surprise. I was astonished to see him in Baghdad, for I have an appointment with him tonight in Samara."[7]

Although the meaning of Maugham's homily is more in line with predestination, the point is made that there is a sense of inevitability in life's drama. Although we may think we are freely going about our business, we are actually under the control of hidden masters. Consider the intricate workings of a finely crafted watch. If the hands of the watch possessed consciousness and self-awareness, they might feel like they were freely moving about the watch face, but we the watchmakers would know better. We know that the watch hands are determined because we know that the spring, cogs, wheels, and various parts all work together to cause the hands of the watch to move. We know that the watch is not a volitional being. It does not freely choose to keep accurate time. If the watch is running slow, we do not assign to it such anthropomorphic traits as indolence. It doesn't want to be late. It simply can't help it, and we take it to a jeweler to determine the cause of the problem. We do not assert that the problem with the watch is an insoluble one due to its volitional nature.

This is the axiom of determinism, the doctrine that every event in the universe has a prior cause, and that all effects are predictable if all causes are known. The free will/determinism problem is an ancient one, but for our purposes we begin in the seventeenth century with the rise of modern philosophy through such philosophers as René Descartes, and the ascent of modern science through such scientists as Isaac Newton. With the advent of the Cartesian/Newtonian mechanistic worldview, philosophers and scientists began to think of the universe and everything in it, including us, as determined in a mechanistic manner. The metaphor of choice, in fact, was that the universe is like a clock. The origin and action of every atom, molecule, cell, organism, person, planet, and star are the effects of some mechanical cause or series of mechanistic causes. This view became so pervasive that it was codified by the French mathematician Marquis de Laplace in what has since become known as *Laplace's demon*: "Let us imagine an Intelligence who would know at a given instant of time all forces acting in nature and the position of all things of which the world consists; let us assume, further, that this Intelligence would be capable of subjecting all these data to mathematical analysis. Then it could derive a result that would embrace in one and the same formula the motion of the largest bodies in the universe and of the lightest atoms. Nothing would be uncertain for this Intelligence. The past and the future would be present to its eyes."[8] Alexander Pope elegantly rhymed the problem this way:

> *Think we, like some weak prince, the Eternal Cause,*
> *Prone for his favourites to reverse his laws?*
> *Shall burning Etna, if a sage requires,*
> *Forget to thunder, and recall her fires?*
> *On air or sea new motions be imprest,*
> *O blameless Bethel, to relieve thy breast?*
> *When the loose mountain trembles from on high,*
> *Shall gravitation cease, if you go by?*[9]

By the twentieth century, philosophers spoke of a "causal net"—a network of causes linked to effects throughout the past and into the future. The causal net encompasses all phenomena, past, present, and future, throughout the cosmos, from atoms to galaxies and everything in between, including us. Without the doctrine of determinism, science could not strive for an ultimate understanding of past events or make predictions about future phenomena. The causal net was cast over the legal profession when attorneys began to use it as a tool in defense of their clients. John Hinckley's case shows how science and the law have each dealt with this problem. From what we have already seen about John Hinckley, it is clear that he was not "normal" in any sense of the word. So what was he, sane or insane? The answer turns out to be neither, and both.

John Hinckley and the Paradox of Moral Determinism

In philosophy there is a well-known fallacy of logic called *post hoc, ergo propter hoc,* literally, "after this, therefore because of this." In cognitive psychology there is a related problem called the "hindsight bias," where it seems that once we know the outcome, there is a sense that "I knew it all along." Much of what we believe about the world seems right only *after the fact*. Before the fact, however, things are not always so clear. Before the FBI's assault on the Waco compound of the Branch Davidians, for example, charging the building with armed guards seemed like the right thing to do. After four FBI agents were shot, it was abundantly clear to everyone that disaster was a foregone conclusion. Monday morning quarterbacking is everyone's favorite hobby. Causality is easy to infer after the effect; it is nearly impossible to know before. That is why the experimental methods of science demand rigid controls over intervening variables that might compound and confuse the results of an experiment.

In the case of John Hinckley, if you did not already know what he did, you would be hard-pressed to find anything in his background that would lead him to commit such an extreme act of violence. Hinckley was the youngest of three children, the son of a stay-at-home mom and a workaholic father prominent in the oil business. His mother described him as clinging and dependent. In reviewing JoAnn and Jack Hinckley's autobiographical book about their son and family, *Breaking Points,* Laura Obolensky wrote in the *New Republic*:

> Perhaps it is fear of what lies outside that makes the interior of the family so rigid and subdued, like life in a well-run bunker. The world of the Hinckleys was the rootless, middle-class Sunbelt culture that nurtures pro-family values, Christian fundamentalism, and occasional mass murderers. Families move frequently, but without compromising their parochialism. Everywhere, people are white, Christian, Republican (JoAnn explains John's egregious prejudices by saying he had "never been around people of other races"). Somewhere outside there are malign elements—minority groups, rock musicians, big government, and the cynical, Godless cosmopolites who dominate the media. Mothers in this culture do not lavish attention on their children, but on their furniture.[10]

Affectless? Yes. Cold and uncaring? Perhaps. Progenitor of an assassin? Hardly.

Upon graduating from high school, Hinckley muddled through two years of college at Texas Tech, in Lubbock, watching television, playing guitar, and finally dropping out in the spring of 1976. Like so many aspiring entertainers before and after him, Hinckley moved to Hollywood, where he saw Jodie Foster's debut in *Taxi Driver,* a movie he watched fifteen times over the next couple of years. In the film, Foster plays a young prostitute rescued from her pimp by a psychotic taxi driver named Travis Bickle, played by Robert De Niro. Hinckley identified with Bickle, adopting the mannerisms ("you talkin' to me?") and dress of the character (who, significantly, contemplated committing a political assassination), even to the point of keeping a diary, wearing army fatigues and boots, and developing an obsession with guns. More importantly, the film spawned in Hinckley an obsession for Foster.

A year later, Hinckley abandoned his musical aspirations and returned to Texas, wandering aimlessly as his depression deepened. A year after that, he bought his first gun and took up target shooting, and by the fall of 1979, he later confessed, he had twice played Russian roulette. Matters took a more sinister turn in the summer of 1980, when he convinced his parents to finance a writing course at Yale University where, not by chance, Foster was enrolled as a student. A distant obsession now turned to physical stalking, as Hinckley left Foster poems and letters, and twice phoned her.

His awkward love unrequited, Hinckley now considered a political assassination to get Foster's attention. His first target of choice was President Jimmy Carter. Hinckley tailed him on the 1980 campaign trail in Washington, D.C., as well as in Columbus and Dayton, Ohio. He later admitted that he couldn't get into "a frame of mind where I could actually carry out the act," so he returned to Yale to make another attempt at winning Foster's love, and once again failed. He was subsequently arrested for carrying handguns in his suitcase through the Nashville airport, but was released without incident.

With his parents' tuition money spent, Hinckley returned home and overdosed (but recovered) on antidepressants. His parents arranged for Hinckley to see a psychiatrist, but there was apparently no hint in their sessions of the level of Hinckley's obsession, or that he was on the brink of attempting suicide or murder. His depression deepening, Hinckley traveled to New York and considered killing himself on the same spot where John Lennon had been assassinated a few weeks earlier by another obsessed young man named Mark David Chapman. On New Year's Eve of 1980, Hinckley recorded a rambling message in which he spoke of not "really" wanting "to hurt" Foster and that he might kill himself if Foster would not return his affections. He finally decided he would take down Reagan, as he carefully explained to Foster in a never-sent letter: "Jodie, I would abandon this idea of getting Reagan in a second if I could only win your heart and live out the rest of my life with you. . . . I will admit to you that the reason I'm going ahead with this attempt now is because I just cannot wait any longer to impress you. I've got to do something now to make you understand, in no uncertain terms, that I am doing this for your sake! By sacrificing my freedom and possibly my life, I hope to change your mind about me. This letter is being written only an hour before I leave for the Hilton Hotel. Jodie, I'm asking you to please look into your heart and at least give me the chance, with this historic deed, to gain your respect and love."

After shooting Reagan, Hinckley was promptly arrested on the spot and transferred to the federal penitentiary in Butner, North Carolina, for a psychiatric evaluation. Although the initial assessment of Hinckley indicated that he was sane, two suicide attempts in his cell and his demand to his attorneys that they get Foster to testify on his behalf indicated that perhaps his mind was not functioning within normal operating parameters. (Foster did provide a videotaped testimony, but not one Hinckley saw as favorable to his cause, which was more focused on winning her love than winning his freedom.)

As the trial got under way, and with the hindsight bias in full display, attorneys, psychiatrists, jurors, and observers searched in vain for "the cause" of Hinckley's actions. Hinckley's attorney, Vince Fuller, attempted to glean from Hinckley's mother some glitch in his upbringing or an action that could have been a clue to the imminent assassination attempt. In fact, just months before the shooting, she had told Hinckley's psychiatrist, Dr. Hopper, "Things are fine." Hinckley's father testified that the last time he saw John he told him, "O.K., you are on your own. Do whatever you want to do." In retrospect, and with the hindsight bias driving him to despair, Jack Hinckley lamented, "I'm sure that it was the greatest mistake in my life. I am the cause of John's tragedy—I forced him out at a time when he simply couldn't cope. I wish to God that I could trade places with him right now." After the fact, anyone can be a Monday morning psychiatrist. Science requires predictions, not postdictions.

In her testimony, Foster said Hinckley's first sets of letters to her were "lover-type letters," but that the later letters were "distress-sounding" and "I gave them to the dean of my college." In a missive dated March 6, 1981, for example, just three weeks before the shooting, Hinckley wrote, "Jodie Foster, love, just wait. I will rescue you very soon. Please cooperate. J. W. H." Asked whether she'd "ever seen a message like that before," Foster replied, "Yes, in the movie *Taxi Driver* the character Travis Bickle sends the character Iris [Foster's role] a rescue letter."

Hinckley's lead psychiatrist for his defense was Dr. William Carpenter. After forty-five hours of conversation with Hinckley, Carpenter concluded that he showed four major symptoms of mental illness: "an incapacity to have an ordinary emotional arousal," "autistic retreat from reality," depression including "suicidal features," and an inability to work or establish social bonds. Hinckley's inability to properly identify with real people, Carpenter explained, led him to emulate *Taxi Driver*'s Bickle. Alone in his college dorm room and playing his guitar, he also identified with John Lennon. When the pop star was assassinated, Hinckley's self-identification was discombobulated. His New Year's Eve monologue, said Carpenter, indicated just how deep his insanity had gone. Here is what Hinckley said:

> John Lennon is dead. The world is over. Forget it. It's just gonna be insanity, if I even make it through the first few days. . . . I still regret having to go on with 1981. . . . I don't know why people wanna live. . . . John Lennon is dead. . . . I still think—I still think about Jodie all the time. That's all I think about really. That, and John Lennon's death. They were sorta binded together. . . .

Hinckley's identification then switched from Lennon back to the fictional Bickle, indicated by his signature in the guest register at the hotel where he stayed after his father refused to let him return home: "J. Travis."

Another psychiatrist for the defense, Dr. David Bear, concurred with Carpenter that Hinckley was psychotic, suffering from Schizophrenia Spectrum Disorder and an extreme reaction to Valium called Paradoxical Rage. For example, Hinckley believed that the character Bickle was talking to him through the film. It was "like he was acting out a movie script," Bear explained. Could he have been faking his symptoms? Not likely, Bear answered, because imposters fake both positive and negative emotions, whereas Hinckley exhibited only negative emotions. The prosecution challenged that the never-sent letter to Foster proved that Hinckley had *planned* to shoot the president, but Bear denied that it was a *rational* plan: "This is so much the heart of the issue—a logical man plans, Hinckley simply reacted, the very opposite of logic. Do I conclude he was rational in plan? My God, my sense of justice says absolutely not." A CAT scan of Hinckley's brain was also introduced into evidence by Bear, who said that widened sulci in his brain were "powerful evidence" of his schizophrenia, because only about 2 percent of the general population show widened sulci, whereas about one-third of schizophrenics do. Finally, Hinckley's score on the Minnesota Multiphasic Personality Inventory indicated that he was near the top of the range for abnormality. According to one expert, only one person out of a million with Hinckley's score would not be suffering from serious mental illness. Case closed.

Or was it? The government's psychiatric team drew a rather different conclusion. While Hinckley may exhibit numerous personality disorders and has some obviously unlikable traits, they argued, he was not insane. His flying about the country, purchasing handguns and ammunition, and plotting the place and timing of Reagan's assassination all indicated that he was rational and organized. How different really was his identification with John Lennon from that of thousands of crazed rock fans? Was it really an "obsession" with Jodie Foster or just an exaggerated form of infatuation exhibited by so many young adults for celebrities? Hinckley, the prosecution argued, was a bored, spoiled, lazy, manipulative rich kid and little more: "Mr. Hinckley's history is clearly indicative of a person who did not function in a usual reasonable manner. However, there is no evidence that he was so impaired that he could not appreciate the wrongfulness of his conduct or conform his conduct to the requirements of the law." Hinckley said as much in a deposition with the prosecution. Basically, he explained, he just wanted to be famous so he could get Foster's attention and affection. "It worked. You know, actually, I accomplished everything I was going for there. Actually, I should feel good because I accomplished everything on a grand scale. . . . I didn't get any big thrill out of killing—I mean shooting—him. I did it for her sake. . . . The movie isn't over yet."

The movie came to an end after eight weeks of evidence and arguments. The jury was instructed by Judge Barrington Parker that the prosecution had the burden of proving beyond a reasonable doubt that Hinckley was not insane and that on the day of the assassination attempt he could appreciate the wrongfulness of his actions. After three days of deliberation the jury concluded that the prosecution had failed to do so, returning the same verdict for all thirteen counts: not guilty by reason of insanity.

The trial captured the paradox of moral determinism. After the verdict, the public was outraged; the jury had acquitted a man whose crime had been witnessed by everyone in the country on national television. Lawmakers promised to launch an investigation into the insanity plea. One reporter sardonically labeled his insanity "dementia suburbia," because Hinckley was from a well-to-do suburban family. Nevertheless, the day after the trial Hinckley was placed in St. Elizabeth's Hospital in Washington, where he has resided ever since, continually denied parole. In 1985 Hinckley returned to court to request grounds privileges and to lift the hospital ban on his access to telephones. His obsession with Foster was finally over, he told the judge. But it wasn't true. Hinckley had previously written a letter to *Time* magazine, claiming, "The most important thing in my life is Jodie Foster's love and admiration. If I can't have them, neither can anyone else. We are a historical couple, like Napoleon and Josephine, and a romantic couple like Romeo and Juliet." The judge denied his request. A search of his room during another hearing on privileges, in 1987, turned up twenty photographs of Foster and numerous writings about her, along with correspondence with serial killer Ted Bundy. Hinckley had even attempted to contact Charles Manson. Even nearly a decade and a half later, his obsession had not diminished. In 2000, shortly after he earned the right to unsupervised furloughs, a search of his room uncovered a book about Foster, and he was once again confined to quarters.

Within a month of the Hinckley verdict, the legal world was awash in debate on the question of moral culpability. The House and Senate held hearings, and Senator Arlen Specter proposed shifting the burden of proof of insanity from the prosecution to the defense. Even Reagan commented, "If you start thinking about even a lot of your friends, you would have to say, 'Gee, if I had to prove they were sane, I would have a hard job.'" Within three years, two-thirds of the states made that shift on the burden of proof. Utah abolished the defense entirely, and eight other states changed the plea to "guilty but mentally ill." In 1984 legislation was passed requiring a defendant to prove that the "severe" mental disease made him "unable to appreciate the nature and quality or the wrongfulness of his acts."

Moral Determinism and the Law

The insanity defense is based on the moral psychological theory that most people most of the time in most circumstances freely choose to follow the law, but that a few people, in certain times and circumstances, and under particular mental duress or disease, cannot make that choice. Therefore, the traditional punishment of prison is inappropriate, because how can you punish someone for an act they did not voluntarily choose to commit?

The legal awareness that there is a paradox of moral determinism dates back to 1843 when an apparently psychotic man named Daniel M'Naghten attempted to assassinate British prime minister Robert Peel, under the paranoid belief that he was being persecuted. Foreshadowing the Hinckley trial, M'Naghten pleaded insanity, while the prosecution argued that his behaviors indicated volition because of the planning needed to execute the attack. Several physicians provided expert testimony that M'Naghten was insane, and the court agreed. The ensuing legal brouhaha (which included Queen Victoria and the House of Lords protesting the outcome of the trial) led to a set of criteria by which jurors were to judge the sanity or insanity of a defendant. These became known as the M'Naghten Test, which was in place in both the United Kingdom and the United States through the early 1960s. In brief, the M'Naghten Test of moral insanity includes a "right-wrong test" in which jurors were to ask themselves two questions: (1) did the defendant know what he was doing when he committed the crime? or (2) did the defendant understand that his actions were wrong? Jurors "ought to be told in all cases that every man is to be presumed to be sane, and to possess a sufficient degree of reason to be responsible for his crimes, until the contrary be proved to their satisfaction; and that, to establish a defense on the ground of insanity, it must be clearly proved that, at the time of the committing of the act, the party accused was laboring under such a defect of reason, from disease of the mind, as not to know the nature and quality of the act he was doing, or, if he did know it, that he did not know he was doing what was wrong."

Over time different jurisdictions modified the M'Naghten Test, some adding an "irresistible impulse" clause where defendants could be acquitted even if they knew an act was wrong but could prove that they could not stop themselves from committing it. The psychological theory behind this addendum was that some forms of mental illness are so powerful that they can cause a person to act against his or her better judgment. It is not simply "I knew this was wrong, but I wanted to do it anyway," but, rather, "I knew this was wrong, but I could not stop myself from doing it anyway." Critics argued that this clause meant anyone could claim irresistible impulses and that the whole point of social laws and moral codes is to curb those impulses. Where do you draw the line between normal resistible and abnormal irresistible impulses?

This conundrum led to the Durham Test, which arose out of the Washington, D.C., 1954 case of *Durham v. United States*. The Durham Test was a modification of the M'Naghten Test, in which jurors were now to ask themselves these two questions: (1) did the defendant have a mental disease or defect? and (2) if so, was the disease or defect the reason for the unlawful act? To return a verdict of not guilty by reason of insanity, both answers had to be in the affirmative. The psychological theory behind the Durham Test was that mental illness was a disease that took control of a person's moral volition. The Durham Test, however, never quite caught on, and by 1972 the D.C. Circuit out of which it originally arose abandoned the test and in its stead adopted the American Law Institute (ALI) Test, a more flexible model designed to recognize degrees of incapacity. Instead of a binary choice of knowing or not knowing the difference between right and wrong, defendants could show degrees of moral and psychological incapacity. Specifically, it stated: "The concept of belief in freedom of the human will and a consequent ability and duty of the normal individual to choose between good and evil is a core concept that is universal and persistent in mature systems of law." The presumption is that good and evil actions are choices made by volitional beings. Thus, "Criminal responsibility is assessed when through free will a man elects to do evil." Free will holds us accountable. Abandon free will in favor of determinism and moral culpability flies out the courtroom window.

By the 1970s almost all federal circuit courts had adopted the ALI Test, and it was this one that was in place for the Hinckley trial. The hue and cry over Hinckley's acquittal, however, reversed the ALI Test precedence when states either abolished it entirely or shifted the burden of proof to the defendant. The Insanity Defense Reform Act of 1984 clearly stated the limits of the insanity plea: "It is an affirmative defense to a prosecution under any federal statute that, at the time of the commission of the acts constituting the offense, the defendant, as a result of a severe mental disease or defect, was unable to appreciate the nature and quality or the wrongfulness of his acts. Mental disease or defect does not otherwise constitute a defense." In other words, the Reform Act eliminated the volitional aspect of the defense, required that the mental disease must be "severe," and replaced "unable to appreciate" with "lacks substantial capacity" in order to clarify the boundaries between a total lack of understanding and partial comprehension. Many states also changed the plea "not guilty by reason of insanity" to "guilty but mentally ill." Perhaps most telling, the American Medical Association cast its vote for the total abolition of the insanity defense. And so the paradox remains mired in legal muck because of the lack of a clear scientific understanding of where to draw the line between sanity and insanity, and between free will and determinism.[11]

Free Will as a Useful Fiction: Is the Free Will/ Determinism Problem an Insoluble One?

Since the time of the ancient Greeks through the present, some of the greatest minds in every generation have grappled with the problem of free will and determinism, and no one to date has proposed a solution that satisfies most people. It could be that this is a really hard problem—on par, say, with celestial mechanics—and our Newton has yet to produce the *Principia* of free will. Perhaps, and this is even more distressing, there is no solution to the problem. Like the question of God's existence, the free will/determinism paradox may be an insoluble one. This may be a "mysterian" mystery, where our brains are sophisticated enough to conceive of the problem but not advanced enough to solve it.[12]

This is in lockstep with the fideist position in theology, where pragmatist philosophers like William James, Charles Peirce, and Miguel de Unamuno argue that it is acceptable to take a leap of faith on issues of extreme importance to human existence, when the evidence is inconclusive one way or the other, and you must choose. That is, just make a choice one way or the other even if uncertainty remains high as to which choice is the correct one. My friend and skeptical colleague Martin Gardner is a fideist and takes this approach to both the God question and the free will/determinism problem. He has chosen God and free will, not because there is better evidence for them, but because they are important issues, the evidence is inconclusive, and it works better for him to believe in God and free will. With the free will problem, Gardner says it "cannot be solved because we do not know exactly how to put the question."[13] Asking the question, Is there free will? is like asking, Why is there something rather than nothing? or What is time? After reviewing all the arguments for free will, Gardner concludes, "Like time, with which it is linked, free will is best left—indeed, I believe we cannot do otherwise—an impenetrable mystery. Ask not how it works because no one on earth can tell you."[14]

If the problem is an insoluble one and thus it is acceptable to choose either free will or determinism, can one have both? The belief that free will can be derived out of a deterministic universe is called *compatibilism,* and it is shared by many, such as philosopher and neurobiologist Owen Flanagan, who argues that the free will/determinism problem is "ill-posed" and that even though "ours is a causal universe . . . no one yet knows the exact range of deterministic and indeterministic causation—assuming the universe contains some of each." Flanagan's solution is creative, if nothing else: "My proposal is this: Change the subject. Stop talking about free will and determinism and talk instead about whether and how we can make sense of the concepts of 'deliberation,' 'choice,' 'reasoning,' 'agency,' and 'accountability.' "[15]

Compatibilist solutions such as these are really pseudosolutions or, more positively, pragmatic solutions. That is, pragmatically speaking, they are true if they work. Along those lines, here is one that works for me; maybe it will work for you: *free will is a useful fiction.* I feel "as if" I have free will, even though I know we live in a determined universe. This fiction is so useful that I act as if I have free will but you don't. You do the same. Since the problem may be an insoluble one, why not act as if you do have free will, gaining the emotional gratification and social benefits that go along with it?

Insolubility and compatibilism are, at best, pseudosolutions, which, for the most part, satisfy no one. Can science help clear up this conundrum? I shall close this chapter with six scientific-based attempts to derive free will out of determinism: the uncertainty principle of quantum indeterminacy, fuzzy logic, neuroscience, genetics, evolutionary theory, and chaos and complexity theory.

Free Will and Indeterminism

One solution to the problem makes an appeal to quantum indeterminacy, a derivative of a field of physics called quantum mechanics. One of the pioneers of quantum mechanics, German physicist Werner Heisenberg, discovered that you cannot determine both the position and the speed of an electron moving about the nucleus of an atom. If you determine where the electron is located, you cannot know its speed. If you determine its speed, you cannot know its position. This became known as the Heisenberg uncertainty principle. Further, when you observe an atom, the "wave function collapses" (a mathematical description made by quantum mechanicists), thus bringing into reality its location. That is, the atom is in a state of uncertainty until it is observed. When observed, the wave function collapses and the state of the atom becomes certain. Finally, there is an additional level of uncertainty in that when atoms decay—as when, say, potassium atoms decay into argon atoms (a process so predictable that it serves as an atomic clock for dating geological events in the earth's history)—it is not possible to know which particular atom will decay. The decay process is, quite literally, uncaused and unpredictable. It is truly indetermined.

From this fact, some philosophers and scientists argue that perhaps these random and indeterministic atomic events associated with quantum mechanics might trigger the random firing of neurons in the brain, leading to indeterminant mental states. Perhaps, they suggest, this is where free will arises. This argument was critiqued by one of the leading quantum physicists, Nobel laureate Murray Gell-Mann, who derisively called it quantum flapdoodle.[16] Quantum effects cancel each other out at the macro level in which everyday events (like human thought) occur. Is it really possible—we might ask rhetorically in an analogy with the uncertainty principle—that the orbit of Mars, like the orbit of an electron, is scattered randomly about the sun until someone observes it, at which point the wave function collapses and the planet appears in one spot? Obviously not, any more than we might think that the moon ceases to exist until it is observed (someone once actually proposed that the moon does not exist until observed). Quantum effects wash out at large scales.

Yet even if it could be established that quantum uncertainties lead to random neuronal firings, that does not produce free will; it just adds another deterministic causal factor, one that is random rather than non-random. This last point was well made by the philosopher Daniel Dennett in his book *Elbow Room: The Varieties of Free Will Worth Wanting,* in which he also argued that if there is too much free will—if people are completely free of all determining or influencing forces—there could be no room for modification of immoral behaviors.[17] Truly free agents could thumb their noses at all attempts to curb their freely chosen actions. Social chaos would be the result. We need a proper balance between determinism and free will, and indeterminism is not freedom.

Free Will, Fuzzy Logic, and Hinckley's Guilt

A universally accepted set of criteria to determine moral (and thus legal) responsibility for a crime has proved to be elusive because of a deep and fundamental difference between science and the law. The law requires unambiguous categories in order to reach a judgment of guilty or not guilty based on the defendant's sanity or insanity. Here again we see the type of Platonic thinking that troubled us over the problem of evil. As with good and evil, "sane" and "insane" are reified things, typologies meant for classification of unchanging entities. Science offers us a solution because it recognizes that there are many shades between such categories, such that a fuzzy logic solution, as we saw in solving the problem of evil, once again proves to be a useful heuristic. Asking if John Hinckley was sane or insane is the binary logic of Aristotle's *A or not-A*. Instead, let us inquire about the quantitative level of Hinckley's sanity or insanity. We can see in Hinckley's background that the shift from sanity to insanity was a fuzzy one, say, .9 sane and .1 insane in his youth, to .8 sane and .2 insane in his teens, to .7 sane and .3 insane during his two years of college, to .6 sane and .4 insane during his year in Hollywood, to .5 sane and .5 insane during the following year of aimless drifting, to .4 sane and .6 insane as he pursued Foster at Yale, to .3 sane and .7 insane as he contemplated assassinating Carter or killing himself, to .2 sane and .8 insane when he hatched the idea to assassinate Reagan, to .1 sane and .9 insane when he penned his final letter to Foster and headed for the Hilton Hotel with his gun in hand and his moral sense fully disengaged. The three photographs of John Hinckley in figure 18 show the slow and gradual (and fuzzy) deterioration of his mind over time, and along with it his capacity for moral reasoning.

A and not-A. Sane and insane. Such psychological states are variable over time and dependent on the context. Hinckley's long-term obsession with Foster became a form of insanity, but his assassination attempt on Reagan was not. He knew exactly what he was doing. He carefully planned it out and he understood that it was morally wrong because he knew doing it would draw enormous public and media attention. The whole point of the assassination attempt was to get Foster's attention, and recall that he said, "It worked. You know, actually, I accomplished everything I was going for there. I did it for her sake." Was Hinckley insane in his obsession over Foster? Was Hinckley insane when he shot Reagan? Approached scientifically, these are separate questions. Hinckley's obsession grew into a form of insanity. Yet he knew what he was doing even as he pulled the trigger, so we should hold him morally accountable for his crime against Reagan. In other words, Hinckley's obsession over Foster yields a verdict of not guilty by reason of insanity, and treatment by mental health professionals is an appropriate response. Hinckley's assassination attempt of Reagan, however, generates a straight guilty verdict, and he should have been punished accordingly with a stiff sentence in a maximum-security prison. The long road that took Hinckley from college dropout in 1976 to the Hilton Hotel in 1981 was gradual enough that he could have reversed his course. As he told *Newsweek* magazine later that year, "The line dividing life and art can be invisible. After seeing enough hypnotizing movies and reading enough magical books, a fantasy life develops which can either be harmless or quite dangerous." That may be the most insightful thing Hinckley ever said. It is not the fantasy life that is the problem. After all, fiction writers are paid to pour out their fantasies. What matters is whether those fantasies are converted into dangerous behaviors.

Even here we can apply the findings of science to a fuzzy logic analysis. Hinckley's unrequited love for Foster was the driving force behind his violence. We know that the fervor of unreciprocated love can become one of the most dangerous of all the passions, overriding reason and

Figure 18. The Fuzzy Deterioration of John Hinckley

Instead of asking whether John Hinckley was sane or insane in a binary choice, a fuzzy logic analysis allows us to assess his state of mind in shades of gray between sanity and insanity, and how it changed over time. One can see the changes in his face in these three photographs. (Courtesy of Associated Press)

rationality. But there are degrees of passions, fractional reactions, and fuzzy responses. Research shows, for example, that 95 percent of all men and women have experienced unrequited love at least once by the age of twenty-five (on either the sending or the receiving end).[18] Most people whose love goes unwanted by another feel rejected, suffer a temporary dip in self-confidence and self-esteem, but quickly move on to find someone who returns their passionate overtures. A few (mostly men, but some women, too) undertake a vigilant campaign to win the heart of their chosen beloved, and occasionally they succeed. Their efforts are assertive, but not aggressive. But when some do not succeed (and by now it is almost entirely men), and they continue the pursuit despite their target's efforts to reject them, charges of stalking and harassment can be filed and convictions won. This terminates almost all remaining attempts.

Almost all. Hinckley's response to Foster's indifference, however, was at the extreme end of the fuzzy scale. But he is still on the scale. His response is still a fractional one within the whole number of human response variability. Between .1 and .9 is still not 0 or 1, and so a scientific approach upholds moral culpability. Even if freedom is diminished, it is not extinguished. Thus we have one solution—fuzzy freedom—to the paradox of moral determinism.

Free Will and Neuroscience

How does the mind work? According to evolutionary psychologists, the mind is like a Swiss Army knife, equipped with specialized tools that evolved in our Paleolithic past to solve specific problems of survival, such as face recognition, language acquisition, mate selection, and cheating detection. In this model the brain is represented as a host of modules, or bundles of neurons, some located in a single spot (as in Broca's area for language), others sprawled out over the cortex. Large modules coordinate inputs from smaller modules, which themselves collate neural events from still smaller neural bundles. This reduction continues all the way down to the single neuron level, where highly selective neurons, sometimes described as "grandmother" neurons, fire only when subjects see someone they know. Caltech neuroscientists Christof Koch and Gabriel Kreiman, in conjunction with UCLA neurosurgeon Itzhak Fried, have even found a single neuron that fires when the subject is shown a photograph of Bill Clinton.[19] The Monica Lewinsky neuron must be closely connected.

What do these modules tell us about how the mind works? For one, experiences that appear to be external may, in fact, be internal. Five centuries ago, for example, demons haunted our world, with incubi and succubi tormenting their victims as they lay asleep in their beds. Two centuries ago spirits haunted our world, with ghosts and ghouls harassing their sufferers all hours of the night. Last century aliens haunted our world, with grays and greens abducting captives out of their beds and whisking them away for probing and prodding. Today people are having out-of-body experiences, floating above their beds, out of their bedrooms, and even off the planet into space. What is going on here? Are these elusive creatures and mysterious phenomena in our world or in our minds?[20] New evidence indicates that they are, in fact, a product of the brain.

Neuroscientist Michael Persinger, in his laboratory at Laurentian University in Sudbury, Canada, for example, can induce all of these experiences in subjects by subjecting their temporal lobes to patterns of magnetic fields. Persinger places on a subject's head a motorcycle helmet specially modified with electromagnets. The subject sits in an easy chair in a soundproof room with eyes covered. The electrical activity generated by the electromagnets produces a magnetic field that stimulates "microseizures" in the temporal lobes of the brain, which, in turn, produce a number of what can best be described as "spiritual" and "supernatural" experiences—the sense of a presence in the room, an out-of-body experience, bizarre distortion of body parts, and even religious feelings. Persinger calls these experiences "temporal lobe transients," or increases and instabilities in neuronal firing patterns in the temporal lobe. Having now studied over 600 subjects in the past decade, Persinger speculates that such transient events may account for psychological states routinely reported as happening outside the mind. These events, he suggests, may be triggered by the stress of a near death experience (caused by an accident or traumatic surgery), high altitudes, fasting, a sudden decrease in oxygen, dramatic changes in blood sugar levels, and other stressful events.[21] I participated as a subject in Persinger's experiment and had a mild out-of-body experience.

Similarly, in 2002, Swiss neuroscientist Olaf Blanke and his colleagues discovered that they could bring about out-of-body experiences through electrical stimulation of the right angular gyrus in the temporal lobe of a forty-three-year-old woman suffering from severe epileptic seizures. In initial mild stimulations she reported "sinking into the bed" or "falling from a height." More intense stimulation led her to "see myself lying in bed, from above, but I only see my legs and lower trunk." Another stimulation induced "an instantaneous feeling of 'lightness' and 'floating' about two meters above the bed, close to the ceiling."[22]

In a related study, researchers Andrew Newberg and Eugene D'Aquili found that when Buddhist monks meditate and Franciscan nuns pray, their brain scans indicate strikingly low activity in the posterior superior parietal lobe, a region of the brain the authors have dubbed the orientation association area (OAA), which orients the body in physical space (people with damage to this area have a difficult time negotiating their way around a house). When the OAA is booted up and running smoothly there is a sharp distinction between self and nonself. When OAA is in sleep mode—as in deep meditation and prayer—that division breaks down, leading to a blurring of the lines between reality and fantasy, between feeling in body and out of body. Perhaps this is what happens to monks who experience a sense of oneness with the universe, or with nuns who feel the presence of God, or with alien abductees floating out of their beds up to the mother ship.[23]

Since our normal experience is of stimuli coming into the brain from the outside, when a part of the brain abnormally generates these illusions, another part of the brain interprets them as external events. Hence, the abnormal is thought to be the paranormal. What these studies show is that mind and spirit are not separate from brain and body. In reality, all experience is mediated by the brain. Further, and more to the point of our discussion on free will and the brain, we now know from recent research in the neurosciences that every brain is wired, and continues to be rewired throughout life, in response to unique genetic, environmental, and historical conditions. Evolutionary psychologists Peggy La Cerra and Roger Bingham, for example, in their book *The Origin of Minds,* argue that our ancestral inheritance is not a set of fixed cognitive tools, but a living "brain/mind-construction system" that exploits pliable brain tissue, changing it with new experiences. The Swiss Army knife, it seems, can design new blades for cutting through new environments.[24] How does it do this?

The mind is an emergent property of billions of individual neurons, each of which is connected to thousands of other neurons that together produce trillions of potential neuronal states. As the individual grows and develops into adulthood the interconnections grow and develop according to individual life experiences. Although we share a common evolutionary ancestry that generated a universal neural architecture, since no life paths are the same, and with trillions of possible permutations of neuronal connections in each brain, the result is that every human mind is unique. There are literally six billion different minds. The foundation of this neural system is what La Cerra and Bingham call the adaptive representational network (ARN), "a network of neurons that memorializes a brief scene in the ongoing movie of your life, linking together your physical and emotional state, the environment you are in, the behavior or thought you generate, and the problem-solving outcome." What they are describing is an autocatalytic (self-generating) feedback loop. New experiences stimulate neurons to grow new synaptic connections. Those new connections are distinctive to every individual mind, which then responds to the environment in an idiosyncratic way, producing a behavioral repertoire of responses. The ARN evolved as an adaptation to help organisms survive in an ever-changing environment. No brain module can do what the ARN does, because modules evolved to solve specific problems, whereas the ARN evolved to solve a range of problems, even those never encountered.

How does this apply to real-world choices? La Cerra and Bingham reinterpret clinical depression in terms of its adaptive response consequences. The symptoms of depression—restlessness, agitation, disturbed sleeping and eating, impaired concentration, and loss of motivation—are not signs of an illness; rather, they represent an adaptive response to do something different in your life. "Because behavior is so enormously expensive energetically, the best thing a person in this situation can do is to stop what he has been doing, reconfigure his life, and try to formulate a more viable trajectory into the future." Why would this intelligence system have evolved? "If you were an ancestral human who was being exploited by another individual or group of individuals, a complete behavior shutdown could abruptly force a renegotiation of the inequitable social relationship." Even in the modern world, depression "serves as a wake-up call, prodding people to abandon dead-end jobs and relationships."[25]

What does all this neuroscience tell us about free will and determinism? Cognitive psychologist Steven Pinker, for one, argues that the brain is wired to "feel" like it is making choices, so we should listen to what our brains are telling us. "The experience of choosing is not a fiction, regardless of how the brain works," he explains. "It is a real neural process, with the obvious function of selecting behavior according to its foreseeable consequences." That is, making choices that lead to behaviors that result in actual consequences for survival and reproduction in our evolutionary history would have led to the evolution of brain mechanisms that give the illusion of free will. "You cannot step outside it or let it go on without you because it *is* you." Even "if the most ironclad form of determinism is real," Pinker concludes, "you could not do anything about it anyway, because your anxiety about determinism, and how you would deal with it, would also be determined."[26] Thus, with such convoluted and complex brains as we possess and living in a world with so many options, our brains evolved a choice-making module that, whether truly free or truly determined, nonetheless makes us feel free.

Free Will and Genetics

In 1985, while racing a bicycle along a lonely rural highway in Arkansas in the 3,000-mile nonstop transcontinental Race Across America, I was asked by ABC television commentator Diana Nyad how it felt to be too far behind the leader of the race to win. I told her that while I would prefer winning I had done everything I could in training, nutrition, equipment, and preparation, and that the only thing I could have done to improve my performance was to pick better parents. When that comment aired months later on *Wide World of Sports,* I called my parents to assure them I only meant that genetics plays a powerful role in athletics. I acquired the comment from renowned sports physiologist Per-Olof Astrand, who told an exercise symposium, "I am convinced that anyone interested in winning Olympic gold medals must select his or her parents very carefully."[27]

From an evolutionary perspective, our parents have been very carefully selected for us—by natural selection. But we are also the products of our parental upbringing, family dynamics, peer groups, community values, teachers and education, preachers and religion, culture and politics, and much more. The science of assigning some portion of our lives to genetics and the remaining portion to the environment has a long and controversial history. The process strikes me as an exercise in futility because of the interactive nature of genes and memes, evolutionary history and cultural history. Such binary thinking, particularly since the completion of the mapping of the human genome, for example, has led to oversimplified claims for a "math gene," a "risk-taking gene," a "promiscuity gene," a "rape gene," or a "smoking gene."[28]

In reality, the story is much more complex, and claims for genetic determinism are greatly exaggerated. Consider as one example among many a gene called D4DR, located on the short arm of the eleventh chromosome. D4DR codes for dopamine receptors, a neurotransmitter released by neurons that, when received by other neurons receptive to its chemical makeup, sets up dopamine pathways throughout the brain that stimulate the organism to be active (or not, if a shortage exists). A complete lack of dopamine, for example, causes patients (or rats) to slip into a virtual catatonic state. High levels of dopamine turn humans schizophrenic and rats frenetic. Dopamine stimulation, in fact, is the basis of the famous experiment where rats pressed a bar to stimulate their so-called pleasure center, which they did until collapsing in exhaustion. This is the fascinating work of geneticist Dean Hamer who, in his quest to find genes for smoking and homosexuality, discovered the gene (or, more precisely, the gene-complex) for a thrill-seeking personality. It turns out that the D4DR gene sequence repeats on chromosome eleven, and while most of us have four to seven copies, some people have two or three, and others have eight, nine, ten, or eleven copies. More copies of D4DR sequence means lower levels of dopamine, which translates into higher novelty-seeking behavior that artificially produces more dopamine (jumping off buildings and out of planes will do the trick). Hamer took 124 people who scored high on a survey that measured their desire to seek novelty and thrills (bungee jumpers and sky divers knock the roof off these tests), then looked at their DNA—specifically, chromosome eleven. He found that people who like to jump off buildings and out of planes had more copies of D4DR sequence than those who prefer knitting and watching grass grow.

When Hamer's research was picked up in the media, headlines declared that scientists had discovered the novelty-seeking gene, implying that perhaps all of our personality traits are genetically coded at a single point on a single chromosome arm. Alas, if only it were that simple—whenever you get that urge to jump off the top of Yosemite's Half Dome, just take a dopamine tablet and you'll prefer to stay on the marked trails. But there is another side to this story. When you actually read the original research, it turns out that Hamer claims to explain no more than 4 percent of novelty-seeking behavior by D4DR sequences. That is, if we say that humans vary by 100 percent in their novelty-seeking behavior—catatonics on one end and X-Game skateboarders careening down hills at 50 mph two inches off the ground on the other—only 4 percent of that variance can be accounted for by D4DR. That's it! As the science writer Matt Ridley explains in his analysis of the research:

> Do you see now how unthreatening it is to talk of genetic influences over behaviour? How ridiculous to get carried away by one "personality gene" among 500? How absurd to think that, even in a future brave new world, somebody might abort a foetus because

> one of its personality genes is not up to scratch—and take the risk that on the next conception she would produce a foetus in which two or three other genes were a kind she does not desire? Do you see now how futile it would be to practise eugenic selection for certain genetic personalities, even if somebody had the power to do so? You would have to check each of 500 genes one by one, deciding in each case to reject those with the "wrong" gene. At the end you would be left with nobody, not even if you started with a million candidates. We are all of us mutants. The best defence against designer babies is to find more genes and swamp people in too much knowledge.[29]

Nature is so intertwined with nurture that to say that a complex human characteristic like personality or intelligence or—to the point of this book—morality is, say, 40 percent genetics and 60 percent environment (to arbitrarily pick two figures) misses something very important: *inheritability of talent does not mean inevitability of success, and vice versa.* We are free to select the optimal environmental conditions that will allow us to rise to the height of our biological potentials. In this sense, athletic success, like any other type of success, may be measured not just against others' performances, but also against the upper ceiling of our own ability. To succeed is to have done one's absolute best. To win is not just to have crossed the finish line first, but also to cross the finish line in the fastest time possible within one's own limits. The closer one comes to reaching the personal upper limit of potential, the greater the achievement, as depicted in the Genetic Range of Potential model in figure 19. Individual "A" may have more absolute talent potential than individual "B," but this does not guarantee relative success. If "B" prepares to the height of his or her upper limit of potential, but "A" slacks off below that mark, inherited talent becomes meaningless. There is not much we can do about selecting our parents, but we can select our environmental conditions to push us to the top of our range of potential.

Free Will and Evolution

Free will, Dennett says, emerges out of our deterministic world from the fact that we evolved a large cortex that allows us to weigh the consequences of the many courses of action available to us, that we are

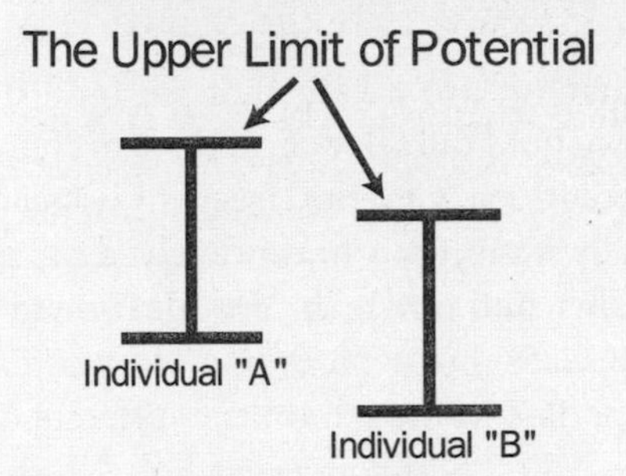

Figure 19: Genetic Range of Potential Model

Human behavior is a function of both genetics and environment, arrayed in a complex and interactive feedback loop. Behaviors are never "fixed" in some absolute sense by genetics; instead, genes code for a range of potential behaviors, which environments then affect. Genetically predisposed behaviors may be affected by environments to be expressed at the low end of the range or the high end of the range, or in between. Individuals may be determined to fall within a given range of potential, but where within that range they end up is a function of environmental determiners as well as self-determination, or free will.

aware that we (and others) make these choices, and that we hold ourselves and them accountable.[30]

In *Freedom Evolves,* Dennett expands on his arguments in *Elbow Room,* adding an evolutionary component to his deduction of free will. Dennett's thesis can be summarized as follows: (1) humans are evolved animals without a soul but with free will; (2) we are the only species with free will because we have a "self," a sense of being self-aware, and are even aware that others are self-aware, because (3) we have symbolic language that allows us to communicate the fact that we are aware and self-aware; and (4) we have extremely complex neural circuitry and many degrees of behavioral freedom (a jellyfish, like a hot-air balloon, for example, has one degree of freedom: up and down; we have many more); and (5) we have a theory of mind about other selves who are also (6) moral animals in the sense of having evolved moral sentiments or feelings of making right or wrong choices as members of a social species, and with symbolic language, we have the representational power to reason with each other about what we ought to do; therefore (7) free will emerges out of our deterministic world from the fact that we can weigh the consequences of the many courses of action available to us, that we are aware that we (and others) make these choices, and that we hold ourselves and them accountable.

In Dennett's evolutionary theory, free will is located in the "self," a metaphor for an adaptation our brains evolved for monitoring what is happening in our own and others' brains. But where is the self located? The answer is not clear, but wherever it is, it is not in one location. Reaction-time experiments that monitor different parts of the brain indicate that there is no "Self-contained You." Instead, "all the work done by the imagined homunculus in the Cartesian Theater has to be broken up and distributed in space and time in the brain."[31] We have a functional "layer" of decision-making power that no other species has (this is not a brain layer, but what Dennett calls "a virtual layer" found "in the micro-details of the brain's anatomy"). For example, "a male baboon can 'ask' a nearby female for some grooming, but neither of them can discuss the likely outcome of compliance with this request, which might have serious consequences for both of them, especially if the male is not the alpha male of the troop. We human beings not only can do things when requested to do them; we can answer inquiries about what we are doing and why. It is this kind of asking, which we can also direct to ourselves, that creates the special category of voluntary actions that sets us apart."[32]

This argument for freedom from evolution brings a fresh perspective to an ancient problem. But is it true? I have my doubts. Although I accept the first six of Dennett's points listed above and agree that he has thoroughly debunked the indeterminism argument, I remain unconvinced that free will can ultimately be derived from determinism in any consistent logical way. The terms are incompatible. What we are left with is a type of free will from ignorance, ignorance of all the determining causes in our lives, such that we are, de facto, free because when we make choices we cannot know all the causal variables. This theory of free will derives from chaos and complexity theory.

Free Will and Chaos and Complexity Theory

There is one more way to get free will, and that is through the complex world of human and social systems. The causal-net theory of determinism means that human behavior is no less caused than other physical or biological phenomena, just more difficult to understand and predict because of the number of elements in the system and the complexity of their interactions. Since no cause or set of causes we select to examine as the determiners of human action can be complete, in terms of human freedom they may be pragmatically considered as conditioning causes, not determining ones. That is, our thoughts and actions are shaped by a myriad of causes—genetic, environmental, and historical. Every individual set of genes is unique (with the exception of identical twins), each environmental setting is matchless, and every historical pathway that each of us has gone down in our individual lives is distinctive. We are, each and every one of us, unique and different from every other of the six billion members of our species. And those conditions are so complex, so interwoven, that no one could possibly know all of the causal variables for themselves or anyone else. Human freedom arises out of this ignorance of causes.

I derived this solution out of a model I developed called the *model of contingent-necessity.*[33] Its primary function is as a tool for the historical sciences, but it can generate another solution to the paradox of moral determinism. By contingency I mean *a conjuncture of events occurring without perceptible design*, and by necessity I mean *constraining circumstances compelling a certain course of action*. Contingencies are the sometimes small, apparently insignificant, and usually unexpected events of life—the kingdom hangs in the balance awaiting the horseshoe nail. Necessities are the large and powerful laws of nature and trends of history—once the kingdom has collapsed, 100,000 horseshoe nails will not save the realm. Leaving either contingency or necessity out of the historical formula, however, is to ignore an important component in the development of historical sequences. The past is constructed by both contingencies and necessities, and therefore it is useful to combine the two into one term that expresses this interrelationship. I call this *contingent-necessity,* taken to mean *a conjuncture of events compelling a certain course of action by constraining prior conditions.*

Randomness and predictability—contingency and necessity—long seen to be opposites on a continuum, are characteristics that vary in the amount of their respective influence and at what time their influence is greatest in the historical sequence. There is available a rich matrix of interactions between early pervasive contingencies and later local necessities, varying over time, in the model of contingent-necessity: *in the development of any historical sequence the role of contingencies in the construction of necessities is accentuated in the early stages and attenuated in the later.* At the beginning of a historical sequence, actions of the individual elements are chaotic, unpredictable, and have a powerful influence on the future development of that sequence. But as the sequence slowly but ineluctably evolves, and the pathways become more worn, the chaotic system self-organizes into an orderly one. The individual elements sort themselves and are sorted into their allotted positions, as dictated by what came before—the conjuncture of events compelling a certain course of action by constraining prior conditions. But aren't both necessities and contingencies caused, and themselves are the causes of effects? And, if so, then isn't all human action caused, and thus determined? We can express the problem this way:

Necessity is omnipotent
Contingency is omnipotent
Humans have free will

If human history is absolutely determined by necessitating forces of any kind, then neither contingency nor free will can exist. If contingency is all-powerful, then there can be no absolutely determining forces, and all history is reduced to just "one damn thing after another." Since it is obvious that there are necessitating forces at work in history, and it is equally obvious that contingencies push and direct historical sequences, then how do we resolve the problem of historical causality and human freedom? Here is a helpful analogy. Atoms moving about in space, like people moving about the environment, are caused, but their collisions (atomic) and encounters (human) happen by a combination of contingencies and necessities. Contingency leads to collisions and encounters; necessity governs speed and direction. Events may occur as a result of accidental causes (a conjuncture of unplanned events), but not by accident, in the sense of being uncaused. An effect, dependent upon the activity of one or more causes, may seem to be produced by accident, but it is really the result of a conjuncture of events compelling a certain course of action by constraining prior conditions. The words *compelling* and *constraining* imply powerful influence but not causal determinism.

Another way to approach the problem is to think of necessities as "what had to be" and contingencies as "what might have been." If history is a product of contingencies and necessities, then necessities (what had to be) imply determinism, while contingencies (what did not have to be) imply, in a way, a type of freedom. If things could have turned out differently because of some small but carefully placed human action, this gives us one more way around determinism. We can make a difference. Our actions matter. And in the rich panoply of causes that determine our actions, we can feel the freedom to choose to make a difference by doing the right thing to change the course of our personal histories or global history.

The number of causes and the complexity of their interactions make the predetermination of human action pragmatically impossible. We can even put a figure on the causal net of the universe to see just how absurd it is to think we can get our minds fully around it. Tulane University theoretical physicist Frank Tipler has calculated that in order for a computer in the far future of the universe to resurrect in a virtual reality every person who ever lived or could have lived, with all causal interactions between themselves and their environment, it would need 10 to the power of 10 to the power of 123 bits (a 1 followed by 10^{123} zeros) of memory. An entity capable of this would be, for all intents and purposes, omniscient and omnipotent, and this is what Tipler calls the Omega Point, or God.[34] Suffice it to say that no computer within the conceivable future will achieve this level of power; likewise no human brain even comes close. Thus, as far as we are concerned, the causal net will always be full of holes. Therefore, in the language of this model: *human freedom is action taken with an ignorance of causes within a conjuncture of events that compels and is compelled to a certain course of action by constraining prior conditions.*

In other words, the enormity of this complexity leads us to feel as if we are acting freely as uncaused causers, even though we are actually causally determined. Since no set of causes we select as the determiners of human action can be complete, the feeling of freedom arises out of this ignorance of causes.

To that extent we may act as if we are free. There is much to gain, little to lose, and personal responsibility follows. I close with William Ernest Henley's powerful poem "Invictus," especially fitting since he wrote it when he was terminally ill and in the context of the nineteenth-century push for scientific determinism, as if to say it ain't so:[35]

Out of the night that covers me,
Black as the pit from pole to pole,
I thank whatever gods may be
For my unconquerable soul.

In the fell clutch of circumstance
I have not winced nor cried aloud.
Under the bludgeonings of chance
My head is bloody, but unbowed.

Beyond this place of wrath and tears
Looms but the Horror of the shade,
And yet the menace of the years
Finds and shall find me unafraid.

It matters not how strait the gate,
How charged with punishments the scroll,
I am the master of my fate:
I am the captain of my soul.

II

A SCIENCE OF PROVISIONAL ETHICS

The aim of ethics is to render scientific—i.e., true, and as far as possible systematic—the apparent cognitions that most men have of the rightness or reasonableness of conduct, whether the conduct be considered as right in itself, or as the means to some end conceived as ultimately reasonable.

—Henry Sidgwick, *The Methods of Ethics*, 1874

5

CAN WE BE GOOD WITHOUT GOD?: SCIENCE, RELIGION, AND MORALITY

The greatest part of morality is of a fixed eternal nature, and will endure when faith shall fail.

—Bernard de Mandeville, *An Inquiry Into the Origin of Moral Virtue,* 1723

On the morning of Tuesday, April 20, 1999, two students in black trench coats killed fourteen of their fellow classmates and a teacher at Columbine High School in Littleton, Colorado, a suburb of Denver. The gunmen, Eric Harris, eighteen, and Dylan Klebold, seventeen, reportedly asked their victims, "Do you believe in God?" and allegedly snuffed out their young lives if they responded in the affirmative. The boys began their attack in the parking lot, picking off students with apparent indiscrimination, then proceeded to a ground-floor cafeteria, moved through school hallways, and ended up in a second-floor library before finally turning their weapons on themselves.

What was the cause of this murderous rampage? By the time I tuned in to CNN that afternoon, "experts" were already proffering theories that included television and movie violence, rock music, morbidly violent computer games, gangs and cults, parental neglect, teenage angst, and revenge for peer ridicule and rejection. Months later there was still no causal consensus. Perhaps if only parents paid more attention to their children, or if school administrators tried to decrease campus racism, or if school counselors could nip student bullying in the bud, or if teachers could check anti-Christian prejudice at the classroom door, or if everyone could learn to love instead of hate. As the documentary film producer Michael Moore wondered in his Academy Award–winning

Figure 20. The Massacre at Columbine High School

Eric Harris, age eighteen, and Dylan Klebold, age seventeen, are captured on a surveillance video moving through the Columbine High School cafeteria on the morning of Tuesday, April 20, 1999, on a mission to kill as many of their fellow students as possible before killing themselves. What would cause people to commit such acts of violence? (Courtesy of Associated Press)

film *Bowling for Columbine,* since the last thing the boys were doing before they went on their shooting rampage was bowling, it is surprising that no one has placed the blame there.

When the Jefferson County Sheriff's Department released its report the following year, it contained more than 10,000 pieces of evidence that included over 4,000 leads and over 5,000 interviews. The attack, they concluded, was driven by indiscriminate hate, was intended to wipe out most of the student body of Columbine High, and was supposed to end in suicide. In Harris's journal, which opened unequivocally with "I hate the fucking world," he railed against everyone from the WB network and slow drivers to racists, minorities, and whites. In his rambling screed, on one page he praised Hitler's efforts to eradicate European Jewry and on the next he obsessed about finding a date for the high school prom. His celebrated "hit list" included targets as risibly ridiculous as Tiger Woods.

The report was a monumental disappointment to those searching for *the* cause of the crime, the magic bullet, the single cause that could be directly addressed through legislation or social action. "I know a lot about both of them," said Kate Battan, the lead investigator for the Jefferson County Sheriff's Department. "This was not about killing jocks or killing blacks or killing Christians. . . . It was about killing everybody." But *why* did the boys kill, she was asked? "Everybody wants a quick answer. They want an easy answer so that they can sleep at night and know this is not going to happen tomorrow at their school. And there is no such thing in this case. There's not an easy answer. I've been working on this nonstop daily since April 20th and I can't tell you why it happened."[1]

That opened the door for wannabe social commentators and ad hoc social scientists to speculate wildly and with no evidence about the deeper cause of Columbine. The violent computer game Doom, for example, was blamed, as when the *New York Times* reported that the boys played "popular computer games in which players stalk their opponents through dungeon-like environments and try to kill them with high-powered weapons." The *Washington Post* described the online gaming world as a "dark, dangerous place." CNN said that the boys "reportedly played computer games often, spending hours trying to kill each other with digital guns and explosive devices." So-called Doom-sayers rallied to its defense, with such comments as this from a Web posting: "Doom has nothing to do with this. I enjoy making Doom more gruesome, I watch movies such as *Evil Dead* 2 and *Terminator* 2, and I listen to . . . [rocker] Rob Zombie, but I don't even want to touch a real gun, bomb, chain saw, or anything." Another fan correctly pointed out that if Doom were the cause of teenage violence, then "surely everyone who played Doom would be running around with guns and other instruments of violence and death." Similarly, a gamer with the log-in name Theoddone33 skeptically observed, "Everyone is always quick to point out murderers that play violent video games, but no one ever thinks of the millions of people that play video games and aren't murderers."[2]

One of the fundamental tenets of science is that a theory should be able to explain the exceptions to its generalizations. This is a problem for the computer-game theory of violence, as it is for the other theories. For example, physician and author Dr. Julian Whitaker blamed the use of prescription drugs: "When I first heard about the Columbine High

School massacre, my initial thought was, 'Lord help us, were they taking Prozac?' Nine days later, it was reported that Eric Harris, one of the shooters, was taking Luvox, which, like Prozac, Zoloft, and Paxil, belongs to the class of drugs known as selective serotonin reuptake inhibitors (SSRIs). In one out of every twenty-five children taking it, Luvox causes mania, 'a psychosis characterized by exalted feelings, delusions of grandeur and overproduction of ideas.' "[3] Whitaker's theory, however, fails to explain why Columbine would be the only case of SSRI-induced mass murder. Another shortcoming can be found in an explanation offered by Dr. Ned Holstein, president of Fathers and Families in Boston, who claimed that fatherless homes cause teen violence: "The strongest predictor of youth violence is not poverty, not race, not inadequate gun laws, not the presence of gangs, not the wearing of camouflage clothing, not portrayals of violence in the media and not lack of midnight basketball. It is the lack of a father in the home."[4] Unfortunately for this theory, both Harris and Klebold were from intact two-parent families.

Initially, much was made of the fact that April 20 is Adolf Hitler's birthday, that Harris had praised Hitler for the "final solution" to the Jewish question, and that both boys had occasionally been seen wearing swastikas. Cults and gangs in general, in fact, were also targeted, such as the "Goths," who wear black and unusual clothing, and the "Trench Coat Mafia," because the boys were known to wear long black dusters as seen in old West photographs. When further investigations failed to turn up any additional links to Nazis, neo-Nazis, or cults of any kind, a gang unit specialist for the Denver Police, Steve Rickard, blamed emotional problems at home: "A lot of times entertainment—music, movies—is the trigger. It's not the cause, necessarily, it's the little push that makes them do something."[5] If not music and movies perhaps, some wondered, the push came from homosexuality. Harris and Klebold had allegedly been called "faggots" by some of the Columbine High jocks, so the rumor mill churned out stories about gays gone mad.[6] The girlfriends of the boys, however, disconfirmed this thesis. Another howler was suggested by the World Socialist Web master David Walsh: "Defenders of capitalism . . . long for a society where profit and loss are the only means of determining the value of any activity or human being. . . . What would such a society, guided only by selfishness and violence, look like? The events in Jonesboro [a shooting tragedy similar to Columbine] give some indication."[7] This theory also fails to account for disconfirmatory evidence, such as teen violence in noncapitalistic countries, or capitalistic countries like Japan where violence of any type is almost unheard of, among both teens and adults.

Ironically, some even identified *insufficient* violence as the cause, violence, that is, in the form of good old-fashioned parental discipline and adult authority. "I feel like the lack of discipline has led to what we are into now, total chaos and disrespect," said Senator Frank Shurden of Henryetta, Oklahoma, who, after Columbine, proposed a bill in the Oklahoma state legislature that would encourage parents to use "ordinary force" such as spanking, paddling, or whipping to discipline their children. (The bill passed in the Senate 36–9, and in the House 96–4.) "Back when I grew up, we got our tails whipped at school, then got it again when we got home. We didn't have shootings." Sorry, Senator Shurden, single anecdotes do not make a science. As for adult authority, onetime presidential candidate Gary Bauer fingered teachers and administrators at Columbine High: "Why did adults in that school feel that they couldn't grab Eric [Harris] and Dylan [Klebold] and say, 'You know, if I see you give the Nazi salute again, if I don't break your arm you're going to be out of this school for the rest of this year.'"[8] Of course, the parents of kids who get their arms broken by teachers and administrators are unlikely to feel that this is an appropriate form of discipline and authority.

Because of the round-the-clock media coverage that Columbine captured, high-profile politicians could not resist tossing in their own extemporized explanations. Former House Speaker and noted conservative Newt Gingrich blamed (who else?) the liberal elite: "I want to say to the elite of this country—the elite news media, the liberal academic elite, the liberal political elite: I accuse you in Littleton, and I accuse you in Kosovo of being afraid to talk about the mess you have made, and being afraid to take responsibility for things you have done, and instead foisting upon the rest of us pathetic banalities because you don't have the courage to look at the world you have created."[9] The chief elite liberal of the day, President Bill Clinton, understandably focused on a different causal vector—Hollywood: "We cannot pretend that there is no impact on our culture and our children that is adverse if there is too much violence coming out of what they see and experience." Hollywood promptly fired back a defending salvo: "If you're looking for violence, what about the evening news?" David Geffen asked rhetorically. "America is bombing Yugoslavia; it's on every day. It's not a movie, it's real."[10]

Guns, of course, were an easy target for Columbine commentators, with noted gun control advocates like Sarah Brady squawking for more legislation. (Brady is the chairperson of the Brady Campaign to Prevent Gun Violence, and the wife of James Brady, who was shot by John Hinckley in his attempt to assassinate Reagan.) The ledger was predictable, with liberals calling for more gun control and conservatives seeking a more sinister evil lurking behind those who wield the guns that kill people.

That evil, we are told, is the lack of morality and religion in public life, especially public schools. On Wednesday, June 16, 1999, barely two months after the Columbine shootings, for example, Congressman Tom DeLay, the Majority Whip, read a letter on the floor of the House of Representatives that reverberated throughout the country and became a flash point of political pundits and radio talk show hosts for months to come. It was written by Addison L. Dawson to the editor of the *San Angelo Standard-Times* (Texas). The letter was originally published in the paper on April 27, one week after the massacre, but DeLay's reading of it led to the mistaken belief that it was written by DeLay, and he has been quoted as its author ever since. No matter, because by reading the letter, DeLay was endorsing Dawson's thesis, which was that guns do not kill people; rather, something else kills people, that something else being broken homes, children's lack of quality time with parents, day care, television sex and violence, computer games, contraception and planned parenthood, abortion, small family size (a direct result of the previous two), short prison sentences for hardened criminals, and, most notably for our discussion (the style of the letter is sarcastic):

> It couldn't have been because our school systems teach the children that they are nothing but glorified apes who have evolutionized out of some primordial soup of mud by teaching evolution as fact and by handing out condoms as if they were candy.
>
> It couldn't have been because we teach our children that there are no laws of morality that transcend us, that everything is relative and that actions don't have consequences. What the heck, the president gets away with it.
>
> Nah, it must have been the guns.[11]

In my opinion, of all the theories an Columbine (as well as other social ills), people deeper than this one: a scientific an with the theory of evolution, implies that th tive basis for morality, no moral principles Archimedean point outside of us from whic world. One minister succinctly summarize the Bible gets it wrong in biology, then w when it talks about morality and salvation."[12] We saw such arguments even before Columbine. The school shootings in Paducah, Kentucky, in December 1997, for example, were blamed on the "godlessness" of the perpetrator, fourteen-year-old Michael Carneal, who opened fire on a school prayer meeting, killing three fellow students and wounding five others. Christian commentators branded Carneal an atheist, and some Christian students at Paducah High claimed that gangs of atheists roamed the hallways targeting Christians for violence. (Subsequently, Carneal's priest, Reverend Paul Donner of St. Paul Lutheran Church, corrected the mistake: "Michael Carneal is a Christian. He's a sinner, yes, but not an atheist."[13] Thank God for that.)

Following Columbine, Christian organizations went into overdrive to push for legislation to bring God back into the "Big House," as Wendy Zoba referenced the spheres of public education and politics in her post-Columbine treatise on what's wrong with America. "Columbine posed a question we weren't prepared to answer and answered a question we did not ask," Zoba reflected. "The question Columbine presents is not what the killers did or did not ask their victims about God but what their deeds ask us about God. If what happened on April 20, 1999, is something we, as a people, cannot abide—as we seem to be concluding—we are forced to confront the follow-up question: Do we need to invite God back into the Big House?" Of course, Zoba is savvy enough to know that posting the Ten Commandments on the walls of Columbine would have done nothing to deter Harris and Klebold in their rampage. "But the sentiment is an expression of a larger truth: There is a God, and he has established a moral order, and we must find a way to make both part of the cultural conversation. How do we heal a nation whose moral fabric has come apart without introducing the language of faith in a higher law?" Zoba answers her own question this way: "Columbine has become the crucible for a larger cultural debate; not about whether Americans believe in God—numerous surveys reveal

—but about whether the God they believe in is relevant. question Harris and Klebold put to their victims when they Do you believe in God?" while pointing a gun to their heads. It e question their victims' responses posed to us. It is the question at has made us all 'Columbine.'"[14]

Good God

Without God are we all "Columbine"? Is a belief in God necessary to right the wrongs of immoral behavior? The 103rd archbishop of Canterbury (St. Augustine was the first, in 597) thinks so. On Friday, May 24, 1996, this spiritual leader of over 70 million Anglicans told 425 civic, business, and religious leaders at the Los Angeles Biltmore Hotel that "secularism" is the cause of much of the West's moral woes. Paradoxically, this was followed by a litany of "unspeakable atrocities against innocent people" committed in the name of religion, as in Bosnia and against the Christian minority in Islamic Sudan. The archbishop—the Most Reverend George L. Carey—told his audience that only faith could stop these atrocities:

> How else can momentum be found for combating the worst excesses of poverty and inequality around the world? How else can we find the self-restraint in the interest of future generations in order to save our environment? How else can we combat the malignant power of exclusive nationalism and racism? All this requires the dynamic power of commitment, faith and love. The privatized morality of "what works for me" will not do.[15]

Agreed, unalloyed self-interested morality will not suffice. But is our only choice between godly morality and godless immorality, as is so often presented by both theologians and the religious virtue peddlers of pop culture? Can we be good without God?

The Grand Inquisitors Say No

The "Grand Inquisitor" is the literary antagonist of Fyodor Dostoyevsky, a Russian socialist who was arrested in 1848 following the political revolutions that threatened the Russian monarchy. He was found guilty of conspiring against the Orthodox Church and the Russian government as part of the Petrashevsky circle, followers of the French socialist Fourier. The sentence included a bizarre mock death sentence and execution, followed by a six-year stint in a penal colony where Dostoyevsky had not "one single being within reach with whom I could exchange a cordial word. I endured cold, hunger, sickness. I suffered from the hard labors and the hatred of my companions" but "the escape into myself . . . did bear its fruits." Among the fruits was his profound religious crisis, which was triggered in Dostoyevsky after he read the Bible. As a result, he eschewed the social and political ideas of his youth and became deeply religious. Subsequent years of turmoil and poverty left him feeling like "a foreigner in a foreign land," and while in exile he composed his greatest work, which was to explore "the problem that has consciously and unconsciously tormented me all my life."

That problem was the existence of God, and the work became *The Brothers Karamazov.* Among the many deep issues addressed in his tome, Dostoyevsky considered the following question: if God does not exist, does anything go? "Nothing is more seductive for man than his freedom of conscience, but nothing is a greater cause of suffering."[16] As the Grand Inquisitor noted, if God granted us freedom to make moral choices, then what is the use of the ancient laws He gave us? "In place of the rigid ancient law, man must hereafter with free heart decide for himself what is good and what is evil, having only Thy image before him as his guide. But didst Thou not know that he would at last reject even Thy image and Thy truth, if he is weighed down with the fearful burden of free choice?"[17]

If God does not exist, then what is the origin of morality? The answer presented in part 1 of this book is that evolution generated the moral sentiments out of a need for a system to maximize the benefits of living in small bands and tribes. Evolution created and culture honed moral principles out of an additional need to curb the passions of the body and mind. And culture, primarily through organized religion, codified those principles into moral rules and precepts. The next logical question to ask, then, and one that is answered in part 2 of this book, is this: can we lead moral lives without recourse to a transcendent being that may or may not exist? Can we construct an ethical system without religion? Most believers and theists answer no. Dostoyevsky's Grand Inquisitor, for example, suggested that this need for a higher source for morality lies at the very foundation of religion: "So long as man remains free he strives for nothing so incessantly and so painfully as to find someone to worship. . . . This craving for community of worship is the chief misery of every man individually and of all humanity from the beginning of time."[18] In his *Casti connubii* of December 31, 1930, Pope Pius XI agreed that without God anything goes: "For the preservation of the moral order neither the laws and sanctions of the temporal power are sufficient nor the beauty of virtue and the expounding of its necessity. A religious authority must enter in to enlighten the mind, to direct the will, and to strengthen human frailty by the aid of divine grace."[19]

Of course, we should not be surprised to find that the leader of the world's largest religious denomination believes that religion is a fundamental necessity for sustained moral behavior. A more contemporary and pop-culture answer is provided by an individual who, you might say, in recent years has become America's Grand Inquisitor. In the 1990s a self-appointed religious moral authority entered the American landscape to enlighten her listeners to the need of divine grace. She is Laura Schlessinger.

On March 19, 1998, I attended a prayer breakfast sponsored by the Glendale, California, Chamber of Commerce, with Dr. Laura as the featured speaker, and, considering the hour (6:00 A.M.), there was a remarkable turnout of approximately 850 people. When she was introduced, it was announced that the week before she had surpassed Rush Limbaugh in number of listeners to become the most popular radio talk show host in America. The syndication of her program set an all-time record for growth (now in excess of 450 stations). Her books are national best-sellers. Her lectures are typically standing room only. During her daily three-hour program, over 65,000 people jam the phone lines, hoping to be one of the lucky few to be able to speak to her. At the breakfast I attended, the title of her lecture was "Can You Be Good Without God?" Her short answer was: "Here and there, but not consistently through all the things that humans have to suffer."

Schlessinger's long answer included an exposition on her personal history, in which she recapped her youth in the "anything goes" 1960s when she was relatively freewheeling and "grew up with no God." Her mother was a "nice Catholic girl from Italy," and her father "a nice Jewish boy from Brooklyn," neither of whom believed in God. So for Schlessinger, and for so many others that decade, anything went morally.

(Well, not quite for Schlessinger, who admitted that her parents did instill some moral principles in her. Something like this argument is often used by believers to explain how and why nonbelievers are moral—even if raised in a nonbelieving household the culture in which they reside is a Judeo-Christian one in which moral precepts and beliefs are inculcated tacitly as part of the general zeitgeist and prevailing milieu.) What is "good" in this system, she explained, is "what I really want to do, what is really turning me on, what is titillating, what is available, what is seductive, what is exciting, what is fun. Without God that is pretty much how we define 'good'—it is a matter of opinion. Your opinion of what is good is probably going to be based on what you were taught, some opportunities that are available, and this magnificent brain that can rationalize anything." Schlessinger explained that over the long haul, however, this morally carefree philosophy was unsatisfying to her—not miserable, mind you, but nothing like a feeling of moral closure or satisfaction. Like most religious converts who describe their conversion in terms of a fulfillment process, Schlessinger explained, "something was missing" from her life. Finally she found God, converted to Judaism, and now has a moral compass that she points at Americans daily from noon to three. "I'm a prophet," she told the *Los Angeles Times*. "This is a very serious show."[20]

Schlessinger's argument was similar to the one she made to me privately several months prior when she resigned from the editorial board of *Skeptic* magazine, a science publication I edit. In 1994 we invited Schlessinger to be on our board because of the public position she bravely took on the recovered memory movement (in which therapists alleged they could extract repressed memories of childhood sexual abuse through suggestive talk therapies, hypnosis, fantasy role playing, and the like). Schlessinger even spoke for us as part of our public science lecture series at Caltech, and she delivered a brilliant exposition on self-reliance, critical thinking, independence of thought, and other attributes admired by freethinkers, humanists, and skeptics. Even after her conversion to orthodox Judaism and a surfeit of critical letters from readers that subsequently came pouring into *Skeptic*'s office, we left her on our board because we do not believe in excluding people based on their religious beliefs. Her later resignation, then, surprised me: "Please remove my name from your Editorial Board list published in each of your *Skeptic* Magazine issues immediately. Science can only describe what; guess at why; but cannot offer ultimate meaning. When man's limited intellect has the arrogance to pretend an ability to analyze God, it's time for me to get off that train." Our follow-up conversation clarified to me that, for Schlessinger, the subject of God's existence was off-limits to science. There is a God. Period. And morality follows.

As she wrote in an opinion editorial in the *Calgary Sun* on September 9, 1997: "There are those who say it is feasible to be moral without God or religion. I think they are all wrong."[21] The bottom line, Schlessinger believes, is that humans are naturally deceitful, innately evil, and inherently bad. "Being good is not natural. Being good requires you to overcome your own self-interest." In short, if we think we can get away with something, we will. Of course, we cannot get away with just anything since we have laws and customs, so we try to get away with what we can, hoping we will not get caught. "Getting caught," says Schlessinger, is the level of morality most people attain, but a belief in God elevates morality to a higher level. If you think you can get away with something, this is when anything goes, and she hears tales of this every day on the radio. But, says Schlessinger, you never get away with anything because God is always watching. He can even see through concrete she explained (in an offhanded one-liner about God knowing you are stealing from a store even if no one can see you). "The notion of God is really, fundamentally, all we have to truly lead us to be good or else we make our own decisions and we become, individually, our own Gods."[22]

That's it. That is the core of Schlessinger's argument. There was nothing about these moral principles being worthy of following in their own right. There was nothing about treating other people as you would like to be treated. There was nothing about human rights or human dignity. For Schlessinger, it comes down to this: you'll be busted by Mr. Big if you sin, so don't.

On the simplest of levels, of course, any of us can be good without God. Most sincere and honest religious folks admit that anyone—even atheists—can *occasionally* be good. Their deeper argument lies in the sustainability of right moral actions across varied circumstances and extended periods of time. Without a religious foundation, they argue, the flesh is weak and the mind is a willing coconspirator in the justification of doing the wrong thing. It is often difficult to do the right thing and, they argue, without that extra transcendental boost from above we fail too often.

Theism, Atheism, and Morality

Events of an evil and primeval nature force us to confront the deepest questions about our moral nature, beginning with the nature of morality if there is no God and the fate of ethics in a secular and scientific society. Are we doomed to destruction if we do not accept the objective value of moral absolutes offered by religion? Do we need religion-based morality as an antidote to the alleged nihilism of a secular and scientific society? Many theologians and religious believers think that we do, and they are often responding to their perception of what it means to embrace science and secularism. Among both social commentators and moral philosophers the consummate example of the result of secular morality is the Holocaust. The Nazi regime, we are told, was a godless atheistic one that led directly and ineluctably to a relativistic morality that justified the brutal murder of millions of people.

The problem with this particular case is that Hitler and the Nazis were not atheists. In *Mein Kampf* Hitler observed that "faith is often the sole foundation of a moral attitude" and that "the various substitutes have not proved so successful from the standpoint of results that they could be regarded as a useful replacement for previous religious creeds." In fact, Hitler argues that an attack against religion "strongly resembles the struggle against the general legal foundations of a state" and "would end in a worthless religious nihilism."[23] Hitler's most famous statement on the subject was made in his Reichstag speech of 1938, when he proclaimed: "I believe today that I am acting in the sense of the Almighty Creator. By warding off the Jews I am fighting for the Lord's work."[24] As for the Third Reich itself, number twenty-four of the original twenty-five points of the German Workers' party proclaimed liberty for all religious denominations "so far as they are not a danger to it and do not militate against the moral feelings of the German race." The party, it was stated, "stands for Positive Christianity."[25] In 1934, Professor Ernst Bergmann penned a twenty-five-point catechism for the core of this new "Positive Christianity" that included, in point number six, this denunciation of atheism and nonbelief: "The German religion is a religion of the people. It has nothing in common with free thought, atheist propaganda, and the breakdown of current religions."[26]

Although it is certainly true that the Lenin-Stalin regime of the Soviet Union was atheistic in principle, sociologists of religion are now discovering that throughout the seventy-five-year-long social experiment of Communism, religious faith remained steadfast, albeit underground and practiced with considerable stealth.[27] As for the rest of the twentieth century, at its beginning the Great War featured God-fearing, Ten Commandment–swearing men who killed other God-fearing, Ten Commandment–swearing men, all in the name of God. By the end of the century, wars, revolutions, and acts of terrorism committed in the name of God were almost nightly news affairs. The fact that the twentieth century was the bloodiest century in human history (by raw numbers only, not by percentage of population casualties) has nothing whatsoever to do with a lack of religious or moral values (which, clearly, were not lacking). Given the killing technologies of modern states (and their correspondingly larger populations) there is little doubt that the crusades, inquisitions, and religious wars of the medieval and early modern periods would have easily produced the vast killing fields of our time. The problem is not a lack of God, religion, or morals. It is the wedding of extremism, fundamentalism, and absolute morality, coupled with the means of murder and access to masses of humanity that results in the wanton destruction we have witnessed in modern times. And it is only fair to ask, what if religion is not the solution but is actually part of the problem? This is not an argument that I am particularly disposed to make, but one can make the observation that if more (and a greater percentage of) Americans believe in God than ever before in history, and if America is going to hell in an immoral handbasket as never before, then at the very least the argument that we cannot be good without God would seem to be gainsaid.

What Would You Do if There Were No God?

Turning from the level of collective politics to that of individual people, what would you do if there were no God? Would you commit robbery, rape, and murder, or would you continue being a good and moral person? Either way the question is a debate stopper. If the answer is that you would soon turn to robbery, rape, or murder, then this is a moral indictment of your character, indicating you are not to be trusted because if, for any reason, you were to turn away from your belief in God (and most people do, at some point in their lives), your true immoral nature would emerge and we would be well advised to steer a wide course around you. If the answer is that you would continue being good and moral, then apparently you can be good without God. QED.

As anyone with any life experience or a sense of history knows, religious people are more than capable of committing sins and crimes, and nonreligious people are more than capable of being moral and trustworthy citizens and friends. (I am not arguing that religious people are more immoral, just that they are not any more moral than nonreligious people.) Think of child-molesting priests, money-scamming televangelists, or flimflam faith healers. At the same time, think of all the people you know who are not religious, yet who daily perform acts of kindness and generosity. Many of your friends are probably either nonbelievers or give little to no thought to religion. Are they robbers, rapists, or murders? Probably not. How then did they come to be moral? Why do they continue to be moral? Personally, it would frighten me to believe that the people I deal with on a day-to-day basis treat me tolerably well only because they are afraid of God and divine retribution. What happens when their belief in God diminishes or departs altogether? Where do their moral principles go then? To me it is a higher level of morality to be good for its own sake than for the consequences it may bring.

How We Can Be Good Without God

An argument could be made that since America is still primarily a Judeo-Christian society even nonbelievers have imbibed these values, regardless of their personal upbringing—that is, atheists are good because of all the good theists around them. Maybe, but as I argued in part 1, religion codified these moral principles for sound reasons that have nothing to do with divine inspiration. The moral sentiments and principles came first, evolving over the course of a hundred thousand years of humans living in a Paleolithic environment. Religion came second, co-opting morality and codifying it to its own end, all of which happened in just the past couple of thousand years. What would happen if we jettisoned religion altogether? Would society collapse into immoral chaos?

No, it would not. And we have a two-centuries-long experiment in the separation of church and state to prove it. When the United States of America was founded, the original framers of the Constitution, heavily influenced by the secular Enlightenment philosophers whose writings over the previous century had laid the philosophical groundwork for a secular ethical and political system, made it clear that regardless of which religion (or even no religion) one professes belief in, certain moral principles hold. These include the right to life, liberty, and the pursuit of happiness, as well as the other rights protected in the Bill of Rights. What the Enlightenment philosophers were arguing, and the U.S. Constitution framers adopted, was the belief that humans have certain rights and values in and of themselves. These rights and values are grounded not in religion, or any other transcendental state or supernatural force, but in themselves. They stand alone. Humans deserve life, liberty, and happiness, not because God said so but because we are human. Period. These rights and values exist because we say they exist, and that is good enough. They are inalienable because we say they are, and that suffices.

Does this secular system work? To answer the question, we have only to compare the levels of life, liberty, and the pursuit of happiness of the citizens of the United States to those of the citizens of other countries, particularly those still ruled by theocracies. The system is not perfect, plenty of people fall through the cracks, rights are abused, lives are unjustly lost, liberties are unfairly trammeled, and too many are not achieving the levels of happiness that they could. But these are relative judgments, relative to what came before and to what exists elsewhere. Like science, secular ethics may be primitive and flawed, yet it is the most precious thing we have.

6

HOW WE ARE MORAL: ABSOLUTE, RELATIVE, AND PROVISIONAL ETHICS

In science, "fact" can only mean "confirmed to such a degree that it would be perverse to withhold provisional assent."

—Stephen Jay Gould, *Hen's Teeth and Horse's Toes*, 1983

One day in 1991 an attractive middle-aged woman was passing through the locker room of a health club on her way to meet a friend for lunch at the snack bar. She was early and there was no one around. Glancing about for her friend her eye was drawn to a shiny object on the floor. Looking closer she discovered that it was a large diamond ring. She vaguely recalled seeing it on someone at the club before, but could not recall to whom it belonged. She picked up the ring and put it in her pocket. When her friend arrived she immediately showed it to her and asked her to accompany her to the front desk so that she would have a witness to verify that she did not steal the ring, but had simply found it. "No one would ever have known that I had the ring," she later recounted. "I could have hocked it for thousands of dollars, but I didn't." Why? Reflecting on the incident, she explained, "One just doesn't do that. My conscience would not allow me to take it. I consider myself an honest person who tries to do the right thing, and in that instance I knew what the right thing to do would be." Why is that the right thing to do? "Because if it were my ring I would hope that someone would do the same for me." Golden rings and golden rules.

That woman was my late mother, and this story is a classic example of the Golden Rule in practice. She treated the owner of that ring as she hoped someone would treat her if she lost her own. I recount the story here not because I think that my mom was some extraordinarily moral person, but because, in fact, as a moral agent I think she was quite ordinary and that most people most of the time in most circumstances would have done the same thing. She told me this story not as a moral homily to impart some extraordinary advice, but to show the ordinary nature of moral reasoning in response to a question I posed to her about the origins of morality: why are you moral? My mother, who had considerable influence on my thinking and moral upbringing, was not a religious person and had no belief in God. It was not something she thought a lot about—she simply did not believe in God and saw no reason to foist a pretense of belief. She did not raise me to be religious or irreligious. The subject almost never came up. Yet she was a decent, moral person, as is my father, and I think my siblings and I are an ordinary moral family. How was she able to be such an ordinarily moral person without believing in an extraordinarily moral being? Without absolute morality, aren't we reduced to accepting an "anything goes" relative morality? No. There is a middle way between absolute morality and relative morality that I call provisional morality.

Absolute Morality

As defined earlier, morality involves right and wrong thoughts and behaviors in context of the rules of a social group, and ethics is the scientific study of and theories about moral thoughts and behaviors in context of the rules of a social group. Thus, we may define *absolute morality* as *an inflexible set of rules for right and wrong thought and behavior derived from a social group's canon of ethics*. The claimed source of that canon may be God, the Bible, the Koran, the state, nature, an ideology, or a philosophy.

An obvious and immediate problem with all systems of absolute morality—known formally in ethical theory as *absolutism*—is that they set themselves up to be the final arbiters of truth, creating two types of people: good and evil, right and wrong, true believers and heretics. This was most succinctly expressed by that sage philosopher Maxwell Smart—agent Eighty-Six on television's *Get Smart* comedy series—who explained to his morally incredulous fellow agent Ninety-Nine: "Don't be silly, Ninety-Nine. We have to shoot, kill, and destroy. We represent everything that's wholesome and good in the world."

Sadly, such black-and-white thinking is not restricted to the little screen. Richard Nixon used such rhetoric for political gain when he admitted, "It may seem melodramatic to say that the U.S. and Russia represent Good and Evil, Light and Darkness, God and the Devil. But if we think of it that way it helps to clarify our perspective of the world struggle." Ronald Reagan was even more histrionic in his proclamation that the Soviet Union was the "evil empire." Most recently, George W. Bush effectively labeled Osama bin Laden and his Al Qaeda operatives as "pure evil."

Most absolute moral systems are religiously based, but not all. Immanuel Kant's *Categorical Imperative,* for example, is a secular rational attempt at an absolute morality. A Categorical Imperative is an unconditional command without exceptions, which Kant contrasted with (by way of rejecting it) a *Hypothetical Imperative,* a conditional command with exceptions. For Kant, if you want to judge the rightness or wrongness of an action, "Act only on that maxim through which you can at the same time will that it should become a universal law."[1] Would we ever want to universalize lying, stealing, or adultery? Of course not. That would put an end to contracts, property, and marriage.

But people do occasionally lie, steal, and commit adultery, and often there are perfectly rational reasons to do so. In the Categorical Imperative we witness a violation of the law of the excluded middle, also known as the either-or fallacy in logic, where options between extremes are excluded by forcing the issue into a binary choice. Here yet another problem is averted with fuzzy logic, where shades of fuzzy probabilities allow us to assign fractional values to moral answers that are more or less likely to be applicable. The world is usually more complex than the two choices typically presented by antagonists who wish to simplify issues for rhetorical sake. A type specimen of a statement of absolute morality can be found in the words of Christian author Francis Schaeffer:

> If there is no absolute moral standard, then one cannot say in a final sense that anything is right or wrong. By absolute we mean that which always applies, that which provides a final or ultimate standard. There must be an absolute if there are to be morals, and there must be an absolute if there are to be real values. If there is no absolute beyond man's ideas, then there is no final appeal to judge between individuals and groups whose moral judgments conflict. We are merely left with conflicting opinions.[2]

Cartoonist Wiley Miller illustrated the concept cleverly in a *Non Sequitur* cartoon (figure 21) in which Moses is admonishing modern moral relativists that God called them "commandments," not "recommendations," because they are absolute and final, no exceptions.

The ultimate fallacy with all forms of absolute morality is that since virtually everyone claims they know what constitutes right versus wrong thought and action, and since effectively all moral systems differ from all others to a greater or lesser degree, then there cannot be a universally accepted absolute morality. In reality, and ironically, it is absolute moralities that leave us with nothing but conflicting opinions and no moral compass. Nowhere is this problem more evident than in religion.

Most ethical systems are absolute, most absolute systems are derived from religious sources, and by far the most popular source of moral precepts and ethical conjectures is religion (making Divine Command Theory one of the most common of all ethical systems). The 2001 *World Christian Encyclopedia,* for example, reports that of the earth's 6.1 billion humans fully 5.1 billion of them, or 84 percent, declare themselves followers of some form of organized religion. Christians dominate at just a shade under 2 billion adherents (with Catholics counting for half of those), with Muslims at 1.1 billion, Hindus at 811 million, Buddhists at 359 million, and ethnoreligionists (animists and others in Asia and Africa primarily) accounting for most of the remaining 265 million. Such overall numbers, however, tell us little. There are, in fact, 10,000 distinct religions of ten general varieties, each one of which can be further subdivided and classified. For example, Christians may be found among an astonishing 33,820 different denominations. The variety of non-Christian religions is also stunning, with worldwide distribution outstripping Christian religions despite the tireless efforts of evangelists to convert as many souls to Christ as possible. One table in the encyclopedia, for example, tracks the number of Christians (69,000) and non-Christians (147,000) by which the world will increase over the next twenty-four hours. Another table reveals the global convert/defector ratio, adjusted for births and deaths, indicating that the sphere of evangelism continues to expand into non-Christian belief space.[3] Given this almost unfathomable level of religious differences, it is obvious that any claim to sole possession of absolute moral truth is fleeting. Clearly they cannot all be right.

Figure 21. Absolute v. Relative Morality

The Ten Commandments are a form of absolute morality. (© 2002 Wiley Miller. Distributed by Univeral Press Syndicate. Reprinted with permission)

Relative Morality

Relative morality is taken to mean *a flexible set of rules for right and wrong thoughts and behaviors derived from how the situation is defined by the social group*. The problem with relative morality—known formally in ethical theory as *relativism*—is that one can justify almost any behavior, implying that all moral actions—from self-sacrifice to human sacrifice—are equal. On a theoretical and scientific level, this is simply not true. On a practical level no one believes this. (Ethical theorists distinguish between *descriptive ethical relativism,* which passes no judgment on whether any of the numerous relative ethical theories are valid or not; and *normative ethical relativism,* which claims that each ethical theory, while relative in value compared to others, is absolutely valid for the culture in which it is practiced.)

When I was a senior in high school in 1971 I became a born-again Christian. I took my commitment seriously enough to enroll at Pepperdine University, a highly regarded Christian institution affiliated with the Church of Christ and nestled in the foothills of Malibu, California, with grand vistas of the Pacific Ocean (okay, so the attraction was not purely academic). There I studied theology and psychology, attended chapel at least twice a week (admittedly attendance was a requirement), wrestled with the relationship between science and religion, and struggled with the normal carnal impulses of youth when they bump up against moral restrictions on their expression. (One student in our dorm, desperately seeking a rationale for what he knew he could not control, actually prayed for God to provide him with an acceptable sexual outlet—read partner—because, he reasoned, he could witness for the Lord better if he were not so distracted by such basic urges.) After graduating from Pepperdine and studying evolutionary biology and experimental psychology in a graduate program at California State University, Fullerton, I turned to science and philosophy for my moral answers, and began to try different ethical systems (not unlike Woody Allen's character in his film *Hannah and Her Sisters,* who examines different religions, for example, coming home one day with a crucifix, a loaf of white bread, and a jar of mayonnaise to try out Catholicism!).

Existentialism initially appealed to me because of its emphasis on moral freedom and individual responsibility. "Existence precedes essence" is a core tenet, meaning that our essence—our being, our very self—is constantly being created by the experiences we choose. We are the authors of our life stories, the architects of our souls. Very few people are innocent victims; rather, we make choices in life that ultimately place us in circumstances in which it might appear we were blameless sufferers but, in fact, most situations are created by the choices we make. Although this puts a rather sizable burden of responsibility for the outcome of your life squarely on your own shoulders, it also means that you can change; you are not stuck where you do not want to be. "Man is a wholly natural creature whose welfare comes solely from his own unaided efforts," wrote one existentialist. To me, existentialism was one of the more optimistic philosophies I examined, but I discovered that I was in a rather small minority in that regard. Most existentialists believe that life is "absurd" because we exist in a meaningless, irrational universe—any attempt to find ultimate meaning can only end in absurdity. Most existentialists seemed to agree with one of the philosophy's founders, Albert Camus, when he lamented, "There is but one serious philosophical problem. That is suicide. Why stay alive in a meaningless universe?"[4] Suicide may be painless (as the *M*A*S*H* theme song croons) but it brings on one major change I found unacceptable.

After existentialism I tried utilitarianism, based on Jeremy Bentham's principle of the "greatest happiness for the greatest number." Specifically, I found his quantitative utilitarianism attractive because of its scientistic approach in attempting a type of hedonic calculus where one can quantify ethical decisions. By "hedonism" Bentham did not mean a simple pleasure principle where, in modern parlance, "if it feels good, do it." In fact, Bentham specified "seven circumstances" by which "the value of a pleasure or a pain is considered":

1. Purity—"The chance it has of not being followed by sensations of the opposite kind."
2. Intensity—The strength, force, or power of the pleasure.
3. Propinquity—The proximity in time or place of the pleasure.
4. Certainty—The sureness of the pleasure.
5. Fecundity—"The chance it has of being followed by sensations of the same kind."
6. Extent—"The number of persons to whom it extends; or (in other words) who are affected by it."
7. Duration—The length of time the pleasure will last.[5]

As a pedagogical heuristic, I once presented the table in figure 22 to my introductory psychology course to draw students into seeing the problem of assigning actual numbers to these seven values (the boxes were blank), in making a rather simple choice between spending money on a good meal, a good date (with the possibility but not certainty of sex), or a good book. The values in the boxes are my own (I was single at the time).

According to Bentham, once the figures are assigned, "Sum up all the values of all the *pleasures* on the one side, and those of all the pains on the other. The balance, if it be on the side of pleasure, will give the good tendency of the act upon the whole, with respect to the interests of that *individual* person; if on the side of pain, the *bad* tendency of it upon the whole."[6] In my example the book wins out over the meal or date. Of course, this is just my opinion, the application of the hedonic calculus to one person. To apply the principle to society as a whole, Bentham says, we must:

> Take an account of the *number* of persons whose interests appear to be concerned; and repeat the above process with respect to each. *Sum up* the numbers expressive of the degrees of *good* tendency, which the act has, with respect to each individual, in regard to

HOW SHOULD
YOU SPEND $25?
(On a scale of 1–10)

HAPPINESS OR PLEASURE	*A GOOD MEAL*	*A GOOD DATE (maybe sex)*	*A GOOD BOOK*
Purity	10	7	2
Intensity	9	8	2
Propinquity	10	8	2
Certainty	9	5	10
Fecundity	0	5	10
Extent	1	2	10
Duration	1	2	10
TOTALS	**41**	**37**	**46**

Figure 22. Jeremy Bentham's Hedonic Calculus

whom the tendency of it is *good* upon the whole: do this again with respect to each individual, in regard to whom the tendency of it is *good* upon the whole: do this again with respect to each individual, in regard to whom the tendency of it is *bad* upon the whole. Take the *balance;* which, if on the side of *pleasure,* will give the general *good tendency* of the act, with respect to the total number or community of individuals concerned; if on the side of pain, the general *evil tendency,* with respect to the same community.[7]

Dismissing the obvious impossibility of doing this on a daily basis and being able to even leave the house, it is clear that you can cook the numbers to make it come out almost any way you like. Doing this on a societal level is simply impossible.

Utilitarianism, particularly in the form of calculating the greatest good for the greatest number as if one were computing an orbital trajectory of a planetary body, is very much grounded in pre-twentieth-century psychological, social, and economic theory that presumed humans (at least Western industrial peoples) to be rational beings who make choice calculations along the lines of a double-entry bookkeeper. (Utilitarians even designated units of pleasure as "hedons" and units of displeasure as "dolors"—in the manner of physicists measuring photons and electrons—and debated among themselves whether we should try to maximize utility or, as *satisficing utilitarians* held, should only try to produce just enough utility to satisfy everyone minimally.) Moral choices, then, were simply a matter of looking at the bottom line.

Thanks to extensive interdisciplinary research by psychologists, sociologists, and economists over the past several decades, however, we now know that humans are emotional and intuitive decision makers subject to the considerable whims of subjective feelings, social trends, mass movements, and base urges. We are rational at times, but we are also irrational, the latter probably a lot more than we care to consider. As we shall see at the end of this chapter, moral reason must be balanced with moral intuition.

These are just a few of the ethical systems that appealed to me, but there are many others for the student of ethics and morality to sample. For example: *consequentialism,* as the name implies, holds that the consequences of an action should determine whether it is right or wrong. *Contractarianism* posits that contractual arrangements between moral agents establish what is right and wrong, where violations of agreements are immoral. *Deontology* claims that one's duty (*deon* is Greek for duty) is the criterion by which actions should be judged as moral or immoral. *Emotivism* holds that moral judgments of right or wrong behavior are a function of the positive or negative feelings evoked by the behavior. *Ethical egoism* (or *psychological egoism*) states that people behave in their own self-interest and thus even apparently altruistic behavior is really motivated by selfish ends. *Moral isolationism,* a form of moral relativism, argues that we ought to be morally concerned only with those in our immediate group, "isolating" those outside our group as not relevant to our moral judgments. *Natural law theory* states that there is a natural order to the human condition, the natural order is good, and therefore the rightness or wrongness of an action should be judged by whether or not it violates the natural order of things. *Nihilism* denies that there is any truth to be discovered, particularly in the moral realm. *Particularity* contrasts with universality and impartiality, holding that we have moral preferences to particular people morally relevant to us. *Pluralism* (an approach very much embraced in this book) holds that there are multiple perspectives that should be considered in evaluating a moral issue, and that no one ethical theory can explain all moral and immoral behavior. *Subjectivism* is an extreme form of relativism, holding that moral values are relative to the individual's sole subjective state alone and cannot even be evaluated in the larger social or cultural context. Encyclopedias of philosophy and morality abound in an alphabet soup of ethical theories and moral labels, and library shelves are sagging with volumes on ethical theories purporting to present the reader with valid and viable criteria of right and wrong human action. What are we to make of all these theories?

Provisional Ethics

If we are going to try to apply the methods of science to thinking about moral issues and ethical systems, here is the problem as I see it: as soon as one makes a moral decision—an action that is deemed right or wrong—it implies that there is a standard of right versus wrong that can be applied in other situations, to other people, in other cultures (in a manner that one might apply the laws of planetary geology to planets other than our own). But if that were the case, then why is that same standard not obvious and in effect in all cultures (as, in the above analogy, that geological forces operate in the same manner on all planets)? Instead, observation reveals many such systems, most of which claim to have found the royal road to Truth and all of whom differ in degrees significant enough that they cannot be reconciled (as if gravity operated on some planets but not others). If there is no absolute moral standard and instead only relative values, can we realistically speak of right and wrong? An action may be wise or unwise, prudent or imprudent, profitable or unprofitable within a given system. But is that the same as right or wrong?

So, both *absolutism* and *relativism* violate clear and obvious observations: there is a wide diversity of ethical theories about right and wrong moral behavior; because of this there are disputes about what constitutes right and wrong both between ethical theories and moral systems as well as within them; we behave both morally and immorally; humans desire a set of moral guidelines to help us determine right and wrong; there are moral principles that most ethical theories and moral systems agree are right and wrong. Any viable ethical theory of morality must account for these observations. Most do not.

In thinking about this problem I asked myself this question: how do we know something is true or right? In science, claims are not true or false, right or wrong in any absolute sense. Instead, we accumulate evidence and assign a probability of truth to a claim. A claim is probably true or probably false, possibly right or possibly wrong. Yet probabilities can be so high or so low that we can act as if they are, in fact, true or false. Stephen Jay Gould put it well: "In science, 'fact' can only mean 'confirmed to such a degree that it would be perverse to withhold provisional assent.'"[8] That is, scientific facts are conclusions confirmed to such an extent it would be reasonable to offer our provisional agreement. Heliocentrism—that the earth goes around the sun and not vice versa—is as factual as it gets in science. That evolution happened is not far behind heliocentrism in its factual certainty. Other theories in science, particularly within the social sciences (where the subjects are so much more complex), are far less certain and so we assign them much lower probabilities of certitude. In a fuzzy logic manner, we might say heliocentrism and evolution are .9 on a factual scale, while political, economic, and psychological theories of human social and individual behavior are much lower on the fuzzy scale, perhaps in the range of .2 to .5. Here the certainties are much fuzzier, and so fuzzy logic is critical to our understanding of how the world works, particularly in assigning fuzzy fractions to the degrees of certainty we hold about those claims. Here we find ourselves in a very familiar area of science known as probabilities and statistics. In the social sciences, for example, we say that we reject the null hypothesis at the .05 level of confidence (where we are 95 percent certain that the effect we found was not due to chance), or at the .01 level of confidence (where we are 99 percent certain), or even at the .0001 level of confidence (where the odds of the effect being due to chance are only one in ten thousand). The point is this: there is a sliding scale from high certainty to high doubt about the factual validity of a particular claim, which is why science traffics in probabilities and statistics in order to express the confidence or lack of confidence a claim or theory engenders.

The same way of thinking has application to morals and ethics. Moral choices in a provisional ethical system might be considered analogous to scientific facts, in being provisionally right or provisionally wrong, provisionally moral or provisionally immoral:

> In *provisional ethics,* moral *or* immoral *means confirmed to such an extent it would be reasonable to offer provisional assent.*

Provisional is an appropriate word here, meaning "conditional, pending confirmation or validation." In provisional ethics it would be reasonable for us to offer our conditional agreement that an action is moral or immoral if the evidence for and the justification of the action is overwhelming. It remains provisional because, as in science, the evidence and justification might change. And, obviously, some moral principles have less evidence and justification for them than others, and therefore they are more provisional and more personal.

Provisional ethics provides a reasonable middle ground between absolute and relative moral systems. Provisional moral principles are applicable for most people in most circumstances most of the time, yet flexible enough to account for the wide diversity of human behavior, culture, and circumstances. What I am getting at is that there are moral principles by which we can construct an ethical theory. These principles are not absolute (no exceptions), nor are they relative (anything goes). They are provisional—true for most people in most circumstances most of the time. And they are objective, in the sense that morality is independent of the individual. Moral sentiments evolved as part of our species; moral principles, therefore, can be seen as transcendent of the individual, making them morally objective. Whenever possible, moral questions should be subjected to scientific and rational scrutiny, much as nature's questions are subjected to scientific and rational scrutiny. But can morality become a science?

Fuzzy Provisionalism

One of the strongest objections to be made against provisional ethics is that if it is not a form of absolute morality, then it must be a form of relative morality, and thus another way to intellectualize one's ego-centered actions. But this is looking at the world through bivariate glasses, a violation of the either-or fallacy, breaking the law of the excluded middle.

Here again, fuzzy logic has direct applications to moral thinking. In the discussion of evil, we saw how fuzzy fractions assigned to evil deeds assisted us in assessing the relative merits or demerits of human actions. Fuzzy logic also helps us see our way through a number of moral conundrums. When does life begin? Binary logic insists on a black-and-white Aristotelian *A or not-A* answer. Most pro-lifers, for example, believe that life begins at conception—before conception not-life, after conception, life. *A or not-A*. With fuzzy morality we can assign a probability to life—before conception 0, the moment of conception .1, one month after conception .2, and so on until birth, when the fetus becomes a 1.0 life-form. *A and not-A*. You don't have to choose between pro-life and pro-choice, themselves bivalent categories still stuck in an Aristotelian world (more on this in the next chapter).

Death may also be assigned in degrees. "If life has a fuzzy boundary, so does death," fuzzy logician Bart Kosko explains. "The medical definition of death changes a little each year. More information, more precision, more fuzz." But isn't someone either dead or alive? *A or not-A*? No. "Fuzzy logic may help us in our fight against death. If you can kill a brain a cell at a time, you can bring it back to life a cell at a time just as you can fix a smashed car a part at a time."[9] *A and not-A*. Birth is fuzzy and provisional and so is death. So is murder. The law is already fuzzy in this regard. There are first-degree murder, second-degree murder, justifiable homicide, self-defense homicide, genocide, infanticide, suicide, crimes of passion, crimes against humanity. *A and not-A*. Complexities and subtleties abound. Nuances rule. Our legal systems have adjusted to this reality; so, too, must our ethical systems. Fuzzy birth. Fuzzy death. Fuzzy murder. Fuzzy ethics.

Moral Intuition and the Captain Kirk Principle

Long before he penned the book that justified laissez-faire capitalism, Adam Smith became the first moral psychologist when he observed: "Nature, when she formed man for society, endowed him with an original desire to please, and an original aversion to offend his brethren. She taught him to feel pleasure in their favorable, and pain in their unfavorable regard." Yet, by the time he published *The Wealth of Nations* in 1776, Smith realized that human motives are not so pure: "It is not from the benevolence of the butcher, the brewer or the baker that we expect our dinner, but from their regard of their own interest. We address ourselves not to their humanity, but to their self-love, and never talk to them of our necessities, but of their advantage."[10]

Is our regard for others or for ourselves? Are we empathetic or egotistic? We are both. But how we can strike a healthy balance between serving self and serving others is not nearly as rationally calculable as we once thought. Intuition plays a major role in human decision making—including and especially moral decision making—and new research is revealing both the powers and the perils of intuition. Consider the following scenario: imagine yourself a contestant on the classic television game show *Let's Make a Deal.* You must choose one of three doors. Behind one of the doors is a brand-new automobile. Behind the other two doors are goats. You choose door number one. Host Monty Hall, who knows what is behind all three doors, shows you what's behind door number two, a goat, then inquires: would you like to keep the door you chose or switch? It's fifty-fifty, so it doesn't matter, right? Most people think so. But their intuitive feeling about this problem is wrong. Here's why: you had a one in three chance to start, but now that Monty has shown you one of the losing doors, you have a two-thirds chance of winning by switching doors. Think of it this way: there are three possibilities for the three doors: (1) good bad bad; (2) bad good bad; (3) bad bad good. In possibility one you lose by switching, but in possibilities two and three you can win by switching. Here is another way to reason around our intuition: there are ten doors; you choose door number one and Monty shows you doors number two through nine, all goats. Now would you switch? Of course you would, because your chances of winning increase from one in ten to nine in ten. This is a counterintuitive problem that drives people batty, including mathematicians and even statisticians.[11]

Intuition is tricky. Gamblers' intuitions, for example, are notoriously flawed (to the profitable delight of casino operators). You are playing the roulette wheel and hit five reds in a row. Should you stay with red because you are on a "hot streak" or should you switch because black is "due"? It doesn't matter because the roulette wheel has no memory, but try telling that to the happy gambler whose pile of chips grows before his eyes. So-called hot streaks in sports are equally misleading. Intuitively, don't we just know that when the Los Angeles Lakers' Kobe Bryant is hot he can't miss? It certainly seems like it, particularly the night he broke the record for the most three-point baskets in a single game, but the findings of a fascinating 1985 study of "hot hands" in basketball by Thomas Gilovich, Robert Vallone, and Amos Tversky—who analyzed every basket shot by the Philadelphia 76ers for an entire season—does not bear out this conclusion. They discovered that the probability of a player hitting a second shot did not increase following an initial successful basket beyond what one would expect by chance and the average shooting percentage of the player. What they found is so counterintuitive that it is jarring to the sensibilities: the number of streaks, or successful baskets in sequence, did not exceed the predictions of a statistical coin-flip model. That is, if you conduct a coin-flipping experiment and record heads or tails, you will encounter streaks. On average and in the long run, you will flip five heads or tails in a row once in every thirty-two sequences of five tosses. Players may feel "hot" when they have games that fall into the high range of chance expectations, but science shows that this intuition is an illusion.[12]

These are just a couple of the countless ways our intuitions about the world lead us astray: we rewrite our past to fit present beliefs and moods, we badly misinterpret the source and meaning of our emotions, we are subject to the hindsight bias where after the fact we surmise that we knew it all along, we succumb to the self-serving bias where we think we are far more important than we really are, we see illusory correlations that do not exist (superstitions), and we fall for the confirmation bias where we look for and find evidence for what we already believe. Our intuitions also lead us to fear the wrong things. Let us return to Adam Smith. According to Smith's theory, our moral sentiments lead us to observe what happens to others, empathize with their pain, then turn to our own self-interest in dreaded anticipation of the same disaster befalling us. The week I wrote this section the ABC television news program *20/20* ran a story about kids who dropped heavy stones off freeway overpasses that smashed through car windows, killing or maiming the passengers within. The producers appealed to the fearful side of our nature by introducing viewers to the hapless victims with mangled faces and shattered lives, evoking our empathy; they then engaged our self-love with the rhetorical question: "could this happen to you?"

Could it? Not likely. In fact, it is so unlikely you would be better off worrying about lightning striking you. Then why do we worry about such matters? Because our moral intuitions have been hijacked by what University of Southern California sociologist Barry Glassner calls a "culture of fear."[13] Who created this culture? Ultimately we did, by buying into the rumors and hearsay that pass for factual data fed to us by the media and other sources. But those factoids and reports had to come from somewhere. Follow the money and those who traffic in fear mongering. Politicians, for example, can win elections by grossly exaggerating (and sometimes outright lying about) crime and drug-use percentages under their opponent's watch. Advocacy groups profit (literally) from fear campaigns that heighten an expectation of doom (to be thwarted just in time, if the donor's contribution is beefy enough). Think of conservatives decrying the demise of the family or liberals proclaiming the destruction of the environment.

Religions play on our fears by hyping up the doom and gloom of this world to make the next world seem all the more appealing. On May 17, 1999, an evangelical Christian friend of mine insisted that we are in the "end times" because the Bible prophesied an increase in immorality and malfeasance. Since everyone knows crime is an epidemic problem in America that worsens by the year ("just look at the recent Columbine shooting," he enthused), the end is nigh. I remember the date because it was the same day the FBI released its findings that we are in the midst of the longest decline in crime rates since the bureau began collecting data in 1930. In other words, we are confronted with the paradox of being more fearful than we have ever been at the same time that things have never been so safe. "In the late 1990s the number of drug users had decreased by half compared to a decade earlier," Glassner explains, yet the "majority of adults rank drug abuse as the greatest danger to America's youth." Ditto the economy, where "the unemployment rate was below 5 percent for the first time in a quarter century. Yet pundits warned of imminent economic disaster."[14] In this century alone modern medicine and social hygiene practices and technologies have nearly doubled our life span and improved our health immeasurably, but Glassner points out that if you tally up the reported disease statistics, out of 280 million Americans, 543 million of us are seriously ill!

How can this be? Benjamin Disraeli had an answer: lies, damn lies, and statistics. We may be good storytellers, but we are lousy statisticians. Glassner shows, for example, that women in their forties believe they have a 1 in 10 chance of dying from breast cancer, but their real lifetime odds are more like 1 in 250. He notes that some "feminists helped popularize the frightful but erroneous statistic that two of three teen mothers had been seduced and abandoned by adult men" when in reality it "is more like one in ten, but some feminists continued to cultivate the scare well after the bogus stat had been definitively debunked."[15] The bigger problem here is the law of large numbers, where million-to-one odds happen 280 times a day in America, and of those the most sensational dozen make the evening news, especially if captured on video. Stay tuned—film at eleven!

Herein lies the problem for our moral sensibilities. We are fed numbers daily that we cannot comprehend about threats to our security we cannot tolerate. Better safe than sorry, right? Not necessarily. Pathological fear takes a dramatic toll on our psyches and wallets. "We waste tens of billions of dollars and person-hours every year," Glassner notes, "on largely mythical hazards like road rage, on prison cells occupied by people who pose little or no danger to others, on programs designed to protect young people from dangers that few of them ever face, on compensation for victims of metaphorical illnesses, and on technology to make airline travel—which is already safer than other means of transportation—safer still."[16]

Of all the institutions feeding our fears, the media takes center stage for sensationalism ("if it bleeds, it leads"). An Emory University study revealed that the leading cause of death in men—heart disease—received the same amount of coverage as the eleventh-ranked vector: homicide. Not surprising, drug use, the lowest-ranking risk factor associated with serious illness and death, received as much attention as the second-ranked risk factor, poor diet and lack of exercise. From 1990 to 1998, America's murder rate decreased by 20 percent while the number of murder stories on network newscasts increased by an incredible 600 percent (and this doesn't count O. J. Simpson stories). The fact is, there is no evidence that secondhand smoke causes cancer or that cell-phone use generates brain tumors; likewise, Gulf War Syndrome appears to be a chimera, television does not cause violence, Satanic cults are phantasmagorical, most recovered memories of childhood abuse are nothing more than false memories planted by bad therapists, silicon breast implants cause nothing more than metastatic litigation, the drug war was lost decades ago, and the drug emperor has no clothes—he's butt naked and it's high time someone said it. We would be well-advised to remember the law of large numbers, and to keep in mind that we have selective memory of the most egregious events and that most of our fears are illusory—the vaporous product of a culture of fear of which we are both creators and victims.[17]

These notable shortcomings to our intuitive instincts aside, however, there is something quite empowering about intuition that cannot be dismissed, especially in the moral realm. In fact, intuition is so ingrained into the human psyche that it cannot be separated from intellect (witness the aforementioned intuitive afflictions). So integrated are intuition and intellect that I have coalesced them into what I call the *Captain Kirk Principle,* from an episode of *Star Trek* entitled "The Enemy Within."[18] Captain James T. Kirk has just beamed up from planet Alpha 177, where magnetic anomalies have caused the transporter to malfunction, splitting Kirk into two beings. One is cool, calculating, and rational. The other is wild, impulsive, and irrational. Rational Kirk must make a command decision to save the landing party now stranded on the planet because of the malfunctioning transporter. (Why they could not just send down a shuttle craft to rescue them is never explained, and thus this episode has contributed to the long list of *Star Trek* bloopers.) Because his intellect and intuition have been split, Kirk is paralyzed with indecision, bemoaning to Dr. McCoy: "I can't survive without him [irrational Kirk]. I don't want to take him back. He's like an animal—a thoughtless, brutal animal. And yet it's me." This psychological battle between intellect and intuition was played out in nearly every episode of *Star Trek* in the characters of the ultrarational Mr. Spock and hyperemotional Dr. McCoy, with Captain Kirk as the near-perfect embodiment of both. Thus, I call this balance the Captain Kirk Principle: *intellect is driven by intuition, intuition is directed by intellect.*[19]

For most scientists, intuition is the bête noire of a rational life, the enemy within to beam away faster than a Vulcan in heat. Yet the Captain Kirk Principle is now finding support from a rich new field of scientific inquiry brilliantly summarized by psychologist David G. Myers, who demonstrates through countless well-documented experiments that intuition—"our capacity for direct knowledge, for immediate insight without observation or reason"[20]—is as much a part of our thinking as analytic logic. Physical intuition, of course, is well known and accepted as part of an athlete's repertoire of talents—Michael Jordan and Tiger Woods come to mind. But there are social, psychological, and moral intuitions as well that operate at a level so fast and subtle that they cannot be considered a product of rational thought. Harvard's Nalini Ambady and Robert Rosenthal, for example, discovered that the evaluations of teachers by students who saw a mere thirty-second video of the teacher were remarkably similar to those of students who had taken the entire course. Even three two-second video clips of the teacher yielded a striking .72 correlation with the course student evaluations![21] How can this be? We have an intuitive sense about people that allows us to make reasonably accurate snap judgments about them.

Research consistently shows how even unattended stimuli can subtly affect us. In one experiment, for example, researchers flashed emotionally positive scenes (a kitten or a romantic couple) or negative scenes (a werewolf or a dead body) for forty-seven milliseconds before subjects viewed slides of people. Although subjects reported seeing only a flash of light for the initial emotionally charged scenes, they gave more positive ratings to people whose photos had been associated with the positive scenes.[22] In other words, something registered somewhere in the brain. That also appears to be the situation in the case of a patient who was unable to recognize her own hand, and when asked to use her thumb and forefinger to estimate the size of an object was unable to do it. Yet when she reached for the object her thumb and forefinger were correctly placed.[23] Another study revealed that stroke patients who have lost a portion of their visual cortex are consciously blind in part of their field of vision. When shown a series of sticks, they report seeing nothing, yet unerringly identify whether the unseen sticks are vertical or horizontal.[24] That's weird.

Intuition especially plays a powerful role in "knowing" other people. The best predictor of how well a psychotherapist will work out for you is your initial reaction in the first five minutes of the first session.[25] The reason for this is because for psychotherapy (talk therapy), research shows that no one modality or style is better than any other. It does not matter what type or how many degrees the therapist has, or what particular school the therapist attended, or whom the therapist trained under. What matters most is how well suited the therapist is for you, and only you can make that judgment, one best made through intuition, not intellect. Similarly, people with dating experience know within minutes whether or not they will want to see a first date again. That assessment is not made through tallying up the pluses and minuses of the date in some intellectual process equivalent to a mental ledger; we don't usually ask for a date's résumé or curriculum vitae before agreeing to a second date. But we do perform something like this in a quick intuitive assessment based on subtle cues—body language, facial expressions, voice tone and volume, wit and humor, politeness, and so forth—all of which can be assessed relatively quickly.

To the extent that lie detection through the observation of body language and facial expressions is accurate (overall not very), women are better at it than men because they are more intuitively sensitive to subtle cues. In experiments in which subjects observe someone either truth telling or lying, although no one is consistently correct in identifying the liar, women are correct significantly more often than men.[26] Women are also superior in discerning which of two people in a photo was the other's supervisor, whether a male-female couple is a genuine romantic relationship or a posed phony one, and when shown a two-second silent video clip of an upset woman's face, women guess more accurately then men whether she is criticizing someone or discussing her divorce.[27] People who are highly skilled in identifying "micromomentary" facial expressions are also more accurate in judging lying. In testing such professionals as psychiatrists, polygraphists, court judges, police officers, and secret service agents on their ability to detect lies, only secret service agents trained to look for subtle cues scored above chance. Most of us are not good at lie detection because we rely too heavily on what people say rather than on what they do. Subjects with damage to the brain that renders them less attentive to speech are more accurate at detecting lies, such as aphasic stroke victims who were able to identify liars 73 percent of the time when focusing on facial expressions (normal subjects did no better than chance). In support of an evolutionary explanation of a moral sense, research shows that we may be hardwired for such intuitive thinking: a patient with damage to parts of his frontal lobe and amygdala (the fear center) is prevented from understanding social relations or detecting cheating, particularly in social contracts, even though cognitively he is otherwise normal.[28] Cheating detection in social relations, such as in the role of gossip in small groups, is a vital part of our evolutionary heritage.

Although most secular theories of morality are rationalist theories, recent research on moral intuition reveals that the Captain Kirk Principle is at work in the moral realm as well. University of Virginia social psychologist Jonathan Haidt, for example, has demonstrated that the mind makes quick and automatic moral judgments similar to how we make aesthetic judgments. We do not reason our way to a moral decision; we jump right in, then later rationalize the quick decision. Our moral intuitions are more emotional than rational. Haidt's "social intuitionist" theory says that moral feelings come first, then the rationalization of those moral feelings. "Could human morality really be run by the moral emotions, while moral reasoning struts about pretending to be in control?" Haidt asks. He answers his own question thusly: "Moral judgment involves quick gut feelings, or affectively laden intuitions, which then trigger moral reasoning."[29] In other words, research supports our usual distinction between morality (thoughts and behaviors about right and wrong) and ethics (theories about moral thoughts and behaviors). In this context, ethics is an expression of emotional moral intuitions aimed at convincing others of the rational validity of our intuitions.

Consider the following moral dilemma and how our moral intuitions respond: you witness a runaway trolley headed for five people. If you throw a switch to derail the trolley, it will save the five but send it down another track to kill one person. Would you do it? Most people say that they would. Rationally, it seems justified: sacrificing one life to save five seems like the logical thing to do. However, consider this minor modification of the moral dilemma: you witness a runaway trolley headed for five people. You can stop the trolley by pushing a person onto the track, killing that one individual but saving five lives in the process. Would you do it? It is the same moral calculation, but most say they would not do it. Why? Princeton University's Joshua Greene believes he has found a reason through brain imaging technology. In presenting these moral dilemmas to subjects and recording what is going on inside their brains as they think about them, the second scenario of pushing the subject onto the tracks triggered the subjects' brains to light up in their emotional areas (normally active when feeling sad and frightened) much more than when they were thinking about the first scenario.[30] The difference in these two scenarios is that in the first one the subject is emotionally detached by being one step removed from the killing process—to save five lives by killing one person, one has only to flip a switch to derail the trolley car. The trolley killed the individual, not the subject. In the second scenario the subject is emotionally involved—to save five lives by killing one person, one has to be directly and viscerally responsible for killing another person. Moral judgment is not calculatingly rational. It is intuitively emotional.

Cognitive biases also play a powerful role in our moral intuitions. The self-serving bias, for example, which dictates that we tend to see ourselves in a more positive light than others actually see us, leads us to think we are more moral than others. National surveys, for instance, show that most businesspeople believe they are more moral than other businesspeople.[31] Even social psychologists who study moral intuition think they are more moral than other social psychologists![32] And we all believe that we will be rewarded for our ethical behavior. A *U.S. News & World Report* study asked Americans who they think is most likely to make it to heaven: 19 percent said O. J. Simpson, 52 percent said former President Bill Clinton, 60 percent said Princess Diana, 65 percent chose Michael Jordan, and, not surprisingly, 79 percent elected Mother Teresa. But the person survey takers thought most likely to go to heaven, at 87 percent, was the survey taker him- or herself![33]

Consistent with these experimental results are studies that show people are more likely to rate themselves superior in "moral goodness" than in "intelligence," and community residents overwhelmingly see themselves as caring more about the environment and other social issues than other members of the community do.[34] In one College Entrance Examination Board survey of 829,000 high school seniors, none rated themselves below average in the category "ability to get along with others," 60 percent rated themselves in the top 10 percent, and 25 percent said they were in the top 1 percent.[35] Likewise, just as behaviors determine perceptions—smokers overestimate the number of people who smoke, for example—moral behaviors determine moral perceptions: liars overestimate the number of lies other people tell. One study found that people who cheat on their spouses and income taxes overestimate the number of others who do so.[36]

Although in science we eschew intuition because of its many perils, we would do well to remember the Captain Kirk Principle that intellect and intuition are complementary, not competitive. Without intellect our intuition may drive us unchecked into emotional chaos. Without intuition we risk failing to resolve complex social dynamics and moral dilemmas, as Dr. McCoy explained to the indecisive rational Kirk: "We all have our darker side—we need it! It's half of what we are. It's not really ugly—it's human. Without the negative side you couldn't be the captain, and you know it! Your strength of command lies mostly in him."

Provisional Morality and Moral Justice: The Best We Can Do

Provisional ethics fits well with the research on moral intuition because how we respond to moral problems depends on a combination of inherited moral sentiments and learned moral rules, the combination of which is often too complex to depend entirely on intellect and reason. There are moral principles that are provisionally true, and we can know and apply these principles best by listening to our moral intuition as well as our moral intellect. The little voice inside should be talking to the little calculator inside.

It cannot be overemphasized that provisional ethics is not relative or situational ethics, nor is it an attempt to eschew moral responsibility or escape moral freedom. As an evolved mechanism of human psychology, the moral sense is transcendent of individuals and groups and belongs to the species. Moral principles, derived from the moral sense, are not absolute, where they apply to all people in all cultures under all circumstances all of the time. Neither are moral principles relative, entirely determined by circumstance, culture, and history. Moral principles are provisionally true—they apply to most people in most cultures in most circumstances most of the time. Although we are all subject to laws of nature and forces of culture and history that shape our thoughts and behaviors, we are free moral agents responsible for our actions because none of us can ever know in its entirety the near-infinite causal net that determines each of our individual lives. Good things and bad things happen to both good and bad people. There is no absolute and ultimate judge to mete out rewards and punishments at some future date beyond the human career on planet Earth. But since moral principles are provisionally true for most people most of the time in most circumstances, there are individual culpability and social justice within human communities that produce feelings of righteousness and guilt and mete out rewards and punishments such that there is at least provisional justice. Provisional ethics leads to provisional justice.

Provisional ethics may not be ultimately satisfying for the moral absolutist, but since there is no justification outside of an omnipotent and omniscient God for such moral absolutism—and there is no convincing scientific evidence that such a God exists—then provisional ethics and provisional justice are the best we can do. If you want more—

if you need some source of moral verification and objectification outside of yourself, your society, and your species—then you are living in the grip of a supernatural illusion. I'm sorry, but you can't get more without eschewing reality. Given the nature of our universe, our world, and our selves, this is the best we can do. Fortunately, it is enough. It leads to a moral humanity because a moral nature is part of human nature. It exists independent and outside of any individual because it belongs to the species. As long as humanity continues so too will morality, provisional though it may be.

7

HOW WE ARE IMMORAL: RIGHT AND WRONG AND HOW TO TELL THE DIFFERENCE

> No man chooses evil because it is evil; he only mistakes it for happiness, which is the good he seeks.
>
> —Mary Wollstonecraft, *Vindication of the Rights of Woman,* 1792

In the 1991 comedy film *City Slickers,* the story of three men turning forty (played by Billy Crystal, Daniel Stern, and Bruno Kirby) who go on a New Mexico cattle drive, there is much discussion about the purpose of life, the meaning of death, and especially sex. Crystal's character Mitch, for example, explains to Phil and Ed (Stern and Kirby, respectively) that "Women need a reason to have sex. Men just need a place." Despite this comedic observation, Mitch has been happily (and faithfully) married to his wife, Barbara, for many years; by contrast, Phil has been in a miserable marriage to his boss's daughter and just got busted by his wife for having an affair with a young checkout clerk; Ed is the consummate playboy bachelor who recently—although somewhat reluctantly—tied the knot with his latest flame. As is often the case among men, the talk turns to sex and infidelity. Ed wants to know if Mitch would cheat on Barbara if he wouldn't get caught. Mitch reminds him of what just happened to Phil. So Ed offers this scenario:

ED: A spaceship lands and the most beautiful woman you ever saw gets out. And all she wants is to have the best sex in the universe with you. And the second it's over she flies away for eternity. No one would ever know. You're telling me you wouldn't do it?

MITCH: No. Because what you're describing actually happened to my cousin Ronald, and his wife did find out about it at the

> beauty parlor. They know everything there! Look, what I'm saying is it wouldn't make it all right if Barbara didn't know. *I'd* know. And I wouldn't like myself. That's all.

Ed persists in pushing Mitch with a cereal analogy, explaining that he has been selecting from a Kellogg's variety pack all his life but now he has to eat the same cereal every day. "And then you wake up one morning, and you're just not hungry anymore." The problem, Ed confesses, is that his wife, Kim, wants to start a family, which means to Ed that he will never have sex with another woman. Poor Phil, whose life appears rather dismal at the moment, can't understand why being married "to this gorgeous twenty-four-year-old underwear model who thinks the sun rises and sets in your pants" is not enough for Ed. Ed retorts: "You don't understand. I don't want to screw around on Kim." Phil admonishes him: "So don't."

Can Ed "just say no" to the temptation of an extramarital affair? Was Phil justified in having an affair since his wife was a ball-busting banshee who regularly refused him sex? Are Mitch's high moral standards the norm or the exception, and is his reason for not accepting the ultimate offer of safe sex—that even if his wife never found out *he* would know and that is reason enough—a higher moral reason than fear of retribution? In this chapter we shall consider how we are immoral by examining a number of principles that help us tell the difference between right and wrong. We will also apply these principles to a number of ethical issues (truth telling and lying, adultery, pornography, abortion, cloning and genetic engineering, and animal rights) to examine where science, fuzzy logic, and provisional ethics can help us resolve them, or at least inform our moral decisions.

The Ask God Principle: Religious Right and Wrong

At the underpinning of all theistic ethical systems is the belief that without God there is no ultimate basis for determining right and wrong. We have already seen the limitations of theistic ethics, but there are two additional questions to consider here: (1) what if the moral issue is not discussed in the sacred writings of the individual's religion? Cloning, stem cell research, and genetic engineering are not discussed in the

Bible, of course, so what are Jews and Christians to think about these very real moral issues? They either have to attempt to infer from ancient biblical writings something that is loosely related to the modern moral issue, or they have to think it through for themselves; (2) what if the moral issue is discussed but is clearly inappropriate or outright wrong in its moral command? With both of these limitations the believer is often forced to selectively read the sacred text, picking and choosing passages without consistency.

Consider, for example, the many Old Testament moral rules that make one blanch with embarrassment for believers. (All biblical passages cited below are from the Revised Standard Version.) For emancipated modern women thinking of adorning themselves in business attire that may resemble men's business wear (or for guys who dig cross dressing), Deut. 22:5 does not look kindly on such behaviors: "A woman shall not wear anything that pertains to a man, nor shall a man put on a woman's garment; for whoever does these things is an abomination to the Lord your God." An even worse abomination is a rebellious child. Deut. 21:18–21 offers this parental moral guideline: "If a man has a stubborn and rebellious son, who will not obey the voice of his father or the voice of his mother, and, though they chastise him, will not give heed to them, then his father and his mother shall take hold of him and bring him out to the elders of his city at the gate of the place where he lives, and they shall say to the elders of his city, 'This our son is stubborn and rebellious, he will not obey our voice; he is a glutton and a drunkard.' Then all the men of the city shall stone him to death with stones; so you shall purge the evil from your midst; and all Israel shall hear, and fear."

If that isn't ridiculous enough, here is the Bible's recommendation on how to deal with women who may or may not have had sex before marriage. According to Deut. 22:13–21: "If any man takes a wife, and goes in to her, and then spurns her, and charges her with shameful conduct, and brings an evil name upon her, saying, 'I took this woman, and when I came near her, I did not find in her the tokens of virginity,' then the father of the young woman and her mother shall take and bring out the tokens of her virginity to the elders of the city in the gate." (For those not accustomed to reading between the biblical lines, the phrase "goes in to her" should be taken literally, and "the tokens of virginity" means the hymen and the blood on the sheet from a virgin's first sexual experience.) If the father of the bride can produce the tokens of virginity, then he "shall spread the garment before the elders of the city. Then the elders of that city shall take the man [the husband] and whip him; and they shall fine him a hundred shekels of silver, and give them to the father of the young woman, because he has brought an evil name upon a virgin of Israel; and she shall be his wife." However, lo to the woman who has dared to have sex before marriage. "But if the thing is true, that the tokens of virginity were not found in the young woman, then they shall bring out the young woman to the door of her father's house, and the men of her city shall stone her to death with stones, because she has wrought folly in Israel by playing the harlot in her father's house; so you shall purge the evil from the midst of you."

Finally, for those of you who have succumbed to the temptations of the flesh at some time in your married life, Deut. 22:22 does not bode well: "If a man is found lying with the wife of another man, both of them shall die, the man who lay with the woman, and the woman; so you shall purge the evil from Israel." Do Jews and Christians really want to legislate biblical morality, especially in light of the revelations of the past couple of decades of the rather low moral character of some of our more prominent religious leaders? And on the legislation question, those on the religious right who are lobbying for the Ten Commandments to be posted in public schools and courthouses should note that the very first one prohibits anyone from believing in any other gods besides Yahweh. (The first commandment is "Thou shall have no other gods before me," a passage indicating that polytheism was commonplace at the time and that Yahweh was, among other things, a jealous god.) That is to say, by posting the Ten Commandments, we are sending the message that any nonbelievers, or believers in any other god, are not welcome in our public schools and courtrooms. Fortunately, the First Amendment of the Bill of Rights prohibits such religious exclusionary practices.

To be fair, not all biblical ethics are this antiquated and extreme. There is much to pick and choose from that is useful for our thinking about moral issues. The problem here is consistency, and selecting ethical guidelines that support our particular personal or social prejudices and preferences. When slavery was the social norm, it was simple for proslavery defenders to point to passages such as those in Exod. 21, which outlines the rules for the proper handling of slaves, for example:

"when you buy a Hebrew slave, he shall serve six years, and in the seventh he shall go out free, for nothing," and "when a man sells his daughter as a slave, she shall not go out as the male slaves do," and, finally, slave families should be kept together, unless the master gave the slave a wife, who then bore him children, in which case the master gets to keep the woman and children when the slave is sold. If you are going to claim the Bible as your primary (or only) code of ethics and proclaim (say) that homosexuality is sinful and wrong because the Bible says so, then to be consistent you should kill rebellious youth, nonvirginal premarried women, and adulterous men and women. Since most today would not endorse that level of consistency, why pick on gays and lesbians but cut some slack for disobedient children, promiscuous women, and adulterous men and women? And why aren't promiscuous men subject to the same punishment as women? The answer is that in that culture, at that time, men legislated and women obeyed. Thankfully, we have moved beyond that culture. But what this means is that we need a new set of morals and an ethical system designed for our time and place, not one scripted for a pastoral/agricultural people who lived 4,000 years ago. The Bible and other sacred texts have wonderfully edifying and sometimes transcendent passages, but we can do better.

The Ask First Principle: Secular Right and Wrong

If we cannot reliably turn to the Bible and other sacred texts to determine moral right and wrong, to whom shall we turn? If we cannot ask God, whom shall we ask? One answer can be found in the first moral principle, the Golden Rule: "Do unto others as you would have them do unto you." The Golden Rule is a derivative of the basic principle of exchange reciprocity and reciprocal altruism, and thus evolved in our Paleolithic ancestors as one of the primary moral sentiments. (If I'm right about this, then it means that religion did not invent the Golden Rule and other moral principles; it co-opted them, then codified them.) In this principle there are two moral agents: the *moral doer*
moral receiver. A moral question arises when the moral doe
tain how the moral receiver will accept and respond to the
question. In its essence this is what the Golden Rule is t
By asking yourself, How would I feel if this were do

are asking, How would others feel if I did it unto them? But the Golden Rule has a severe limitation to it: what if the moral recipient thinks differently from the moral doer? What if you would not mind having action X done unto you, but someone else would mind it? Most men, for example, are much more receptive toward unsolicited offers of sex than are women. Most men, then, in considering whether to approach a woman with an offer of unsolicited sex, should not ask themselves how they would feel if the roles were reversed. We need to take the Golden Rule one step further, through what I call the *ask first principle.*

There is one surefire test to find out whether an action is right or wrong: ask first. The moral agent should ask the moral recipient whether the behavior in question is moral or immoral. If you aren't sure that the potential recipient of your action will react in the same manner you would react to the moral behavior in question, then ask. Consider an easy test of the ask first principle—adultery. If you want to know if having an extramarital affair is moral or immoral, ask first the potentially affected moral recipient—your spouse: "Honey, is it okay if I sleep with someone else?" You will receive your moral answer swiftly and without equivocation. In this example, as with so many others, you do not actually have to ask the question to know the answer. The thought experiment alone should give you a strong sense of what is right and wrong.

Such moral thought experiments are at the heart of moral reasoning. For this process you can monitor your own sense of guilt and other emotions as a guideline. Imagine, in the above example, how you would feel if your partner had sex with someone else. I mean, *literally* imagine it. For a few people, perhaps, their marriages and relationships are so dead that such fantasies have no effect—truly a sign that the emotional attachment has been severed. For most people, however, imagining their partners having sexual relations with others is extremely emotionally disruptive. There is, in fact, solid scientific research on this subject. Evolutionary psychologist David Buss wired up subjects to monitor their pulse, blood pressure, breathing rate, and perspiration (the basic measurements used by the polygraph, or lie detector). He then asked them to imagine their significant other having sex with someone else. For most of Buss's subjects, their heart rate and blood pressure went through the roof, their breathing became rapid and forced, and their bodies perspired profusely.[1] It is not a big leap of the imagination for the moral doer to project that response onto the moral receiver to get an answer to the moral question.

The Happiness Principle: Personal Right and Wrong

In addition to asking the moral receiver, what other criteria might we use to judge the rightness or wrongness of an action? For millennia, philosophers and observers of human behavior have noted that we have a tendency to seek pleasure and avoid pain. Pleasure and pain encompass many things, from pure physical to pure ethereal states. We may find pleasure in a kiss or an idea. We may experience pain in a slap or an insult. Happiness is a good synonym for pleasure, and unhappiness is a good synonym for pain, and thus we may state that one of the fundamental drives of human nature is that we all strive for greater levels of happiness and avoid greater levels of unhappiness, however these may be personally defined. Happiness and unhappiness, then, are emotions that evolved as part of the suite of emotions that make up the human psyche.

As we have seen, humans have a host of moral and immoral passions, including being selfish and selfless, competitive and cooperative, nasty and nice. It is natural and normal to try to increase our own happiness by whatever means available, even if that means being selfish, competitive, and nasty. Fortunately, evolution created both sets of passions, so that we also seek to increase our own happiness by being selfless, cooperative, and nice. *The happiness principle states that it is a higher moral principle to always seek happiness with someone else's happiness in mind, and never seek happiness when it leads to someone else's unhappiness.* My colleague, social scientist and moral philosopher Jay Stuart Snelson, expressed this sentiment well in his "win-win principle": "Always seek gain through the gain of others, and never seek gain through the forced or fraudulent loss of others."[2]

This is not always easy to do. There is a tension in the human condition between these competing motives, and as often as not the darker side of our humanity emerges. The moral animal struggles with the immoral animal within. Whether the moral or immoral animal wins in any given situation depends on a host of circumstances and conditions. Since we have within us both moral and immoral sentiments, and we have the capacity to think rationally and override our baser instincts, and we have the freedom to choose to do so, the core of morality is choosing to do the right thing by acting morally and applying the happiness principle.

So, for any given moral question, one may begin by asking the moral receiver how he or she would respond, then ask yourself if the action in question will likely lead to greater or lesser levels of happiness for yourself and the moral receiver. These two moral principles dovetail, because the moral receiver is, presumably, seeking greater levels of happiness; thus, by asking first what you should do, you will also receive feedback on how the moral receiver's happiness will be affected by your actions.

The Liberty Principle: Social and Political Right and Wrong

In addition to asking the moral receiver how he or she might respond to a moral action and considering how that action might lead to your own and the moral receiver's happiness or unhappiness, there is an even higher moral level toward which we can strive: the freedom and autonomy of yourself and the moral receiver, or what we shall simply refer to here as liberty. Liberty is the freedom to pursue happiness and the autonomy to make decisions and act on them in order to achieve that happiness. *The liberty principle states that it is a higher moral principle to always seek liberty with someone else's liberty in mind, and never seek liberty when it leads to someone else's loss of liberty.* The liberty principle is grounded in history and anchored in modern enlightenment values.

In prehistoric bands and tribes, liberty was limited to the actions and interactions of individuals within their families, extended families, and tiny communities. Liberty as a political concept was nonexistent, because there was no politics. Society was mostly a loose confederation of individuals—families and extended families—within a slightly larger community. The primary purpose of these communities was to resolve conflicts within the band or tribe, to secure food and natural resources, and to protect against other bands and tribes. As bands and tribes coalesced into chiefdoms and states, and populations grew from hundreds to thousands and tens of thousands, political organizations were needed because the informal methods of conflict resolution that worked so well in smaller populations broke down among the much larger populations, and the small skirmishes between bands and tribes grew into much larger and costlier wars between chiefdoms and states.

Because chiefdoms and states require revenue to support a bureaucratic infrastructure and bureaucracies are not designed to be revenue-generating organizations, the individual members of the chiefdom or state must relinquish some percentage of their productive labor. Today this is done through taxes, duties, levies, tolls, excises, and various other financial assessments. Where there is no money, or a limited supply of cash flow, the barter system may be employed, such as in feudalism, where peasants gave over a portion of their agricultural products to the land-owning lord, and/or a fraction of their time to military service in defense of the castle, manor, or realm. Here, and elsewhere, some freedom and autonomy is exchanged for security and resources, and this may lead to an increase in overall liberty and the general good of the chiefdom or state. It is at this point—roughly 3,000 to 5,000 years ago when bands and tribes evolved into chiefdoms and states—that the concept of civil and political liberty was born. Here we can turn to the Bio-Cultural Evolutionary Pyramid, figure 6 on page 48, to see where and how that transition was made. It is at the bio-cultural transitional boundary between the community and the society, where social status and recognition lead to social justice and security, and where the drive of reciprocal altruism gives rise to indirect and blind altruism, that liberty emerges. This is the birth of liberty, the principle that when individual members of the community exchange freedom and autonomy for resources and security, in the long run their overall liberty increases. For example, exchanging a portion of my earnings for food that someone else produces allows me the freedom to pursue non-food-producing activities. Ideally, the exchange of some freedom and autonomy for resources and security leads to other forms of freedom and autonomy. Unfortunately, that is not always the case.

For many millennia the concept of liberty for all members of the state lay dormant, suppressed by the selfish and competitive drives of the political and religious leaders who held the reins of power. Even the occasional enlightened societies that set up quasi-representative bodies to protect the interests of the citizens at large restricted liberty to a narrow class of land-owning or power-wielding males. Only in the last couple of centuries have we witnessed the worldwide spread of liberty as a concept that applies to all peoples everywhere, regardless of their rank or social and political status in the power hierarchy. Liberty has yet to achieve worldwide status, particularly among those states dominated by theocracies that encourage intolerance and dictate that only some people deserve liberty, but the overall trend since the early modern period has been to grant greater liberty for more people. Although there are still setbacks, and periodically violations of liberties disrupt the overall historical flow from less to more liberty for all, the general trajectory of increasing liberty for all humans continues.

The Moderation Principle: Extremism Is No Virtue, Moderation Is No Vice

On July 16, 1964, in his speech accepting the Republican presidential nomination, Barry Goldwater gave voice to one of the most memorable one-liners in the history of politicking: "Extremism in the defense of liberty is no vice. Moderation in the pursuit of justice is no virtue." For most human endeavors, however, Goldwater is wrong. Extremism is almost always a vice that generates countless unintended consequences. Extremism too often leads to violence, terrorism, and even war. From 9/11 to the ongoing Arab-Israeli conflict, and from the bombing of the federal building in Oklahoma City to the blowing up of abortion clinics, the principles of happiness and liberty are violated in the most ultimate fashion.

The opposite of extremism is moderation. The *moderation principle states that when innocent people die, extremism in the defense of anything is no virtue, and moderation in the protection of everything is no vice.* The moral principles behind the moderation principle are happiness and liberty. If you are killing people in the name of anything, you are seeking happiness and liberty at the ultimate expense of someone else's happiness and liberty.

Provisional Ethics Put to the Test

Real-world problems are the true test of any theory of morality. Other than for intellectual recreation, what good is an ethical theory without moral application? In the remainder of this chapter we shall examine how provisional ethics and a science of morality—particularly the ask first principle, the happiness principle, and the liberty principle—might be applied to a number of such ethical issues, including truth telling and lying, adultery, pornography, abortion, cloning and genetic engineering, and animal rights.

Provisional Morality and Truth Telling and Lying

We begin with a relatively easy moral issue: truth telling and lying. We all agree that truth telling is vital for trust in human relations, so in a binary system of morality, truth telling is right and lying is wrong. Life, however, like nature, is never this simple. It turns out that all of us lie every day, but most of the lies are so-called little white lies, where we might exaggerate our accomplishments, or they might be lies of omission, where information is omitted to spare someone's feelings. Such lies are usually amoral. Fuzzy logic better represents life—and lies—than does binary logic. In the case of truth telling and lying, fuzzy provisional ethics allows us to nuance our thinking on such moral issues.

Little white lies, for example, since they are commonplace and mostly harmless, might be ranked a .1 or a .2 lie. Lies of omission might perhaps be catalogued as .3 or .4 lies. Lies of commission—intentionally providing false information—might be classified as .5 or .6 lies. Big lies—lies in the range from .7 to .9—are getting much more serious, and thus can be seen as more immoral than little white lies.

When in doubt as to whether a lie is moral, immoral, or amoral, you can ask yourself how the moral receiver might feel if he or she found out you lied, and whether the moral receiver's happiness and liberty increased or decreased as a result of the lie you told. When telling a lie, most of us, most of the time, do so to increase our own happiness or liberty, or to avoid anticipated unhappiness or loss of liberty if someone else knows the truth. Thus, the moral thing to do is to never tell a lie if it leads to someone else's unhappiness or loss of liberty. Of course, there are circumstances when telling the truth might lead to someone else's unhappiness or loss of liberty. For example, if an abusive husband inquires whether you are harboring his fearful wife, it would be immoral for you to answer in the affirmative if you are, because the truth might lead to the abuse or death of the wife. Here we must be cautious since it would be easy to rationalize a lie when, in fact, telling the truth is usually the right thing to do.

Provisional Morality and Adultery

There is a lighthearted biblical story told by ministers and rabbis about Moses and the Ten Commandments. In this revisionist spin, the great prophet descends from the mountaintop with the divinely chiseled tablets, announcing to his people, "I have good news and bad news. The good news is that God kept the number of commandments down to a manageable ten. The bad news is that God left in the one about adultery."

Well, what if He had not left that commandment in? Would that remove adultery from the list of immoral acts? Would it make it amoral, or even moral? Divine Command Theory, a narrower version of theistic ethics, implies that it would. For Judaism, Christianity, and Islam, if the monotheistic God of Abraham does not decree an act immoral, the implication is that it is not, at least as far as the Bible goes (all three religions have a long and honorable tradition of biblical commentary and moral discourse in which the sages of each generation have produced extrabiblical works on all matters moral). As to whether there is a God or not, however, we need to go beyond divine command as our guide. Eighteenth-century theistic German mathematician Gottfried Leibniz explained it this way: "In saying, therefore, that things are not good according to any standard of goodness, but simply by the will of God, it seems to me that one destroys, without realizing it, all the love of God and all his glory; for why praise him for what he has done, if he would be equally praiseworthy in doing the contrary?"[3]

Religions and societies have long struggled to regulate human sexual behavior, one of the strongest of all the passions. In some cases, it has been outlawed, even in modern America. In Rolling Hills, California, for example, Section 9 of the municipal code, on "immoral conduct," specifically states that "any person occupying, using or being present in any bed, room, automobile, structure or public place with a person of the opposite sex, to whom he or she is not married, for the purpose of having sexual intercourse with such person" is guilty of a crime punishable by up to three months in jail.[4] Such laws sound arcane to us today, but many states once had them on the books because marriage is considered a contract, the violation of which leads to third-party damages. But since most states have moved to no-fault divorce laws, adultery as a cause of divorce fell into disuse, along with the antiadultery laws. (In April 2003, for example, the city council of Rolling Hills voted to repeal the antiadultery statute.) Although adultery may retain its status as a sin, its designation as a crime has proved ineffective as a behavioral curb. If two consenting adults wish to engage in unsanctioned sexual behavior, there is little church or state can do to stop them. Behavioral restraint needs to come from within.

Here, again, we can look to evolutionary theory for a deeper understanding of why we do what we do unto others. In provisional ethics, adultery is provisionally immoral because of the disruption it causes to the natural mating condition of our species. We evolved as pair-bonded primates for whom monogamy, or at least serial monogamy (a sequence of monogamous marriages), is the norm. Adultery is a violation of a monogamous relationship and, as we shall see, there is copious scientific data (and loads of anecdotal examples) showing how destructive adulterous behavior is to a monogamous relationship. In fact, one of the reasons that serial monogamy (and not just monogamy) best describes the mating behavior of our species is that adultery typically destroys a relationship, forcing couples to split up and start over with someone new. Thus, in contrasting provisional ethics with theistic ethics, adultery is immoral because of its destructive consequences no matter what God or the patriarchs said about it. And evolutionary theory provides a deeper reason for adultery's immoral nature that is transcendent because it belongs to the species. If there is a God, and if He does condemn adultery as an immoral act, it is because evolution made it immoral.

According to evolutionary psychologist David Buss, sexual betrayals are primarily a biologically driven phenomenon (although they may be accentuated or attenuated by culture), encoded over eons of Paleolithic cuckolding. Buss argues that there are differences between men and women in this tendency and that these differences cut across cultures, and thus are primarily driven by our genes. He cites a study by Russell Clark and Elaine Hatfield in which college students were approached by an attractive member of the opposite sex who asked one of three questions:

1. "Would you go out on a date with me tonight?"
2. "Would you go back to my apartment with me tonight?"
3. "Would you sleep with me tonight?

The results were revealing, to say the least. For women, 50 percent agreed to the date, 6 percent agreed to return to the apartment, and none agreed to have sex. By contrast, for men, 50 percent agreed to the date, 69 percent agreed to the apartment, and 75 percent agreed to the sex. Yet, even with such a basic and simple drive as sex, Buss admits that genes are only part of the story: "Desires represent only one set of causes of actual mating behavior. Individuals cannot always make decisions that correspond precisely to their desires—people can't always get what they want. Mates possessing all of the desired qualities are scarce and often unavailable. Competition is keen for the limited supply of desirable mates; members of the same sex constrain access. Parents can wield influence. And members of the opposite sex exert preferences that further restrict access."[5]

As for the act of adultery itself, its evolutionary benefits are obvious. For the male, depositing one's genes in more places increases the probability of this form of genetic immortality. For the female, it is a chance to trade up for better genes and higher social status. Its evolutionary hazards, however, are equally obvious. For the male, revenge by the adulterous woman's husband can be extremely dangerous—a significant percentage of homicides involve love triangles. And while getting caught by one's own wife is not likely to result in death, it can result in loss of contact with children, loss of family and security, and risk of sexual retaliation, thus decreasing the odds of one's mate bearing one's own offspring. For the female, being discovered by the adulterous man's wife involves little physical risk, but getting caught by one's own husband can and often does lead to extreme physical abuse and occasionally even death.

Beyond the evolutionary implications, there are the sociocultural problems, such as the risk of sexually transmitted diseases, family and extended-family rejection, social ostracization from one's community, and the like. It would be difficult to justify adultery as a moral act from either an evolutionary perspective or a cultural one, for either the individual or society. Extreme exceptions come to mind, of course, such as a woman whose husband is in a long-term coma and she finds solace through intimacy with another man. But such "lifeboat" cases are so rare as to fall outside the purview of provisional ethics. And one occasionally hears about "open marriages" in which both partners allegedly agree to tolerate extramarital affairs, but such arrangements appear to be desired more by one partner than the other and typically end in divorce when one partner becomes attached to and falls in love with the paramour.

In provisional ethics, then, most moral actors, especially those who are directly or indirectly involved in or affected by the affair—one's mate, the mate of the adulterous partner, the families of both adulterous parties, the community, and the society—would likely offer their provisional assent that adultery is an immoral act for most people, in most circumstances, most of the time, because it causes considerably more harm than good. Thus, overall and in the long run, the adulterous act leads to disastrous consequences and a decrease in liberty and happiness for most parties involved, and thus cannot in most circumstances be justified. If there is any doubt about this, before you set out on an adulterous adventure, ask first your partner.

Provisional Morality and Pornography

On July 26, 1991, Paul Reubens, better known as the affably paedomorphic Pee-wee Herman, was arrested for indecency in the South Trail Cinema, an X-rated theater in Sarasota, Florida. It seems that the actor, then star of his own syndicated children's television series, which followed his wildly popular film *Pee-wee's Big Adventure,* was enjoying *Nancy Nurse* a bit too carnally for a public venue.

The event quickly turned into a media feeding frenzy and exploitative fodder for stand-up comics, including Reubens himself who, as host of a subsequent television awards show, quipped, "Anyone heard any good jokes lately?" The Pee-wee porn affair was much ado about nothing. Most wondered why he did not simply indulge his passions in the privacy of a hotel room or his own home, where such materials are readily available and confidentially enjoyed. Unfortunately, the image of some guy masturbating to an erotic film in a public theater is what far too many people conjure up in their imagination when they hear the word *pornography.*

Simply defined, *pornography* constitutes images in the form of films, videos, photographs, literature, and other materials that enhance sexual arousal. Determining which images specifically constitute pornography is so fraught with moral and legal complications that it led D. H. Lawrence to comment, "What is pornography to one man is the laughter of genius to another," and Supreme Court Justice Stewart to famously pronounce that although he could not define pornography, "I know it when I see it."[6] Unlike all the other primates (with the exception of bonobo chimpanzees), whose sex lives are largely governed by seasonal periods of receptivity (primarily when the female is "in heat"), humans evolved with the desire to engage in sexual behavior at any time of the year. So who would object to enhancing the arousal of a perfectly natural and exceptionally pleasurable activity designed to propagate the species and bring new life into the world?

Plenty of people, as a matter of fact, do object to such supplemental activities, and they have been largely successful in transmogrifying the meaning of the word *pornography* into a lewd and smutty activity conducted by sandal-wearing, tree-hugging, left-leaning, liberal, pinko, godless communists, homosexuals, and perverts of all stripes for the purpose of preventing this great nation from returning to its roots as a Christian country and to subvert its foundation in the puritanical ethic whose greatest fear is that somewhere, someone out there is enjoying carnal pleasure. Pornography, we are told, is immoral—a sin of the mind only slightly less violative than such sins of the body as adultery, masturbation, homosexuality, and premarital sex. Even President Jimmy Carter famously (and shamefully) confessed that he "lusted in his heart."

Let's see how provisional ethics applies to pornography, examining three types: *mental pornography, positive pornography,* and *negative pornography.* In essence, I shall argue that mental pornography and positive pornography are not immoral because most people in most circumstances most of the time are not harmed by them and, in fact, may find much pleasure in them, both individually and in their relationships. There is, however, some evidence that negative pornography (pornography that depicts harm or violence against women, such as pleasure in being raped) is harmful to at least some people in some circumstances some of the time and may therefore be considered provisionally immoral. Let's examine what science can tell us about the effects of pornography of these three types.

Mental pornography. Stripped of its pejorative connotations and seen for what it really is—images that enhance sexual arousal—the simplest form of pornography is the sexual images in our imaginations. Mental pornography, or what Havelock Ellis called "autoeroticism," is one of the most ubiquitous of all sexual activities. I do not know if sexual fantasy itself evolved, providing some selective advantage to individuals who had them versus those who did not, or if autoeroticism is just a spandrel—a by-product of some other evolutionary adaptation. But certainly the ability to fantasize in general did evolve as a useful by-product of a large cerebral cortex, and no doubt this ability did provide a selective advantage (imagining the positive outcome of a hunt or the negative consequences of a fight). Sexual fantasies are probably a contingent free ride that comes with having a large brain capable of fantasizing about other scenarios in life. Since social relations between humans are so important, and because sex is so intimately intertwined with how we feel about and interact with other members of our group, then it would not surprise me if it turned out that fantasizing about sexual relations with others did ultimately serve some functional purpose in our evolutionary history.

Western religion has typically prohibited sexual fantasies. Consider this medieval church punishment in the form of penances for erotic fantasies among church leaders of ascending stripes: twenty-five days for a deacon, thirty days for a monk, forty days for a priest, and fifty days for a bishop. (I guess the pope is not only infallible, but also unimaginative.) How can something so harmless to others (this assumes negative sexual fantasies are not expressed behaviorally—more on this below in the discussion of negative pornography) and yet so fun and fulfilling to the individual be immoral? Science sees it rather differently. Erotic fantasies may serve a variety of personal functions, including the fact that sometimes it is a lot easier to just fantasize about a sexual encounter than it is to actually invest the time, energy, and money, and to risk rejection, failure, disease, social ostracism, or the possibility of an unsatisfactory experience in an actual sexual encounter. Some of the best sex any of us have ever experienced is the sex in our minds. That mental sex may be informed by actual sexual experiences—usually the most enjoyable ones we have had with a partner who is especially important to us—but it remains safely ensconced in the private domain within our skulls.

Therefore, from a provisional ethics perspective, it would be reasonable for us to offer our provisional assent that mental pornography in the form of positive sexual fantasies is not immoral because the evidence confirms that almost everyone has them, they provide numerous benefits, they harm no one else, and thus they are justified if so desired by the individual or couple (sharing your sexual fantasies with your partner, particularly if they are positive and about that partner, can be very stimulating and provocative).

Positive Pornography. I define positive pornography as images that enhance sexual arousal by depicting individuals or couples in non-harmful and nonexploitative sexual situations. (I do not consider sado-masochism [S&M] harmful as long as the partners involved in an S&M encounter are willing participants.) Films, videos, photographs, and literature that depict individuals masturbating or couples engaging in consensual sex and that are viewed by either individuals alone or couples together for personal enjoyment represent pornography in a positive mode. So-called soft-porn films that leave something to the imagination and depict sex as a romantic and loving expression of affection between two people are fine examples of positive pornography. So too is the body of erotica literature in its higher form by authors such as the French diarist and novelist Anaïs Nin. In fact, erotica is a synonym for positive pornography and is a term already in the lexicon that carries positive connotations. Let pornography describe negative pornography. Let erotica describe mental and positive pornography. Erotica is literary, highbrow, graceful, elegant, and, most of all, sensual—the very essence of a positive sexual experience.

Pulp fiction romance novels that portray lovemaking in crass terms, such as describing a man's "throbbing pole of love," and so-called hard-core porn films that leave nothing to the imagination in graphically revealing cunnilingus and fellatio, vaginal and anal penetration, ejaculation, multiple partners, and spontaneous sex with strangers in unlikely venues tend to be preferred much more by men than by women. Contrast these images with the following passage from Anaïs Nin's book of erotica *Little Birds,* in a short story about a young woman married to an older man who delays ultimate intimacy several nights to "woo her slowly and lingeringly, until she was prepared and in the mood." After several nights of teasing kisses and caresses,

> he discovered the trembling sensibility under the arm, at the nascence of the breasts, the vibrations that ran between the nipples and the sex, and between the sex mouth and the lips, all the mysterious links that roused and stirred places other than the one being kissed, currents running from the roots of the hair to the roots of the spine. Each place he kissed he worshiped with adoring words,

> observing the dimples at the end of her back, the firmness of her buttocks, the extreme arch of her back, which threw her buttocks outwards. . . . He encircled her ankles with his fingers, lingered over her feet, which were perfect like her hands, stroked over and over again the smooth statuesque lines of her neck, lost himself in her long heavy hair.[7]

Pages later, the lovers finally embrace in full intimacy. This is positive pornography at its finest, and research shows that it is very effective in sexually arousing both men and women. Physiological research, for example, shows that penile erection, vaginal vasocongestion, blood pressure, and genital temperature all increase in response to exposure to positive pornographic material (such arousal effects can be also be generated through the imagination alone).[8]

There are cases, of course, when positive pornography can become negative (in a manner different from negative pornography to be discussed below), and that is when pornography becomes an alternative, rather than a supplement, to a satisfying sexual relationship with your partner. Here we are well advised to follow Aristotle's golden mean: *all things in moderation.* If the viewing of pornography becomes so addictive and compelling that it replaces sex with your partner, and your partner then becomes dissatisfied with this arrangement, then such pornography is no longer positive. We should always remember that, by definition, pornography is supposed to *enhance* sexual arousal, not replace it. When in doubt, ask your partner.

On the flip side, positive pornography may be a useful substitute for sex when you are between relationships or, for whatever reason, you do not desire a sexual relationship with your partner (and your partner is not frustrated by this substitution). Sex with yourself is safe, and pornography can be a positive enhancement of the self-sexual experience.

Negative Pornography. I define negative pornography as images that enhance sexual arousal by depicting sex as violent, abusive, or exploitative, and especially those that imply or show women being seduced and raped against their will and then enjoying the experience as it unfolds. Here we enter the darker realm of rape and the relationship of pornography to this especially malevolent act.

One argument made against pornography is that it leads men to rape women. Indeed, attacks on pornography often begin here and come not just from the conservative right but from the liberal left as well (mainly from extreme feminists). Catharine MacKinnon, for example, describes all pornography as "the celebration, the promotion, the authorization and the legitimization of rape, sexual harassment, battery and the abuse of children." Andrea Dworkin defines pornography as "the material means of sexualizing inequality."[9] As a blanket causal variable in the study of why men rape, however, there is no evidence to support this claim. Indeed, if only it were as simple as eliminating pornography in order to eliminate rape; but it is not so, as evidenced by the fact that rape has been a tragic part of human history millennia before pornography of any sort made its appearance on the cultural landscape.

With pornography and rape we need to make an important causal distinction: although some rapists have watched and enjoyed pornography (as noted by critics in citing serial rapist and killer Ted Bundy's remark just before his execution, "You are going to kill me . . . but out there are many, many more people who are addicted to pornography, and you are doing nothing about that"), by far the vast majority of men who have watched and enjoyed pornography have never raped. In a review of seven studies on the relationship of pornography to sex offenders of all types, Berl Kutchinsky concluded: "Sex offenders are, as a rule, not more acquainted with pornography or more sexually aroused by such material than are other males—in fact, such differences tend to be in the opposite direction."[10] Indeed, an extensive study of rapists and their backgrounds revealed that instead of being driven to rape by the hypersexuality allegedly produced by pornography, rapists tend to come from sexually repressed environments in which sex was rarely or never discussed, nudity was forbidden, and sexuality was portrayed as sinful. By contrast, nonrapists were more likely than rapists to have experienced pornography while growing up and to have been raised in a family environment in which sex was openly discussed and not shamed into quiescence.[11]

Several correlational studies were equivocal on the relationship between pornography and rape. A 1986 study investigated the relationship between exposure to sexually explicit material and attitudes toward rape in 115 men, finding that only exposure to coercive or violent sexual themes was related to more traditional attitudes about women as submissive and inferior; but contrary to predictions, subjects with greater exposure to general and nonviolent sexual materials held more liberal and egalitarian attitudes toward women. A 1991 study based on data from the Uniform Crime Reports, circulation data from three sexually oriented magazines, and the Standard Metropolitan Statistical Areas found no relationship between pornography and rape. This same study did find, however, that population size, proportion of young adults, percentage divorced, and population change were all significant predictors of rape. Finally, in an extensive cross-cultural study of rape in four countries (Denmark, Sweden, Germany, and the United States), there was no correlation between the availability of pornography (type not specified) and increased sexual violence.[12]

Interestingly, a number of studies point to a possible catharsis effect for pornography, with most citing Denmark and Japan as examples. In the 1960s Denmark experienced a surge in pornography. Instead of taking draconian measures to stop it, the government lifted all bans on pornography. Subsequently, there was a dramatic drop in sex crimes. In Japan, levels of pornography are as high or higher than in America, while rates of sex crimes are fourteen times lower than in the United States (34.5 rapes per 100,000 in the United States versus 2.4 per 100,000 in Japan). "Japanese view the availability of such stimuli as a cathartic valve," wrote the researchers who conducted the study. "It is presumed to provide vicarious satisfaction of a socially unacceptable behavior. In a culture that endorses strict codes of behavior and highly defined roles, the depiction of rape also provides a context in which Japanese men can vicariously abandon all of the explicit signposts of good behavior."[13] Of course, this is not a recipe for subjecting potential rapists to pornography, but at the very least this evidence shows that whatever the cause of rape, it is clearly not pornography by itself.

On the other hand, negative pornography as I have defined it, particularly pornography that depicts a reluctant women who subsequently succumbs to the pressures of her would-be lover and in the end enjoys the experience, may elicit in male viewers inappropriate sexual behavior toward unwilling females. A number of studies show a strong positive correlation between such pornographic scenarios and subjects' self-reported probability of raping a woman.[14] A corroborative study on nonpornographic but aggressive material found an equally positive correlation between portrayed aggression toward women and actual aggression toward women.[15] According to Indiana University psychologist Dolf Zillmann, what generates or increases aggression toward women are not specific sexual or aggressive acts toward women per se, but the overall degree of excitation within the film itself. But this varies considerably among individuals; pornographic and aggressive films appear to have the greatest effect on individuals with limited social and sexual experience. "Persons with limited sexual socialization experience in particular have been found to respond negatively to erotica. Such persons . . . appear to be especially vulnerable to behaving aggressively after exposure to erotica, even to comparatively mild erotica—innocuous as their stimuli may seem to others." Similarly, W. A. Fisher and D. Byrne found that pornography had a greater effect on people whose attitudes toward sex were negative.[16]

A particularly important finding made by Neil Malamuth and James Check was that rapists who report that they are more likely to rape if they think they would not get caught show greater excitatory response to pornography than do nonrapists and men who report that they would not rape even if they would not get caught.[17] Here again we see the reverse causal relationship between pornography and rape. Rapists may be stimulated by pornography, but people who are stimulated by pornography do not become rapists. Interestingly, pornography that shows a woman being sexually seduced against her will, and showing disgust in response, decreased the arousal rating among rapists and potential rapists, in contrast to pornography that supports the myth that women like to be raped.[18]

As with most ethical issues and moral dilemmas, rarely are matters black and white in the world of pornography and erotica. Instead, there are shades of erotic gray. As with other forms of fuzzy morality, assigning fuzzy fractions to shade the world into erotic degrees is much more useful. Watching a stimulating erotica video once in a while is surely no sin, especially if it is not meant as a replacement for intimacy with one's partner, so we might assign it a .1. When the experience of pornography gets to the point of being a daily ritual, is done for masturbation purposes only, and replaces intimacy with one's partner who finds this substitution violative of the relationship, then it might be appropriate to rank that form of pornography as a .9 act of immorality. One can assign fractions in between these extremes according to preference and consequence.

All the scientific studies and reasoned arguments about positive pornography's harmlessness (or even benefits), however, do not amount to a hill of Viagra beans if your partner finds it offensive, or feels repulsed or replaced instead of aroused. Here is another place to apply the ask first principle. Because sex is such a personal matter, and because there is so much variation in what individuals find sexually stimulating, asking your partner first is the simplest and surest way to find out what constitutes acceptable pornography.

Provisional Morality and Abortion

There has been, arguably, no more contentious moral issue of the past half century than abortion, where morality, politics, and science are confoundedly conflated. Moral issues are personal. Political issues are social. Scientific issues are factual. Herein lies confusion. Pro-choicers believe that whether a woman decides to abort a fetus or not is a personal moral issue in which the rights of the mother take precedence over the rights of the fetus.[19] Pro-lifers want to make it a political moral issue in which the rights of the fetus take precedence over the rights of the mother so that society determines what a woman can or cannot do with her body and her fetus.[20] Can science help settle this dispute?

When pro-lifers and pro-choicers square off to debate, they are oftentimes talking at cross-purposes. Pro-lifers speak of the "murder" of innocent fetuses and attack their debate opponents on the grounds that murder is wrong, as if pro-choicers accept murder as moral. In fact, pro-lifers and pro-choicers all agree that murder is immoral. What they disagree about is whether aborting a fetus constitutes murder. This apparent moral question is actually a factual question, because abortion can only be considered murder if it means taking the life of a human being, and when a fetus becomes a human is a question that is difficult to resolve, as Supreme Court Justice Harry A. Blackmun noted in his decision in the 7–2 majority ruling in the 1973 *Roe v. Wade* case: "When those trained in the respective disciplines of medicine, philosophy and theology are unable to arrive at any consensus, the judiciary, at this point in the development of man's knowledge, is not in a position to speculate as to the answer."[21]

The problem is more one of logic than of knowledge. Moral and political decisions are grounded in binary logic in which unambiguous yeses and noes determine Truth. Science is grounded in fuzzy logic in which ambiguous probabilities determine provisional truths. In provisional ethics, moral choices correspond to scientific facts in being provisionally right or wrong, where moral or immoral means confirmed to such an extent that it would be reasonable to offer provisional assent. It remains provisional because, as in science, the evidence might change. Here is how provisional ethics may inform a decision in the abortion debate: most pro-lifers believe that human life begins at conception—before conception not-life, after conception, life. Binary logic. Binary life. With fuzzy logic we can assign a probability to human life—before conception 0, the moment of conception .1, multicellular blastocyst .2, one-month-old embryo .3, two-month-old fetus .4, and so on until birth, when the fetus becomes a 1.0 human life-form. Fuzzy logic. Fuzzy life.

The process does not sound very romantic, but from a scientific perspective, life is a continuum from sperm and egg, to zygote, to blastocyst, to embryo, to fetus, to newborn infant.[22] Neither egg nor sperm is a human individual, nor is the zygote or blastocyst because they might split to become twins or develop into less than one individual and naturally abort.[23] The eight-week-old fetus has recognizable human features such as face, hands, and feet, but neuronal synaptic connections are still being made, so thinking is not possible. Only after eight weeks do embryos begin to show primitive response movements, but between eight and twenty-four weeks (six months) the fetus could not exist on its own because such critical organs as the lungs and kidneys do not mature before that time. For example, air sac development sufficient for gas exchange does not occur until at least twenty-three weeks after gestation, and often later.[24]

Not until twenty-eight weeks, at 77 percent of full-term development, does the fetus acquire sufficient neocortical complexity to exhibit some of the cognitive capacities typically found in newborns. Fetus EEG recordings with the characteristics of an adult EEG appear at approximately thirty weeks, or 83 percent of full-term development.[25] In other words, the capacity for human thought does not exist until just weeks before birth. Of all the characteristics used to define what it means to be human, the capacity to think is provisionally agreed upon by most scientists to be the most important.[26] By this criterion, since virtually no abortions are performed after the second trimester, and before then there is no scientific evidence that the fetus is a thinking human individual, it is reasonable for us to provisionally agree that abortion is not murder and to offer our provisional assent that abortions within the first two trimesters are not immoral because the evidence confirms that during this time the fetus is not a fully functioning human being. Therefore, although one may oppose abortion on a personal level, there is no scientific justification to shift the abortion issue from a personal and moral one to a social and political one.

One objection to this line of reasoning is that science and technology have so fuzzified the boundaries between what were once reasonably discrete categories (even in my fuzzy analysis) that it becomes difficult to justify precisely where to draw the line. Unborn babies are now being treated as patients, with complex surgeries being performed in the womb for such maladies as spina bifida (an opening in the spine through which the spinal cord dangerously emerges), congenital diaphragmatic hernia (the fetus's abdominal organs merge into the chest), and congenital cystic adenomatoid malformation (cysts in the fetus's lungs). Fetuses that would have been aborted before are now being saved, and they are treated medically as little people.[27] On the other end of the spectrum, there are adults whose cognitive capacities are so severely retarded through brain damage that they cannot think at all. They may even lie comatose, completely brain-dead, and yet still retain rights as humans.

The response to this argument is that most fetal surgery is done well into the third trimester, when abortions are rarely performed anyway. And brain-damaged adults already retain rights as humans, so their rights cannot be taken away. We retain hope that they may come out of the coma and regain their thinking ability, so we might think of them as potential humans, in the same sense as it is sometimes argued that a fetus is a potential human. Is a potential human, however, a full rights-bearing person? That depends on how we define personhood. Pro-lifers argue that the genome is entirely in place shortly after conception when the two half-genomes from both parents combine. From that moment on the little clump of cells is a prospective human, a pending member of our species. That is true as far as it goes, but it does not go very far. Human eggs and sperm are potential humans, but no one would consider monthly menstruation or male masturbation to be murder. The counter to this is that the uniting of egg and sperm is a dramatically significant step. Agreed, but a lot of things have to happen during the

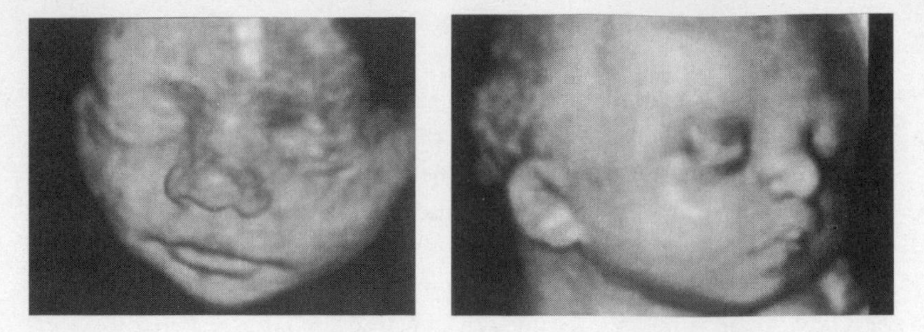

Figure 23. Drawing the Line

The law requires distinct categories—black or white, life or death, right or wrong—whereas science traffics in shades of gray and fuzzy fractions. When does life begin? The law demands that we pick an arbitrary point. Science assigns a fuzzy probability to life—before conception 0, the moment of conception .1, multicellular blastocyst .2, one-month-old embryo .3, two-month-old fetus .4, and so on until birth. Not until twenty-eight weeks—between these two images of a fetus at twenty-three weeks (*left*) and thirty-two weeks (*right*)—does the fetus acquire sufficient neocortical complexity to exhibit some of the cognitive capacities found in newborns. (Courtesy of GE Medical Systems)

nine-month gestation period for the potential human to become an actual human and, unfortunately, there are a lot of things that go wrong that lead to natural abortions, another normal process that no one moralizes about. Potentiality does not equal actuality, and moral rules and principles must be applied first and foremost to actual persons, not potential persons. Given the choice between granting rights to an actual person or a potential person, it is more tenable to choose the former. Herein we find another important distinction to make in the abortion debate, and that is the difference between a human and a person.

A human is a member of the species *Homo sapiens*. A person is a member of a social group or society with legal rights and responsibilities and with moral value. Even if one could justify a fetus as being a human (even if only a potential human), that still does not make it a person. What makes it a person is the granting of legal rights and responsibilities and moral value by the rules governing that society. Pro-lifers are encouraged by changes in the law in many states that grant personhood rights to the unborn, in cases where a pregnant woman has been murdered and the fetus dies as well. No less than twenty-eight states now criminalize harm to a fetus, and many more are moving to pass the Unborn Victims of Violence Act, which in 2003 was renamed Laci and Connor's Law after victims of a notorious murder allegedly committed by Scott Peterson, who was charged with double homicide in the killing of both his wife and unborn child.[28] Legally, if killing a mother and her fetus is double murder, then killing a fetus by itself is single murder. This would make abortion a crime of murder.

Here again, we face moral inconsistencies of deep significance. Where do we draw the line? If our society were to grant personhood rights to the unborn, then it would be logical to do so for nonhuman persons as well, such as our closely related primate cousins, the chimpanzees, gorillas, and orangutans. Since, as I shall argue at the end of this chapter, this is a scientifically justifiable thing to do, then why not grant them to the unborn? After all, we grant rights to adult humans who are so low functioning as to be less "human" than both an eight-month-old fetus and a chimpanzee.

This is a very knotty problem to unravel. We cannot first ask a fetus if it would like to be aborted or not; we can, however, run that thought experiment by imagining ourselves in the position of an unborn potential human who would be granted personhood rights and dignity value upon birth. Presumably, most of us would choose life. By the ask first principle, we would have to conclude that abortion is immoral. However, although asking the unborn can never be more than a thought experiment, there is someone we can ask first, and that is the pregnant woman, who is both a human and a person who already has all the rights, privileges, and moral dignity values bestowed upon her by society. Given the choice between asking the fetus in a thought experiment and actually asking the woman what she thinks should be done, it is logical to give the moral nod to the woman. Given the choice between the potential rights of the fetus and the actual rights of the woman, it makes more sense to go with what already exists in fact over what might exist in potential.

Finally, we can turn to the liberty principle and consider the historical treatment of women. The trend over the past several centuries has been to grant greater freedom, autonomy, and self-determination to minorities, children, and women. Modern civilizations have systematically outlawed slavery, freed children from the burden of excessive labor, and granted women the same rights and privileges as those given to men. We have done so under the principle of liberty: expanding freedom and autonomy to as many members of our species as possible. One of the most important sources of freedom and autonomy for women has been control over their bodies, especially in relation to reproduction. Social and political advances, coupled with scientific and technological discoveries and inventions, have increasingly provided women with greater amounts of reproductive autonomy and control, which, in turn, leads to the overall increase of liberty for women in general. To take away an important source of reproductive control from women by outlawing abortion would be a significant step backward in the historical trajectory of liberty. Thus, given the choice between increasing the liberty of an adult person and the liberty of an unborn fetus, it makes more sense—historically, legally, logically, and morally—to grant that liberty to the adult person, the woman.

This is not to claim that abortion is moral, only that it is not immoral. This brings us back to where we began in making a distinction between individual morality and political morality. If abortion is not murder, then it is not immoral, from a social/political point of view. Or, if a woman decides that even though having a child may burden her physically and financially it is still more important to her to grant life and liberty to her unborn child, that is her choice to make, not the state's. In the end, abortion remains a personal moral choice.

Provisional Morality and Cloning and Genetic Engineering

On December 27, 2002, Dr. Brigitte Boisselier, the scientific director of Clonaid—an organization associated with the Raelians, who believe that life was seeded on Earth by aliens and that cloning is the next step toward immortality—announced at a press conference that a thirty-one-year-old American woman had given birth to the world's first clone, whom they nicknamed, appropriately, "Eve."[29] The story was a bust, although the media feeding frenzy over it generated millions of dollars worth of free publicity for a hitherto obscure fringe cult.

My skepticism is not directed toward the Raelians, however, because whether they succeed or not is superfluous since it will soon become apparent whether cloning is possible or if medical complications will make it impractical as another form of fertility enhancement. That is, the current moral dilemma may be displaced by a scientific factual matter of whether cloning works or does not work. If cloning works as a viable reproductive technology and generator of usable stem cells, then it will be used somewhere by someone regardless of legislative restrictions. If cloning does not work—that is, if it generates genetic monstrosities and nonviable stem cells—it will most likely fall into disuse from lack of interest. Reproductive physicians and their patients will choose other, more viable, technologies, stem-cell researchers and genetic engineers will apply more reliable means to achieve their scientific goals, and madmen will not want to reproduce themselves if they think that their genetic doppelgänger will join Dr. Frankenstein's monstrosity in the Arctic hinterlands.

Given the current scientific limitations on cloning, how can science inform this moral debate? By debunking three fundamental myths about cloning: the *Identical Personhood Myth*, the *Playing God Myth*, and the *Human Rights and Dignity Myth*.

The Identical Personhood Myth is well represented by Jeremy Rifkin, the king of genetic Luddites: "It's a horrendous crime to make a Xerox of someone. You're putting a human into a genetic straitjacket. For the first time, we've taken the principles of industrial design—quality control, predictability—and applied them to a human being."[30] The argument is heard in religious circles as well. The Catholic theologian Albert Moraczewski (echoing the pope's 1987 *Donum Vitae*) proclaimed that cloning would "jeopardize the personal and unique identity of the clone (or clones) as well as the person whose genome was thus duplicated." What about twins? Don't they jeopardize each other's unique identity? No, Moraczewski explained, since they are not the "source or maker of the other," meaning only God can do that.[31]

Baloney. These cloning critics have the argument bass ackwards. Because they tend to be environmental determinists, they should be arguing: "Clone all you like, you'll never produce another you because environment matters as much as heredity." Even proponents of the position that behavior has a significant genetic component to it argue not for genetic determinism, but for gene-environment interactionism. This interactionism starts when genes code for proteins, which generate biochemical reactions, which regulate physiological changes, which govern biological systems, which impact neurological actions, which induce psychological states, which cause behaviors; these behaviors, in turn, interact with the environment, which changes the behaviors, which influence psychological states, which alter neurological actions, which transform biological systems, which modify physiological changes, which transfigure biochemical reactions. This all happens in a complex interactive feedback loop between genes and environment throughout development and into adulthood. The best scientific evidence to date indicates that roughly half the variance between us is accounted for by genetics, the rest by environment. Because it is impossible to duplicate the near-infinite number of environmental permutations that go into producing an individual human being, cloning is no threat to unique personhood. Psychologist and twins expert Nancy Segal cautions that "Genetic influence does not mean that behaviors are fixed, but the ease, immediacy and magnitude of behavioral change vary from trait to trait, and from person to person." It is from that variance that unique personhood, even between identical twins, emerges.[32]

The Playing God Myth has numerous promoters, the latest being Stanley M. Hauerwas, a professor of theological ethics at Duke University, who responded to the news that the Raelians had achieved human cloning with this unequivocal denouncement: "The very attempt to clone a human being is evil. The assumption that we must do what we can do is fueled by the Promethean desire to be our own creators."[33] In support of this myth he is not alone. A 1997 *Time*/CNN poll, conducted after Dolly the cloned sheep was revealed to the world, found that 74 percent of Americans answered yes to the question, "Is it against God's will to clone human beings?"[34] President Clinton, not the most religious president we have ever had, nevertheless threw his spiritual hat into the fear-of-cloning ring with this statement: "Any discovery that touches upon human creation is not simply a matter of scientific inquiry, it is a matter of morality and spirituality as well." Even before the National Bioethics Advisory Commission—established by Clinton to present to him all the ethical considerations surrounding cloning—submitted its report to the White House, Clinton instituted a ban on federal funding related to research on the cloning of humans and asked that the private sector do the same. Shortly thereafter, Clinton held another press conference urging Congress to ban human cloning altogether (not just research funds): "Personally, I believe that human cloning raises deep concerns, given our cherished concepts of faith and humanity."[35] Most religions are against cloning on similar argumentative grounds. The Christian Life Commission of the Southern Baptist Convention, for example, on March 6, 1997, passed a resolution that called on Congress to "make human cloning unlawful" and for "all nations of the world to make efforts to prevent the cloning of any human being." Similarly, Fred Rosner, a Jewish bioethicist, wrote that cloning can be considered as "encroaching on the Creator's domain."[36]

Balderdash. Cloning scientists don't want to play God any more than fertility doctors do. What's godly about in-vitro fertilization, embryo transfer, and other fully sanctioned birth enhancement technologies? Absolutely nothing. Yet we cheerfully accept these advances because we are accustomed to them. In fact, most of us are alive because of medical technologies and social hygiene practices that have doubled the average life span in this century. What's godly or natural about heart-lung transplants, triple bypass surgeries, vaccinations, or radiation treatment? Nothing. The mass hysteria and moral panic surrounding cloning is nothing more than the historically common rejection of new technologies, coupled with the additional angst produced when the sphere of science expands too quickly into the space of religion. The editorial cartoon in figure 25 captures this fear well, in which the cloning of God Himself spells the end of monotheism.

The Human Rights and Dignity Myth is embodied in the Roman Catholic Church's official statement against cloning, based on the belief that it denies "the dignity of human procreation and of the conjugal union,"[37] as well as in a Sunni Muslim cleric's demand that "science must be regulated by firm laws to preserve humanity and its dignity."[38] Members of Congress, assigned to deal more with legalities than moralities, have decreed that cloning violates the rights of the unborn. *Bunkum.* Clones will be no more alike than twins raised in separate environments, and no one is suggesting that twins do not have rights or dignity or that twinning should be banned.

In his 1950 science fiction novel *I, Robot,* Isaac Asimov presented the "Three Laws of Robotics": "1. A robot may not injure a human being, or, through inaction, allow a human being to come to harm. 2. A robot must obey the orders given it by human beings except where such orders would conflict with the First Law. 3. A robot must protect its

Figure 24. Cloning God
(Courtesy of Hilary Price)

own existence as long as such protection does not conflict with the First or Second Law."[39] Given the irrational fears people express today about cloning that parallel those surrounding robotics half a century ago, I would like to propose the "Three Laws of Cloning" that address these misunderstandings:

1. A human clone is a human being no less unique in its personhood than an identical twin.
2. A human clone is a human being with all the rights and privileges that accompany this legal and moral status.
3. A human clone is a human being to be accorded the dignity and respect due any member of our species.

Instead of restricting or banning cloning, I propose that we adopt the Three Laws of Cloning, the principles of which are already incorporated in the laws and language of the U.S. Constitution. Most of the horror-laden scenarios proposed by moralists are already addressed by law—a clone, like a twin, is a human being, and you cannot harvest the tissues or organs of a twin. A clone, like a twin, is a person no less than any other human.

Cloning is going to happen whether it is banned or not, so why not err on the side of liberty and allow scientists to freely explore the possibilities—not to play God, but to do science. The soul of science is found in courageous thought and creative experiment, not in restrictive fear and prohibitions. For science to progress, it must be given the opportunity to succeed or fail. Let's run the cloning experiment and see what happens.

Provisional Morality and Animal Rights

At the top of the Bio-Cultural Evolutionary Pyramid (figure 6 on page 48) is the biosphere, which includes all life on earth, a love of which Ed Wilson calls "biophilia." The sentiment behind biophilia, if there is one, I call bioaltruism. Bioaltruism appears to be almost entirely the product of culture, not evolution, and as such it must be learned. In the long run, bioaltruism may be the most important moral sentiment that will save our species, along with other species upon which we depend, from extinction.

Moving up the BCE pyramid from the family to the extended family to the community to the society, the liberty principle has been expanded to encompass more members of our species. Eventually we will achieve complete species altruism and belongingness, in which all members of our species are considered equal in terms of liberty rights. Achieving liberty for all human animals has been a difficult struggle against tyrannies of many types. It has been only a century and a half since slavery was outlawed. Only over the past century have women begun to approach anything like the liberty rights enjoyed by men. And it has been less than half a century since minorities began to see some progress in gaining civil liberties (gays, for example, have the legal right to marry in only three countries and are still fighting to gain rights for such matters as adoption, inheritance, and insurance benefits). What hope, then, can there be for *nonhuman* animals? Animal rights activists must often feel like King Sisyphus, condemned for eternity to roll the boulder of freedom up the hill of oppression, only to have it roll down again just before cresting the top. If far too many people still think that women and minorities do not deserve full liberty rights, what hope can there be for chimps and dolphins?

At the heart of the animal rights debate is how we should treat nonhuman animals. Clearly, when we are talking about "animal rights," no one is proposing that animals have the right to vote or the right to a public education. Animal rights are much more basic. Some, in fact, are already in place. In many Western countries, for example, it is a crime to treat animals inhumanely. Humans can be tried and convicted for the crime of cruelty to animals, although this is usually restricted to horrific abuses of our favorite pets, such as dogs, cats, and horses (the Animal Planet television network even has a series on animal-abuse busters who drive around, camera crews in tow, looking for abusers). Modifying the legal system to include some animals under its protective umbrella is going to require breaking through a number of substantial barriers, including economic, religious, legal, and psychological.[40] On a positive note, these are the same obstacles faced by women and minorities who, in time, managed to leap the hurdles.

Economically, the trade in animals, as it was in slavery in nineteenth-century America, is so extensive that if animal rights were suddenly instituted just for all mammals, the economy would suddenly grind to a disastrous halt. The blood and fat from cows, to pick just one example, goes toward the production of adhesives, contraceptive jellies, cosmetics, cough syrup, crayons, detergents, dyes, fabric softeners, fertilizer, fire extinguisher foam, ink, jet engine lubricants, lubricants, plastics, shaving cream, soaps, textiles, and countless medical products.[41] And this does not even include the food industry. How many of us are willing to give up eating juicy tenderloin steaks? We are the dominant carnivore species on the planet. We cannot simply grant all nonhuman mammals the same rights as human mammals without considerable economic consequences. *Religiously,* the Bible says that God bestowed upon one species (us) dominion over all other species. Of course, for centuries God and the Bible were invoked by slave owners and oppressors of women. Christians enslaved Africans, Muslims enslaved Africans, and Jews enslaved gentiles, Africans, and other Jews. One justification for the enslavement of black Africans was that they were subhuman and thus not full rights-bearing members of the human species.[42] Similarly, for centuries religion laid the foundation for the control of women, by fathers, husbands, and patriarchal society at large.[43] If religion can justify the treatment of blacks and women as subhuman, we should not be surprised about the religious attitudes toward nonhumans. *Legally,* animals are, in essence, things—products and property for humans to buy and sell. The legal treatment of nonhuman animals as things means that we think of them in substantially different ways than we think about human animals. *Psychologically,* it is the same attitude found in the proslavery movement of centuries past, as the slave historian David Brion Davis explains: "Today it is difficult to understand why slavery was accepted from pre-biblical times in virtually every culture and not seriously challenged until the late 1700s. But the institution was so basic that genuine anti-slavery attitudes required a profound shift in moral perception."[44] That shift could not come about until there was a psychological shift toward including all people as members of the species. For the animal rights movement to succeed, there must be a psychological shift from speciesism toward bioism.

Fuzzy Animal Rights

Arguably, one of the most extreme religions in the world is Jainism, a form of Hinduism whose members hold such a deep reverence for life that it drives them to sweep the ground before them to avoid squashing insects. This is the reductio ad absurdum of the animal rights movement—since there is no good place to draw the line, then we should grant all animals all rights, including the mosquitoes buzzing about our heads. This position is so ridiculous that, thankfully, almost no one embraces it. However, this is the logical opposite of the system of binary rights that we presently hold—all rights for us, no rights for them. But what constitutes "us" and "them"? There is a logic, however binary and restrictive, to speciesism. Although species are nonstatic and malleable over evolutionary time scales, they are static and fixed entities on historical time scales. I shall never forget memorizing Ernst Mayr's definition of a species in my first course in evolutionary theory: "Species are groups of actually or potentially interbreeding natural populations which are reproductively isolated from other such groups."[45] Since humans are reproductively isolated from all other species (that is, we cannot interbreed with them and produce viable offspring), the species level is a clear and distinctive place to "draw the line."

In 1955, the French novelist Vercors penned a science fiction story entitled *You Shall Know Them,* in which a scientist impregnates a female chimpanzee, kills the offspring, then turns himself in for murder. During his trial the fuzzy boundaries between human and nonhuman primates are explored, demonstrating how difficult it is to justify any particular line between us and them.[46] Something like this scenario almost played itself out in our own evolutionary history. Considerable fossil evidence now reveals that not long ago—within the past 100,000 years—there were several species of hominids living simultaneously and reproductively isolated in the same geographic regions of the globe, most notably *Homo erectus* in Asia, *Homo neanderthalensis* in Europe, and Cro-Magnons, or anatomically modern humans.[47] If the Neanderthals had not gone extinct about 35,000 years ago, and instead were living among us in Europe (where they flourished for 200,000 years before our arrival), would they be granted rights? Their brains were as large as if not larger than ours, but they showed little cultural progress and they may or may not have had language. If we were able to control them, imagine the justifications for Neanderthal slavery and slave labor: "Neanderthals to the mines!"

Still, that is not what happened, and here we are, the dominant primate species, the last one left standing at the end of the Pleistocene. So, yes, we are the only hominids around and we are reproductively isolated from all other species. But once we expand our thinking to include, well, thinking, the lines blur with a handful of other species. This is fuzzy logic applied to animal rights. Fuzzy logic allows us to expand beyond our fuzzy species boundary to include in our circle of liberty rights a tiny cadre of other big-brained, intelligent, emotional mammals, including the great apes (chimpanzees, bonobos, gorillas, and orangutans) and what we might correspondingly call the great marine mammals (dolphins and whales). Fuzzy logic is a very effective way of thinking about animal rights, so I was not surprised to encounter it in Steven M. Wise's brilliant analysis of the animal rights debate, *Drawing the Line: Science and the Case for Animal Rights*.[48] My fuzzy scale is not as inclusive as Wise's (see figure 26), which includes African elephants and African grey parrots. I will admit to considerable sympathy, even empathy for elephants, especially when seeing them grieve over the murder of one of their brethren by poachers.[49] But I sense that I am over-anthropomorphizing; I also retain some skepticism about some of the language and cognitive studies conducted with chimps, gorillas, and parrots, particularly because of the tendency for their scientist handlers to exaggerate and anthropomorphize their abilities. Nevertheless, there is now a sizable body of evidence that makes a potent case that in terms of evolutionary closeness, cognitive abilities, and emotional capacities, the great apes are too near to us to withhold from them certain basic liberty rights; and although dolphins and whales are genetically more distant from us than the great apes, their large brains, convoluted cortexes, socially complex groups, and

Figure 25. Equal Rights for All Hominids?

What if *Australopithecus afarensis* and other hominids had not gone extinct? Would we grant them equal rights, or would we enslave them? (Illustration by Michael Rothman. Reproduced by permission)

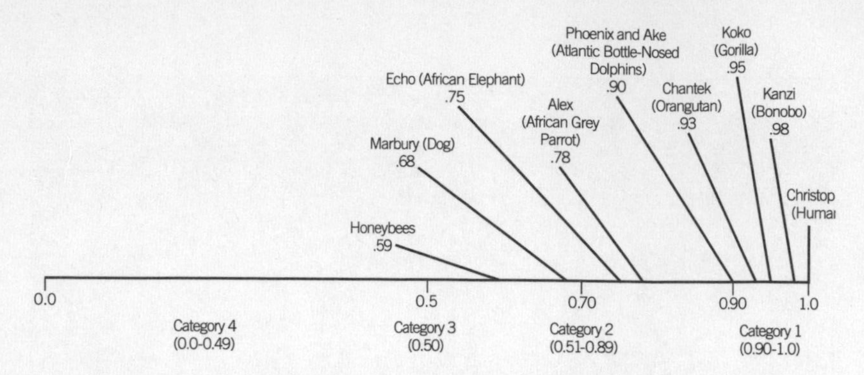

Figure 26. The Fuzzy Logic of Animal Rights

In *Drawing the Line*, attorney and animal rights activist Steven M. Wise applies fuzzy logic to arrive at a scientific basis for the granting of basic liberty rights to some species. In Category 1 he includes species "who clearly possess sufficient autonomy for basic liberty rights," including the great apes (although not on the chart, he includes chimpanzees). In Category 2 he includes species who might qualify for basic legal rights, depending on what other criteria we might consider. In Category 3 he includes species for which we do not have enough knowledge to determine what rights they should have, and Category 4 includes those species who lack sufficient autonomy for basic liberty rights. (From Steven Wise's *Science and the Case for Animal Rights*, 2002. Courtesy of Steven M. Wise)

apparent symbolic languages put them in the same fuzzy set to receive basic liberty rights.[50]

As Wise notes, for example, the mind of Koko the gorilla is on par with the mind of a six-year-old human. Koko has a sense of self, in that she can pass the mirror self-recognition test in which a red dot is clandestinely placed on her forehead and she notices that something is different. Koko can pass an "object permanence" test in which she can remember the shape and location of objects, as well as the conservation of liquid, in which she realizes that the quantity does not change when it is poured into a different-sized container. Koko imitates the actions of humans and other gorillas, has learned hundreds of symbolic language signs with which she can answer questions and attempt to deceive her handlers, and has even attempted to teach language signs to other gorillas. This implies that Koko has a "theory of mind"; that is, she not only has a sense of self, she realizes that other gorillas and humans have the same sense of self. Most tellingly (in terms of how many people judge an animal's moral worth), Koko has scored between seventy and ninety-five on a standard human child intelligence test.[51] And there is evidence that the minds of a chimpanzee named Washoe, a bonobo named Kanzi, and an orangutan named Chantek are functionally equivalent to that of Koko, bringing all of the great apes into the same cognitive set.[52]

The case for dolphins is also compelling. Two extensively studied bottle-nosed dolphins, named Phoenix and Ake, exhibit evidence of symbolic communication, and other dolphins studied in marine laboratories have passed the mirror self-recognition test, which means they too have a sense of self.[53] Dolphin brains make for an interesting comparison. Although the brains of gorillas (average 500 cubic centimeters, or cc), chimpanzees (400 cc), bonobos (340 cc), and orangutans (335 cc) are considerably smaller than ours (1440 cc), dolphins' brains (1700 cc) are slightly larger than ours, even when adjusted for body size and weight (figure 29). More importantly, the surface area of a dolphin's cortex—where the higher centers of learning, memory, and cognition are located—is enormous, averaging 3700 cc squared compared to 2300 cc squared for humans. Although this is a little misleading—the thickness of the dolphin's cortex is roughly half that of humans—when absolute cortical material is compared, dolphins still average an impressive 560 cc compared to our 660 cc. Dolphin brains are also asymmetrical, which some neuroscientists believe is related to intelligence and language ability.[54]

Provisional Animal Rights: The Moderation Principle

For the animal rights movement to succeed, the moderation principle must be applied. Torching scientific laboratories that use white rats for subjects is no virtue; in fact, it is immoral, illegal, and idiotic. Human property rights, liberties, and freedoms are destroyed in the extremist name of animal liberation. The end result is that the hearts of people are hardened against granting any animals any rights, and the animals are not liberated. Here, extremism is a vice. Why worry about lower mammals when higher mammals can't get a fair hearing in the court of public opinion? Let's take this movement one step at a time, justifying each claim with a mountain of scientific evidence and legal precedence. We have that for apes and dolphins. Based on their evolutionary close-

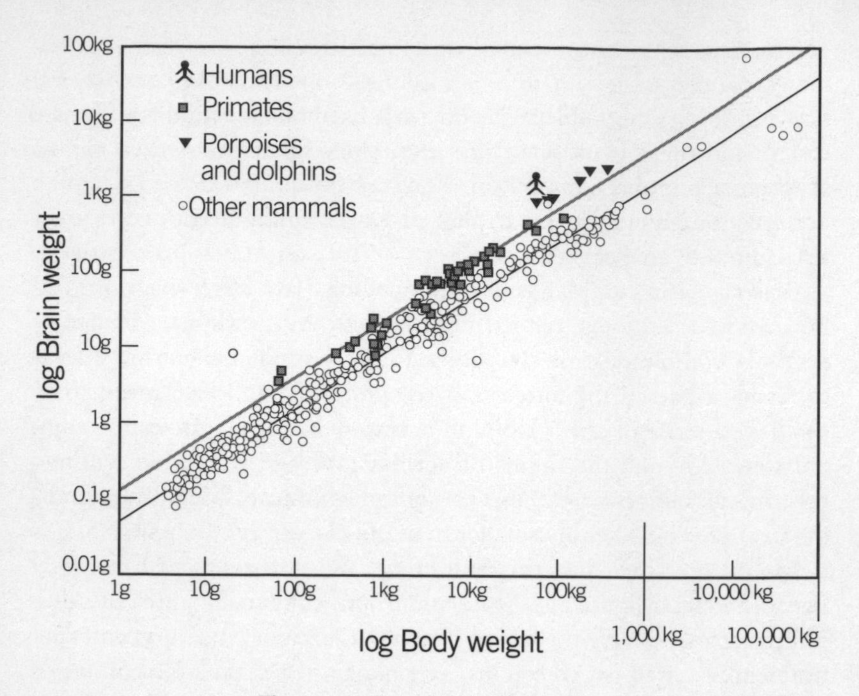

Figure 27. Comparative Brain Size

Comparing the brain sizes of humans, primates, dolphins and porpoises, and other mammals reveals that there is a direct relationship between brain weight and body weight—in general, bigger bodies take bigger brains to operate. However, when controlling for body size, there is still a powerful trend showing that humans, primates, dolphins, and porpoises have larger brains for their body weight than all other mammals. In addition to their capacity for symbolic language and their ability to pass the mirror self-recognition test for self-awareness, this is powerful data in support of granting these species basic liberty rights. (From John M. Allman's *Evolving Brains*, 1999. Courtesy of John Allman)

ness, cognitive abilities, and emotional capacities, it is reasonable to offer our provisional assent to extend basic liberty rights to them. If we can win for them basic liberty rights, then we can worry about monkeys, elephants, dogs, and parrots. What rights? We can begin with the most basic rights granted by the U.S. Constitution: life, liberty, and the pursuit of happiness.

Applying the ask first principle is difficult with nonhuman animals, of course, but as a thought experiment, it is a reasonable means of approximating what an animal might think about what you are contemplating doing to it. (Although I'd bet dollars to doughnuts that if you taught Washoe, Kanzi, Koko, and Chantek the sign for electric shock, and then asked them with sign language if they would like to receive one as part of an experiment, you would very quickly receive an answer.) We cannot ask a chimpanzee how she might feel if we lock her up in a small, cold stainless steel cage for the rest of her life in order to inject her with various human diseases, but we can observe her nonverbal communication, and, for most chimps, it is abundantly clear that they are none too pleased about such arrangements (as in the title of Wise's first book, *Rattling the Cage*). In regard to the happiness principle and the liberty principle, as a first step toward a higher moral principle, *we should never seek happiness and liberty when it leads to a great ape's or a great marine mammal's unhappiness and loss of liberty.* For chimpanzees, bonobos, orangutans, and gorillas, this may mean nothing more than the freedom to go hunting and foraging for food in an open natural environment and to be protected from poachers who kill them for the "bush meat" market. For dolphins, this may just mean the liberty to swim in social bands throughout the planet's oceans, free from tuna fishermen who drown them in their extensive fishing nets. It's that simple. Don't kill 'em. Don't eat 'em. Don't wear 'em. Don't cage 'em. Just let 'em be.

To many animal rights activists, this proposal will seem cowardly in its timidity. To many anti–animal rights activists, this proposal will seem ridiculous in its grandiosity. It is, I think, the moderation principle in practice. If we adopt the historian's stock in trade, deep time, I think it is a provisionally moderate moral step up the Bio-Cultural Evolutionary Pyramid, expanding our sentiments to include, first, more people and then more species into our circle of sympathies that will ultimately lead us toward the bioaltruism that will one day save our biosphere and, as a consequence, our species.

Cicero's Warning

For all of humanity it has been a long journey on the evolutionary and historical pathway to where we stand today, on the brink of triumph or disaster, survival or extinction. Which road we take depends on which moral choices we make. Since we have the capacity for both moral and immoral actions, and the freedom to choose, our destiny lies within.

In the first century B.C., the Roman statesman Cicero remarked, "Although physicians frequently know their patients will die of a given disease, they never tell them so. To warn of an evil is justified only if, along with the warning, there is a way of escape." As we shall see in the last chapter, there is an escape from our immoral disease.

8

RISE ABOVE: TOLERANCE, FREEDOM, AND THE PROSPECTS FOR HUMANITY

> Nature, Mr. Alnutt, is what we were put in this world to rise above.
>
> —Katharine Hepburn to Humphrey Bogart in *The African Queen,* 1951

In an episode of the original series of *Star Trek,* the Enterprise crew encounters an alien civilization playing a war game that is just about to escalate into a full-scale conflict. Captain James T. Kirk tells the alien leaders to "just say no" to war. Though instinctive, war may be resisted. There can be peace.

> ALIEN: There can be no peace. Don't you see? We've admitted it to ourselves. We are a killer species. It's instinctive. It's the same with you.
>
> KIRK: All right, it's instinctive. But the instinctive can be fought. We are human beings with the blood of a million savage years on our hands, but we can stop it. We can admit that we are killers but we're not going to kill . . . today. That's all it takes—knowing that you're not going to kill . . . today.[1]

Are we, by nature, a killer species? The scientific evidence we have surveyed in this book answers in the affirmative. Built into our makeup is the capacity for aggression, violence, and war. Providentially, the scientific evidence also points to a solution. The rise of state societies, and with them the development of codified moral systems—initially through religious organizations and subsequently through secular institutions—have truly "civilized" our species, accentuating the moral and attenuating the immoral sentiments. We have done well, but we can do better. Our moral and immoral natures are in delicate balance. Too many immoralities confront us daily, even in the most civilized of cultures. Homicides and genocides, wars and revolutions, rape and domestic violence make their appearances too frequently on the nightly news. We still retain the power to trigger our own extinction, whether through ecocide with the mass destruction of our environments or through genocide from nuclear, chemical, and biological weapons. If nature put us into this world, how best shall we rise above it? In this final chapter on the prospects for humanity, we shall examine the evidence that our species is on a long evolutionary trajectory toward greater amity toward members of our own group, a long historical path toward more tolerance and inclusivity, and a long political path toward more liberties for more people in more places, whether they are members of our group or not. Out of this analysis arise two recommendations, one on individual tolerance and the other on social and political freedoms.

The Domesticated Primate

UCLA evolutionary biologist Jared Diamond once classified humans as the "third chimpanzee."[2] Genetically we are very similar, and when it comes to high levels of between-group aggression, we also resemble our ape cousins. But in comparing ourselves to other species of great apes, there is hope for humanity. In looking at within-group levels of aggression, it turns out that we are much more like bonobos, who are well known for their peaceful and loving ways. Once classified as "pygmy chimpanzees" because of their diminutive skulls, bonobos are now placed in their own grouping, primarily based on these dramatic behavioral differences from chimpanzees. We are much more peaceful than chimpanzees with our fellow in-group members, and our prodigious sexuality is far more like that of bonobos than chimpanzees. Although humans still exhibit within-group violence—as witnessed on the nightly news—compared to chimpanzees, who almost daily exhibit uncontrolled outbursts of male on male aggression, and male on female violence, statistically we are closer to bonobos than to chimpanzees. Domestic violence among humans, for example, while still at unacceptable levels, is nevertheless significantly less than that seen in chimpanzee families.[3]

What are we to make of this contrast between humans as within-group bonobos and humans as between-group chimpanzees? Anthropologist Richard Wrangham proffers a plausible theory that as a result of selection pressures for greater within-group peacefulness, humans and bonobos have gone down a different behavioral evolutionary path than chimpanzees. That difference may be witnessed in morphology. The "pygmy chimpanzee" moniker was given to bonobos because, compared to chimpanzees, their skull, jaws, and teeth are much reduced in size, even while their other body parts are quite similar. It has been observed that when artificially selecting for docility in wild animals, along with far less aggression, breeders also cause a suite of other changes, including and especially a reduction in skull, jaw, and teeth size. This is called pleiotropy, in which a single gene may affect a series of traits. Selecting for one set of traits (for example, nonaggression) may generate other unintended changes (for example, reduced skull, jaw, and teeth size). The most famous study on selective breeding for domesticity in wild animals was begun in 1959 by the eminent Russian geneticist Dmitri Belyaev at the Institute of Cytology and Genetics in Siberia (and continues today by Lyudmila N. Trut). Silver foxes (*Vulpes vulpes*) were bred for friendliness toward humans (defined by a series of criteria, from the animal allowing itself to be approached, hand fed, petted, to it actively seeking to establish human contact). In only thirty-five generations (remarkably short on an evolutionary time scale), the researchers were able to produce tail-wagging, hand-licking, peaceful foxes. What they also fashioned were foxes with significantly smaller skulls, jaws, and teeth than their wild ancestors.

Similar changes can be seen in comparing domesticated dogs to their ancestral wild wolves. Mitochondrial DNA sequence evidence places the ancestor of every dog on earth, including New World dogs, to a single population of Asian wolves active roughly 15,000 years ago. In addition to smaller skulls, jaws, and teeth, domesticated dogs evolved a set of social-cognitive abilities that enable them to read human communicative signals indicating the location of hidden food. These abilities are not shared by their wild forebears nor by any of the other great apes. Specifically, domesticated dogs were able to pick the right container of concealed food when the experimenter looked at, tapped, or pointed to it, whereas wolves, chimpanzees, and other primates are unable to do so. Such nonverbal communication skills are vital for both domesticated dogs and domesticated humans.[4] There were additional pleiotropic effects in the foxes. After several generations of breeding, for example, the foxes, like their canine cousins, began exhibiting floppy ears, curly tails, and striking color patches on their fur, including a star-shaped pattern on the face similar to those found in many breeds of dogs.

What is going on here? The Russian scientists believe that in selecting for docility, they inadvertently selected for paedomorphism, or the retention of juvenile features into adulthood, such as floppy ears (found in wild pups but not in wild adults), the delayed onset of the fear response to unknown stimuli, and lower levels of aggression. It seems that the selection process led to a significant decrease in levels of stress-related hormones such as corticosteroids, which are produced by the adrenal glands during the flight or fight response, as well as a significant increase in levels of serotonin, thought to play a leading role in the inhibition of aggression. Curiously, in selecting only for docility, the Russian scientists were also able to accomplish what no breeder had done before—increase the length of the breeding season.[5]

Wrangham suggests that over the past 20,000 years, as humans became more sedentary and their populations grew, there was selection pressure for less within-group aggression, and this effect can be seen in the reduced size of our skulls, jaws, and teeth (compared to our immediate hominid ancestors), our year-round breeding season and more bonobo-like prodigious sexuality, and our paedomorphism (see figure 28). Wrangham also cites research that shows how area 13 in the limbic frontal cortex in humans, believed to mediate the inhibition of aggressive behavior, more closely resembles in size the equivalent area in bonobos' brains than it does that in chimpanzees' brains.[6]

An additional test of this hypothesis is to compare serotonin levels in humans with those in chimpanzees and bonobos. The prediction is that human serotonin levels should more closely match those in bonobos. This comparison has never been made, but as Robert Sapolsky, a Stanford University biologist who studies aggression in primates, told me: "Overall, the literature shows that low levels of serotonin, or serotonin breakdown products in the cerebral spinal fluid, is a predictor of aggression, impulsivity, disinhibition, and so on. It probably has much to do with indirectly inhibitory roles that serotonin plays in the frontal cortex." Paul Zak, an expert on the effects of oxytocin on cooperation and the control of aggression, agreed with Sapolsky's assessment, and

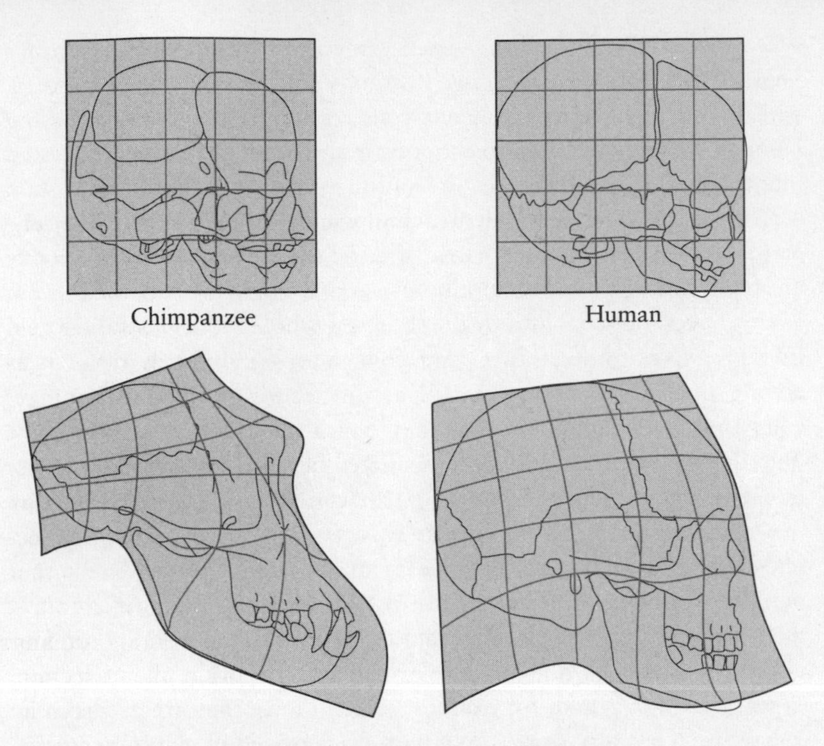

Figure 28. The Paedomorphic Primate

Paedomorphy is the retention of juvenile features into adulthood. Humans, like bonobos and domesticated dogs and foxes, show paedomorphy of the skull, jaw, and teeth. This illustration shows the growth of a chimpanzee skull (*left*) from juvenile (*top*) to adult (*bottom*), compared to the growth of a human skull (*right*), from juvenile (*top*) to adult (*bottom*). The chimpanzee skull matures very differently than the human skull. We are the paedomorphic primate. (From John M. Allman, *Evolving Brains,* 1999. Courtesy of John M. Allman)

added: "Oxytocin (OT) is a feel-good hormone and we find that it guides subjects' decisions even when they are unable to articulate why they are acting in a trusting or trustworthy matter. I think the same 'sense' of right and wrong in moral dilemmas may utilize OT." As for the difference between chimpanzees and bonobos, Zak suggested that "there is evidence that serotonin (5HT) increases the release of OT. I would speculate that bonobos have higher OT (due to the frequency of sexual activity and touching) and therefore generally higher 5HT and less aggression (though I have not seen such a study done)." He also noted that in his lab they have been studying what he called oxytocin's "ugly cousin," "the neuroactive hormone arginine vasopressin (AVP), which is strongly related to reactive aggression in mammals, including humans." Where does all this happen in the brain? Sapolsky suggested that the inhibition of aggression is ultimately controlled by the frontal cortex, of which we have plenty, chimps have some, and the baboons he studies on the Serengeti Plain of Africa have very little. "Baboons are far less disciplined than chimps, and when you map their brain anatomy you notice that they don't have a whole lot of frontal cortical function. Even though there are tremendous individual differences among the baboons, they're still at this neurological disadvantage, compared to the apes, and thus they typically blow it at just the right time. They could be scheming these incredible coalitions, but at the last moment, one decides to slash his partner in the ass instead of the guy they're going after, just because he can get away with it for three seconds. Their whole world is three seconds long." To the extent that morality is linked to impulse control and the delay of gratification, if there is a moral module in the brain, it is either in the frontal cortex or directly linked to it and is heavily mediated by brain chemistry and experience. Zak noted, for example, that in mice that are predisposed to have high levels of oxytocin, if they are deprived of maternal nurturing, "a large proportion of brain areas with OT receptors atrophy."[7]

These data fit well with the theory on the evolution of the moral sentiments presented in the first half of this book. The increased size of our brains, particularly the frontal cortex, gave humans a cognitive advantage over other primates that allowed us to better control our impulsive emotions and delay gratification, as well as form social coalitions and plan strategies. Between-group competition (with other hominid and primate groups) for limited resources led to the selection for such immoral sentiments as competitiveness and selfishness, but at the same time it led to within-group cooperation and selflessness that enhanced the fitness level of individual members of the group. The result is within-group amity and between-group enmity (figure 29).[8]

This evolutionary scenario bodes well for our species. Even though the behavioral potential for between-group violence exists in our primate brains, we also harbor the seeds of peaceful coexistence. We cannot increase the size of our frontal cortex. However, we can learn to be moral. (Children do it—the frontal cortex is not fully developed until

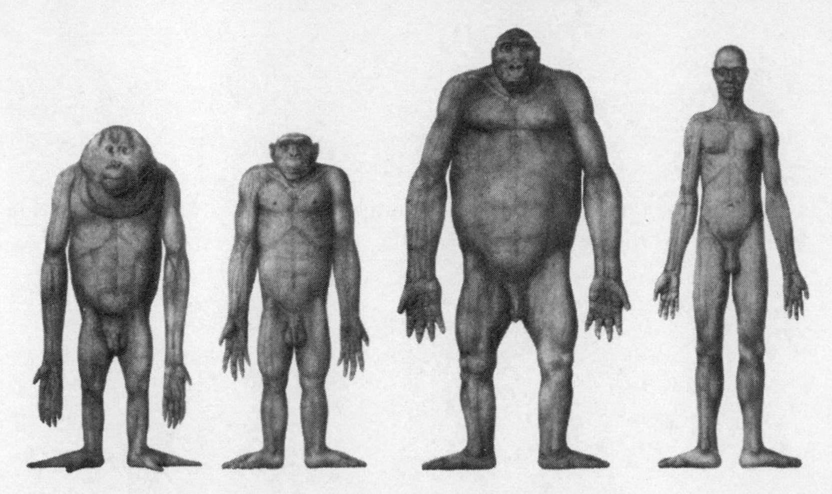

Figure 29. The Domesticated Primates

An orangutan, bonobo, gorilla, and human are pictured. In measures of between-group violence, humans are more like aggressive and territorial chimpanzees. In measures of within-group violence, humans are more like peaceful and loving bonobos. Between-group competition for limited resources led to the selection for within-group cooperation and selflessness and between-group competitiveness and selfishness. (From Adolph Schultz, *The Life of Primates,* 1969. Courtesy of Orion Publishing Group, Inc.)

well into the teenage years, which is why children are, in a manner of speaking, premoral animals, why teens seem so impulsive, and why parents discipline their children in an attempt to shape a moral sense into them.) We can also practice being moral. If making love makes more oxytocin, and more oxytocin means less aggression, then there very well may be a neurological correlate to the cliché, "make love, not war" (figures 30 and 31). We can also learn to think differently about ourselves—perceiving the entire human population as a single group, for example, would eliminate between-group aggression. Although recent history is not encouraging, the long-term trend over the past half millennium has been toward greater inclusiveness and more liberties, for more people, in more places. There are two ways to reinforce the continuation of this positive trend: (1) individual action to accentuate tolerance and attenuate intolerance of individual differences between people; (2) political action to expand whom we tolerate, both legally and morally, to be members of our in-group.

Figure 30. Making War, Not Love

When it comes to between-group levels of violence, humans behave much like chimpanzees, with males fanning out into surrounding environments in search of food and other resources that often result in seek-and-destroy missions against other groups. (*Top:*) Two groups of New Guinea hunter-gatherers square off for war. (Photograph by Robert Gardiner. Courtesy of the Peabody Museum) (*Bottom:*) A group of U.S. soldiers in Vietnam sweeps through rice paddies in an army photograph entitled "Mission—Search and Destroy the Vietcong." (Photograph by SFC Jack H. Yamaguchi)

Figure 31. Making Love, Not War

Bonobos have much lower levels of within-group violence and much higher levels of sexual contact. Conflict resolution among bonobos often involves sexual and erotic contact. Although humans are like chimpanzees in our high levels of between-group aggression, we are more like bonobos in our low levels of within-group aggression and high levels of sexuality and eroticism. (Photograph by Frans de Waal. Courtesy of Frans de Waal)

From Intolerance to Tolerance: How the Mind Works

Since the founding of the Skeptics Society and *Skeptic* magazine in 1991, I have been asked countless times how and why I lost my faith. Although my conversion to Christianity was sudden and dramatic, as is often the case for those who are not inculcated into a religion in childhood, my "de-conversion" was gradual and evolutionary. The scales did not suddenly fall from my eyes. Paul did not morph back into Saul.

Rather, there was a slow but systematic displacement of one worldview and way of thinking by another: genesis and exodus myths by cosmology and evolution theories; faith by reason; final truths by provisional probabilities; trust by verification; authority by empiricism; and religious supernaturalism by scientific naturalism.[9]

Intellectually, I found little substance in the so-called scientific proofs of God's existence or the philosophical arguments dating back to the Middle Ages that form the foundation of Christian apologetics. Although the scientific explanations for the origins of our universe, our world, and ourselves—Big Bang cosmology, historical geology, and evolutionary theory—were not wholly encompassing and had plenty of gaps, they seemed to me to have a higher probability of being true, by orders of magnitude, than the origin myths offered by religion. Anthropologists and social psychologists elucidate the fact that most aspects of religion, including and especially myths of origins and morality, are culturally determined and socially constructed. In my studies there came a point when it seemed to me absurd to even ask if these stories were true. Attempting to calculate how many pairs of animals could have fit on Noah's ark seems like a pointless exercise: if God is omnipotent, He could fit as many animals on the ark as He liked. The point of the Noachian flood story is about destruction and redemption, not how Noah kept the predators away from the prey or on which deck the dinosaurs were housed. In this sense, creationists have butchered the Bible in their attempt to squeeze the square peg of religion into the round hole of science.

Emotionally, over time I found increasingly less in religion that appealed to me. The definitive nature of religious answers struck me as contrived, particularly when contrasted with the provisional nature of answers in science. For my temperament, the uncertainty of science was one of the perks of the job. Now anyone, including me, could join in the search for answers and participate personally in the adventure of exploring our world and our selves. Since there was no Archimedean point of objective observation, it meant that I was on a journey of discovery with the rest of humanity. There are no privileged priests in science.

Morally, there were aspects of religion that I found more than a little troublesome. When I became a born-again Christian, the moral complexities and subtleties of life—that I was only just beginning to explore and comprehend—suddenly vanished in the clarity of the absolute and final answers to moral dilemmas. Or so I thought. My first inkling of a problem came the day after my conversion, when a passionately religious friend reprimanded me for choosing the wrong faith. He told me I was still doomed if I did not switch to his church, the Jehovah's Witnesses. As it turned out, this was not an isolated instance. The more faiths I examined, the more aware I became of the fact that they all think they alone are right.

Religious Absolutism and Intolerance

The belief that one's faith is the only true religion too often leads to a disturbing level of intolerance, and this intolerance includes the assumption that nonbelievers cannot be as moral as believers. The Bible reinforces this idea (Ps. 14:1): "The fool has said in his heart, 'There is no God.' They are corrupt, they have done abominable works, there is none who does good." The most famous pronouncement along these lines came during a news conference on August 27, 1987, by Vice President George Bush, who was making a stop at Chicago's O'Hare Airport on the 1988 presidential campaign trail. After he explained that "faith in God is important to me," a reporter inquired, "Surely you recognize the equal citizenship and patriotism of Americans who are atheists?" Bush replied: "No, I don't know that atheists should be considered as citizens, nor should they be considered patriots. This is one nation under God."[10] This conclusion, however, is not born out in the data from the scientific study of religion and morality. While individual religious believers may be exceptionally moral and tolerant people, and while religion may inspire in some individuals extraordinary morality and tolerance, religion does not necessarily foster these desirable traits.

According to a 1997 survey conducted by the University of Ohio, for example, intolerance among Christian activists is relatively high, especially when it comes to the perceived moral degeneration of America. In fact, 99 percent agreed that "moral decay is the cause of America's problems." What is the perceived source of this moral decay? One-third of the Christian activists who responded to the survey listed the American Civil Liberties Union as the most dangerous group in America, with gay rights groups coming in a close second. About 80 percent stated that members of the ACLU and gay rights groups "should not be allowed to: make a public speech, run for public office, demonstrate in public, or operate legally." Nearly half, 44 percent, declared that such "dangerous" people "should not be allowed to teach in public schools." Most threatening to civil liberties and the separation of church and state, more than half (52 percent) agreed with this statement: "Christians should take dominion over all aspects of society." No less than 91 percent believe that "God works through politics and election returns," and 89 percent think that "the U.S. has prospered when it obeyed God" and that "Clergy and churches should be involved in politics." A vast majority (75 percent) agreed that, "if enough people were brought to Christ, social ills would take care of themselves."[11]

Nothing fuels religious extremism more than the belief that one has found the absolute moral truth. Islamic terrorism, for example, has gradually shifted from secular motives in the 1960s to religious motives today. One study found that in 1980 there were only two out of sixty-four militant Islamic groups whose mission was religiously based. In 1995 that figure had climbed to nearly half.[12] It is a type of fuel that can lead to what Clay Farris Naff, executive director of the Center for the Advancement of Rational Solutions in Lincoln, Nebraska, cleverly calls the "neuron bomb," after its cold-war counterpart, the "neutron bomb," which was designed to kill people while leaving buildings and infrastructure intact. A schematic of the neuron bomb looks like this:

Arming Device: Belief that God's enemies must be defeated or destroyed
Concealment: Can be implanted in any human mind
Cost: Practically nothing
Explosive Materials: Anything at hand
Destructive Potential: Unlimited[13]

Salman Rushdie minced no words in his analysis of the problems between India and Pakistan, two religiously based political systems poised intermittently on the brink of nuclear holocaust: "The political discourse matters, and explains a good deal. But there's something beneath it, something we don't want to look in the face: namely, that in India, as elsewhere in our darkening world, religion is the poison in the blood. So India's problem turns out to be the world's problem. What happened in India has happened in God's name. The problem's name is God."[14]

To be more accurate, India's problem—and the world's—is *extremism* in the name of God, even in the industrial and democratic West. "All faiths that come out of the biblical tradition—Judaism, Christianity and Islam—have the tendency to believe that they have the exclusive truth," writes Rabbi David Hartman of the Shalom Hartman Institute in Jerusalem. "When the Taliban wiped out the Buddhist statues, that's what they were saying. But others have said it too."[15] Others such as Cardinal Ratzinger, a representative of the Vatican, who proclaimed in August 2000: "With the coming of the Saviour Jesus Christ, God has willed that the Church founded by Him be the instrument for the salvation of all humanity. This truth of faith . . . rules out, in a radical way . . . the belief that 'one religion is as good as another.' "[16] Religious extremism in America is particularly potent when gathered together under the umbrella of militia groups. An example can be seen in the life of Eric Robert Rudolph, the American terrorist charged in the 1996 Atlanta Olympics bombing that killed one person and wounded over a hundred others, as well as in the bombing of a gay nightclub and two abortion clinics. When he was captured in May 2003, it was reported that he was a member of the Christian Identity movement (an extremist group that believes Jews are satanic and blacks are subhuman), was known for his anti-Semitic and racist views, and that in his seven-year evasion of the FBI and law enforcement agencies, he probably had help from militia groups as well as local townspeople in Murphy, North Carolina, many of whom apparently share his views.[17]

Not only is there no evidence that a lack of religiosity leads to less moral behavior, a number of studies actually support the opposite conclusion. In 1934 Abraham Franzblau found a negative correlation between acceptance of religious beliefs and three different measures of honesty. As religiosity increased, honesty decreased.[18] In 1950 Murray Ross conducted a survey among 2,000 associates of the YMCA and discovered that agnostics and atheists were more likely to express their willingness to aid the poor than those who rated themselves as deeply religious.[19] In 1969 sociologists Travis Hirschi and Rodney Stark reported no difference in the self-reported likelihood to commit crimes between children who attended church regularly and those who did not.[20] In 1975 Ronald Smith, Gregory Wheeler, and Edward Diener discovered that college-age students in religious schools were no less likely to cheat on a test than their atheist and agnostic counterparts in nonreligious schools.[21] Finally, David Wulff's comprehensive survey of correlational studies on the psychology of religion revealed that there is a consistent positive correlation between "religious affiliation, church attendance, doctrinal orthodoxy, rated importance of religion, and so on" with "ethnocentrism, authoritarianism, dogmatism, social distance, rigidity, intolerance of ambiguity, and specific forms of prejudice, especially against Jews and blacks."[22] The conclusion is clear: not only does religion not necessarily make one more moral, it can lead to greater intolerance, racism, sexism, and the erosion of other values cherished in a free and democratic society.

Since I am a nonbeliever, I might reasonably be accused of a biased selection of data to make my case. Consider, then, the results of the well-known religious pollster George Barna, a born-again Christian, discussed in his 1996 *Index of Leading Spiritual Indicators*. Based on interviews with nearly 4,000 adult Americans, Barna found "Born-again Christians continue to have a higher likelihood of getting divorced than do non-Christians," and "Atheists are less likely to get divorced than are born-again Christians." This seems counterintuitive, yet Barna found that the current divorce rate for born-again Christians is 27 percent, while it is only 24 percent for non-Christians. In addition, the baby boomers—that generation often criticized for sexual indulgence and moral relativism—have a lower rate of divorce (34 percent) than the preceding generation (portrayed in popular culture as the idealized 1950s Ozzie and Harriet family), whose rate hovers at 37 percent. Five years later, in a 2001 survey, Barna found that "33 percent of all born again individuals who have been married have gone through a divorce, which is statistically identical to the 34 percent incidence among non-born again adults."[23] My point in this divorce data dump is to counter the claim—heard all too often in our culture—that one cannot be *as* good without God. That is simply scientifically and statistically false.

Even everyday intolerances are easily derived from such moral certainty. On March 25, 1998, for example, the Reverend Reggie White, better known for his bone-breaking hits as an all-star linebacker for the Super Bowl champion Green Bay Packers, presented the Wisconsin State Assembly with his theory of race and religion: "Homosexuality is a decision, it's not a race. People from all different ethnic backgrounds live in this lifestyle. But people from all different ethnic backgrounds also are liars and cheaters and malicious and back-stabbing." God, says White, granted each race with special gifts. Blacks, for example, are good at worship and celebration: "If you go to a black church, you see people jumping up and down because they really get into it." Whites are good at organization: "You guys do a good job of building businesses and things of that nature and you know how to tap into money." Latinos "were gifted in family structure and you see a Hispanic person, and they can put 20, 30 people in one home." Asians "can turn a television into a watch." Native Americans are "gifted in spirituality."[24] Right. God created blacks to make merry, whites to make money, Asians to make televisions, Latinos to make babies, and Indians to make rain.

In a similar vein, pro-lifer Randall Terry, founder of Operation Rescue, succinctly summarized the absolute intolerance that is possible with absolute morality: "Let a wave of intolerance wash over you. . . . Yes, hate is good. . . . Our goal is a Christian nation. . . . We are called by God to conquer this country. . . . We don't want pluralism." Putting an exclamation point on Terry's philosophy is the abortion clinic bomber John Brockhoeft: "I'm a very narrow-minded, intolerant, reactionary, bible-thumping fundamentalist . . . a zealot and fanatic. . . . The reason the United States was once a great nation, besides being blessed by God, is because she was founded on truth, justice, and narrowmindedness."[25]

There is a simple historical problem with this theory. According to the 2001 *New Historical Atlas of Religion in America,* it is a myth that America was once a Christian nation that has since lapsed into secular debauchery.[26] Most revealing are the historical maps and charts that track the changing demographics of American religion. Conservative pundits who worry about the moral fiber of America and proclaim that we need to return to the good old days when America was a Christian nation should look closely at the graph in figure 32, which shows that church membership among the U.S. population over the past century and a half has increased from 25 percent to 65 percent. If America is going to hell in an immoral handbasket, it is happening when church membership is at an all-time high and a greater percentage of Americans (90–95 percent) than ever before proclaim belief in God.

Absolute morality leads logically to absolute intolerance. Once it is determined that one has the absolute and final answers to moral

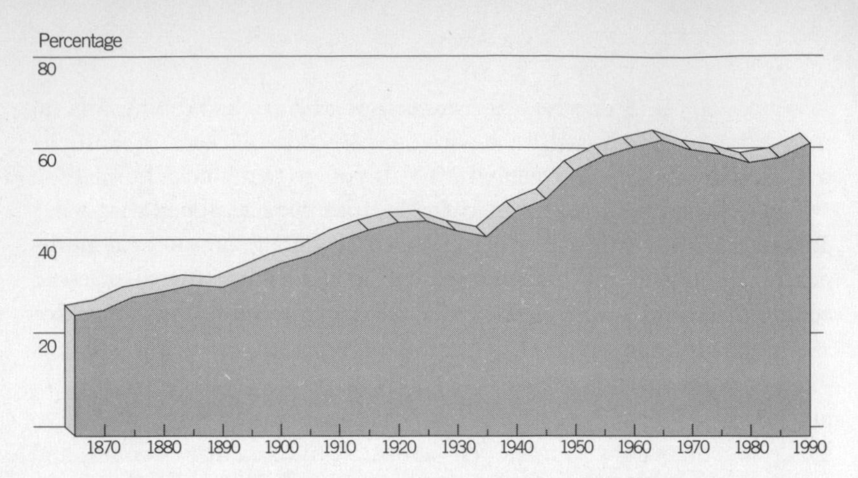

Figure 32. The Myth of Early America as a Christian Nation

Contrary to what modern conservatives claim, America has never been a more religious nation than it is today. Over the past century, church membership has climbed from 25 percent to 65 percent of the American population. (From Edwin S. Gaustad, Philip L. Barlow, and Richard Dishno, *New Historical Atlas of Religion in America,* 2001, figure 4.16)

questions, why be tolerant of those who refuse to accept the Truth? Religiously based moral systems apply this principle in spades. From the medieval Crusades and the Spanish Inquisition to the Holocaust and Bosnia, history is rife with examples of intolerance. In the name of God, religious people have sanctioned slavery, anti-Semitism, racism, homophobia, torture, genocide, ethnic cleansing, and war. In the name of their religion, people have burned women accused of witchcraft. The events of September 11, 2001, are a potent example of religious extremism gone political.

Secular Absolutism and Intolerance

Despite this ripping indictment of religion, it would be intolerant of me not to reiterate that religion per se is not the problem; it is religious extremism and, by extension, extremism of any stripe. So it is only reasonable to note that this generalization is true of secular moral systems as well. Extreme atheist ideologies, most notably the various Marxist regimes throughout the twentieth century, have generated their share of purges and pogroms in the name of an ideological god. But also during this time, a number of secular ethical systems have been proposed by philosophers in a quest to move beyond religion, while avoiding the relativism endemic to so many nonreligious ethical theories, and the intolerance generated by Marxism.

In *A Theory of Justice*, for example, John Rawls argued that there are certain moral principles that are absolute and above cultural modification (and thus are inalienable): "In a just society the liberties of equal citizenship are taken as settled; the rights secured by justice are not subject to political bargaining or to the calculus of social interests." Where those rights came from without reference to a transcendental power and how they can be secured without political power are challenges Rawls has had to answer against his critics. Rawls is perhaps best known for his tenet that justice, or fairness, be defined by conceiving of an "original position" in which you would not know what your status in life would be (for example, which race, class, or religion you would be born into) before making a political decision. Thus, your choice is made from a "veil of ignorance" of your "original position." When you do this you soon learn (and feel) just how unfair the world is and how many people are advantaged or disadvantaged in life to no credit or fault of their own but just by dint of birth and upbringing. To remedy this problem, Rawls says that "society must give more attention to those with fewer native assets and to those born into less favorable social positions. The idea is to redress the bias of contingencies in the direction of equality."[27]

At first blush this does seem, well, fair and just. After all, it's not my fault that my parents were not as well off as other parents who sent their children to exclusive private schools, while I floundered through mediocre public schools. Shouldn't I be compensated somehow? What about poorer black kids attending inner city public schools that fall far below mediocrity? Shouldn't redress be made for this bias? Unfortunately, when you carry this line of reasoning out through all of its consequences, it becomes absurd in the extreme. What qualifies as a native asset? Height, looks, and intelligence? Social psychologists have found that men in excess of six feet in height will earn more money than men under six feet tall, that better-looking people receive more attention from teachers and more breaks on the job, and that people with an IQ in excess of 145 puts them at the top of the game in both native ability and earning power. Should people receive compensation for handicaps of height, looks, and smarts? Of course not. It seems fairly obvious that this trend toward equality would lead to extensive and draconian governmental interventions. Instead of justice for all, we would end up with freedom for none.

A more reasonable alternative, it seems to me, is the libertarianism of Robert Nozick's *Anarchy, State, and Utopia.* Ironically, Nozick makes an argument similar to Rawls's concept, but he arrives at a radically different solution about the role of the state. "Individuals have rights, and there are things no person or group may do to them (without violating their rights). So strong and far-reaching are these rights that they raise the question of what, if anything, the state and its officials may do." As little as possible, Nozick concludes: "A minimal state, limited to the narrow functions of protection against force, theft, fraud, enforcement of contracts, and so on, is justified; that any more extensive state will violate persons' rights not to be forced to do certain things, and is unjustified; and that the minimal state is inspiring as well as right."[28] Inspiring and right. It doesn't get any better than that. But here I may not be purely objective.

In fact, I must plea a mea culpa for enthusiastically embracing what has to be one of the most paradoxical forms of secular absolute moral systems ever devised—Ayn Rand's *Objectivism.* Throughout my youthful forays into divers ethical systems, I clung to a core of philosophy laid down in her magnum opus, *Atlas Shrugged,* a novel many people devour for its ideals of personal responsibility, rugged individualism, and free-market economics. Objectivism is based on four central tenets: 1. *Metaphysics*: Objective Reality; 2. *Epistemology*: Reason; 3. *Ethics*: Self-interest; 4. *Politics*: Capitalism.[29] Thinking for oneself is primary, and behaving morally leads to success and happiness. Rand's moral hero is John Galt, the Atlas who shrugged when his vision of the world failed to take hold. Galt is the "Prometheus who changed his mind. After centuries of being torn by vultures in payment for having brought to men the fire of the gods, he broke his chains and he withdrew his fire—until the day when men withdraw their vultures." When the vultures (read Big Government and Big Religion) finally withdrew their restrictions, Galt (read Rand) exhorted the heroes left standing: "The world you desired can be won, it exists, it is real, it is possible, it's yours.[30]

The problem with Objectivism is its contention that absolute knowledge and final truths are attainable. For Objectivists, once a principle has been discovered through reason to be True, there is no further cause for disputation. If you disagree with the principle, then too bad for you—the principle is True anyway. This is more like theology than it is philosophy. Whatever it is, it is not science. In Rand's circle, such absolutism led to the same end that all absolute moral systems experience if they are carried out to their logical extreme: a bipolarization of people into true believers and heretics, with acceptance of the former and excommunication of the latter. Nathaniel Branden, Rand's chief lieutenant, who began as a true believer and ended up an excommunicated heretic, explained the "implicit premises" to which "everyone in our circle subscribed," including: "Ayn Rand is the greatest human being who has ever lived. *Atlas Shrugged* is the greatest human achievement in the history of the world. Ayn Rand, by virtue of her philosophical genius, is the supreme arbiter in any issue pertaining to what is rational, moral, or appropriate to man's life on earth. Once one is acquainted with Ayn Rand and/or her work, the measure of one's virtue is intrinsically tied to the position one takes regarding her and/or it. No one can be a good Objectivist who does not admire what Ayn Rand admires and condemn what Ayn Rand condemns."[31]

Absolute morality generates absolute intolerance. And the problem is endemic to all absolute systems of thought, from religious to nonreligious, from libertarian to communist. One would think, for example, that Objectivists would embrace all libertarians. But no, like the Baptists and Anabaptists who warred over whether baptism should be implemented at birth or in adulthood (with the Anabaptists opting for the latter), some of Rand's biggest battles were fought not with socialists, but with fellow libertarians. Barbara Branden recalled a dinner catastrophe that resulted from the first meeting between Rand, the libertarian economist Henry Hazlitt, and Ludwig von Mises, the greatest intellectual defender of free-market economics of the twentieth century. "The evening was a disaster. It was the first time Ayn had discussed moral philosophy in depth with either of the two men. 'My impression,' she was to say, 'was that von Mises did not care to consider moral issues, and Henry was seriously committed to altruism. . . . We argued quite violently. At one point von Mises lost his patience and screamed at me.' "[32] Economist and Nobel laureate Milton Friedman, one of the fountainheads of libertarianism, recalled an incident at the first meeting of the Mont Pelerin Society in 1947, at which was gathered a veritable who's who of free-market economists (including

Friedrich von Hayek, Fritz Machlup, George Stigler, Frank Knight, Henry Hazlitt, and Ludwig von Mises). "One afternoon, the discussion was on the distribution of income, taxes, progressive taxes, and so on. In the middle of that discussion von Mises got up and said, 'You're all a bunch of socialists,' and stomped out of the room."[33]

Intolerance is just as prevalent at the other end of the political and economic spectrum. Economist Murray Rothbard, who avers that the Libertarian party was founded in his living room, compared the intolerance of communists with that of Randian Libertarians, with a corresponding result that due to the inability of followers to properly toe the party line, at any given time both groups had more ex-members than members: "an ideological cult can adopt the same features as a more overtly religious cult, even when the ideology is explicitly atheistic and anti-religious." Intolerance is enforced through ideological straitjackets: "Communists preserve their members from the dangerous practice of thinking on their own by keeping them in constant activity together with other Communists . . . of the major Communist defectors in the United States, almost all defected only after a period of enforced isolation." The same was true in the Rand circle. "Every night one of the top Randians lectured to different members expounding various aspects of the 'party line': on basics, on psychology, fiction, sex, thinking, art, economics, or philosophy. Failure to attend these lectures was a matter of serious concern in the movement."[34] Loyal followers often found themselves outcast heretics for the minutest of infractions, such as listening to the "wrong" music or not properly denouncing an irrational idea. Moral absolutism leads to moral absurdities, turning acolytes into apostates.

Judge Not, That Ye Be Not Judged

One of the reasons that Christianity succeeded was its tolerance for diversity and openness to all comers. With the success of Christianity, the within-group intolerant morality of the Old Testament gave way to a more tolerant morality of the New Testament. Contrast Rand's Old Testament–style morality with that of Jesus on the matter of moral judgment. Here is Rand's position:

> The precept: "Judge not, that ye be not judged" . . . is an abdication of moral responsibility: it is a moral blank check one gives to

> others in exchange for a moral blank check one expects for oneself. There is no escape from the fact that men have to make choices; so long as men have to make choices, there is no escape from moral values; so long as moral values are at stake, no moral neutrality is possible. To abstain from condemning a torturer, is to become an accessory to the torture and murder of his victims. The moral principle to adopt . . . is: "Judge, and be prepared to be judged."[35]

Actually, what Jesus said in full (in Matt. 7:1–5) was:

> Judge not, that ye be not judged. For with what judgment ye judge, ye shall be judged: and with what measure ye mete, it shall be measured to you again. And why beholdest thou the mote that is in thy brother's eye, but considerest not the beam that is in thine own eye? Or how wilt thou say to thy brother, Let me pull out the mote out of thine eye; and, behold, a beam is in thine own eye? Thou hypocrite, first cast out the beam out of thine own eye; and then shalt thou see clearly to cast out the mote out of thy brother's eye.

The principle Jesus extols is not moral neutrality or a moral blank check, but a warning against self-righteous severity and a rush to judgment, as explained in the Talmudic collection of commentary on Jewish custom and law called the Mishnah: "Do not judge your fellow until you are in his position" (Aboth 2:5); "When you judge any man weight the scales in his favor" (Aboth 1:6). Jesus wants us to be cautious, not to cross the line between legitimate and hypocritical moral judgment. The "mote" and "beam" metaphor is purposeful hyperbole. The man who lacks virtue feels morally smug in judging the virtue of his neighbor. The "hypocrite" is the critic who disguises his own failings by focusing attention on the failings of others. Perhaps Jesus is offering insight into human psychology where, for example, the adulterer is obsessed with judging other peoples' sexual offenses, the homophobe secretly wonders about his own sexuality, or the liar suspects others of excessive falsehoods.[36]

Methodological Individualism and Moral Tolerance

Why should absolutism necessarily lead to intolerance? Is it just that people who prefer absolute systems of morality tend to be intolerant by temperament, or is there something built into the systems themselves that leads to intolerant attitudes and behavior? An answer can be found in the difference between the binary logic of absolute morality and the fuzzy logic of provisional morality. The basis of most ethical systems is Aristotelian binary logic: black or white, right or wrong, moral or immoral. Ayn Rand well represents this position: "There are two sides to every issue: one side is right and the other is wrong, but the middle is always evil. The man who is wrong still retains some respect for truth, if only by accepting the responsibility of choice. But the man in the middle is the knave who blanks out the truth in order to pretend that no choice or values exist."[37]

Nonsense on stilts. Philosophy often only tells us the way the world should be. Science tells us how it really is, and science reveals a very fuzzy world with multiple shades of gray. Since the basis of provisional ethics is evolutionary theory, it seems fitting to turn to a Darwinian principle that leads to a more tolerant moral guide for our fuzzy world—*methodological individualism.* It assumes that only individual phenomena have a basis in reality—there are no pure Platonic essences, no fixed Aristotelian types. In the natural world, for example, there is no such thing as an immutable species fixed in the mind of some divine Gepetto. There are only individual organisms classified into types we call species. (These species, while temporarily stable in form and function, harbor the seeds of change or extinction. On a human time scale they appear relatively stable, but on a geological time scale species change.) Analogously, there are no fixed "species" of pure good or evil, only individual organisms of good and evil acts. Thus, if we want to understand morality and immorality, we must study individuals who express moral and immoral behaviors. That is, the target of our investigation should be the individual and individual human action in all its wondrous variety.

Consider the work of an entomologist and evolutionary biologist whose specialty was gall wasps and whose methodology was evolutionary individualism. After earning a doctorate from Harvard and landing a post at the University of Indiana, he spent the next twenty years of his career logging tens of thousands of miles and collecting some 300,000 specimens of gall wasps, publishing the results of his research in two large monographs in 1930 and 1936: "In the intensive and extensive measurement of tens of thousands of small insects . . . I have made some attempt to secure the specific data and the quantity of data on which scientific scholarship must be based. During the past two years, as a result of a convergence of circumstances, I have found myself confronted with material on variation in certain types of human behavior." The convergence of circumstances was this: in 1938 his university wanted to offer a course on marriage, a euphemism at the time for sex education. The entomologist was asked to serve as chairman of the committee to regulate the course and to give three lectures on the biology of sex. Thorough scientist that he was, he went to the library and found virtually nothing on human sexuality. So he began to research the subject himself. A student had scrawled a graffito on the title page of Harvard's only copy of the 1936 wasp monograph: "Why don't you write about something more interesting, Al?" Al was Alfred Kinsey, the pioneer in the scientific study of human sexuality.

Kinsey undertook to collect his own data on a massive scale. One colleague described his need "to devour life, to gulp life, to look, and experiment and record"; Kinsey explained, "The technique we are using in this study is definitely the same as the technique in the gall wasp study."[38] As an entomologist, before he would hazard even a cautious conclusion about a particular group or species, Kinsey collected thousands of individual insects. He was no less thorough in his study of human sexuality. He began his research with personal interviews in his office, but as the process became too unwieldy, he developed a sizable staff and procured a separate research office and private grants to support a longitudinal study. By the time he published his results, Kinsey had collected data on more than 18,000 people, far outstripping all other studies done on any type of human behavior.

The reason for such exhaustiveness was that Kinsey realized the unique individuality of all living organisms, from wasps to humans. A taxonomist's generalizations of species, genera, and even higher categories, Kinsey explained, "are too often descriptions of unique individuals and structures of particular individuals that are not quite like anything that any other investigator will ever find." Not just entomologists but psychologists as well are equally guilty of such hasty generalizations: "A mouse in a maze, today, is taken as a sample of all individuals, of all species of mice under all sorts of conditions, yesterday, today, and tomorrow." Worse still, these collective conclusions are even extrapolated to humans: "A half dozen dogs, pedigrees unknown and breeds unnamed, are reported upon as 'dogs'—meaning all kinds of dogs—if, indeed, the conclusions are not explicitly or at least implicitly applied to you, to your cousins, and to all other kinds and descriptions of humans."[39]

If wasps showed so much variation, how much more might humans? In his 1948 book *Sexual Behavior in the Human Male,* Kinsey concluded: "Given the range of variation . . . the clinician can determine the averageness or uniqueness of any particular person, and comprehend the extent to which generalizations developed for the whole group may be applied to any particular case"; such individualist thinking helps "in the understanding of particular individuals by showing their relation to the remainder of the group."[40] Methodological individualism showed that even for such two seemingly dichotomous categories—heterosexual or homosexual—not everyone could be easily classified. "The histories which have been available in the present study make it apparent that the heterosexuality or homosexuality of many individuals is not an all-or-none proposition." One can be both simultaneously, or neither temporarily. One can start as heterosexual and become homosexual, or vice versa. And the percentage of time spent in either state varies considerably among individuals in the population. "For instance," Kinsey observed, "there are some who engage in both heterosexual and homosexual activities in the same year, or in the same month or week, or even in the same day"; therefore, he concluded, "one is not warranted in recognizing merely two types of individuals, heterosexual and homosexual, and that the characterization of the homosexual as a third sex fails to describe any actuality."[41] Extrapolating this methodology to taxonomy in general, Kinsey deduced the uniqueness of individuals (in a powerful statement tucked away amid countless tables and graphs):

> Males do not represent two discrete populations, heterosexual and homosexual. The world is not to be divided into sheep and goats. Not all things are black nor all things white. It is a fundamental of taxonomy that nature rarely deals with discrete categories. Only the human mind invents categories and tries to force facts into separate pigeon-holes. The living world is a continuum in each and every one of its aspects. The sooner we learn this concerning human sexual behavior the sooner we shall reach a sound understanding of the realities of sex.[42]

Approaching human behavior—including the holy grail of moral behavior—from the perspective of methodological individualism leads to moral tolerance. If variation and uniqueness are the norm, then what form of morality can possibly envelop all human actions? For human sexuality alone, Kinsey measured 250 different items for each of over 10,000 people: "Endless recombinations of these characters in different individuals swell the possibilities to something which is, for all essential purposes, infinity."[43] At the end of his 1948 volume on males, Kinsey concluded that there is virtually no evidence for "the existence of such a thing as innate perversity, even among those individuals whose sexual activities society has been least inclined to accept." On the contrary, he demonstrated through vast statistical tables and in-depth analysis that the evidence leads us to conclude that "most human sexual activities would become comprehensible to most individuals, if they could know the background of each other individual's behavior."[44]

Variation, Kinsey concluded, is the basis of both biological and cultural evolution. You cannot categorize humans as either tall or short, blond or brunette, black or white. "Dichotomous variation is the exception and continuous variation is the rule, among men as well as among insects." Likewise for behavior, we identify right and wrong, "without allowance for the endlessly varied types of behavior that are possible between the extreme right and the extreme wrong." That being the case, the hope for cultural evolution, like that of biological evolution, depends on the recognition of variation and individualism: "These individual differences are the materials out of which nature achieves progress, evolution in the organic world. In the differences between men lie the hopes of a changing society."[45]

This extension of his scientific analysis into the realm of morality, coupled with his exposé of what humans do behind closed doors, brought Kinsey much wrath and taught him more about WASPs than wasps. In a "Last Statement" dictated two weeks before his death, Kinsey noted with some bitterness the human foible of bias that seems to enter into the evaluation of human moral behavior. He bemoaned the fact that his strongest detractors were his fellow scientists, who had found difficulty "in facing facts of human sexual behavior with anything like objectivity." One prominent scientist with a powerful political position in Washington, D.C., said unequivocally: "I do not like Kinsey, I do not like the Kinsey project, I do not like anything about the Kinsey study of sexual behavior."[46] Even Kinsey's colleagues at Indiana

University held reservations about the publication of Kinsey's data. One department head recommended complete censorship, while another proposed delaying publication until the material was screened by the Department of Public Relations.

Such reactions are not surprising considering the political climate of 1950s McCarthy-era America. Protestant ethics forbidding sexual activity outside of heterosexual marriage were bumping up against the realities of human nature. As Kinsey noted, men's and women's sexual drives do not arrest themselves while awaiting the delayed marriages of modern culture. For the modern man, for example, "The society in which he lives condemns nearly all forms of sexual outlet except that legalized by marriage, but the economic system in which he finds himself imposes a delay in marriage of something like seven to twelve or more years. The teachers of morals blithely advise him to sublimate his physiologic reactions, though the record indicates that not more than two per cent of the unmarried males completely achieve that theoretical ideal."[47]

For such statements, among others, Kinsey was labeled a communist and moral subversive. The Indianapolis Roman Catholic Archdiocese claimed that Kinsey's books "pave the way for people to believe in communism and to act like Communists." A Bloomington newspaper headline read: KINSEY'S SEX BOOKS LABELED "RED" TAINTED. The publication *Christianity and Crisis* denounced the Kinsey report as "animalistic."

Most telling was the fact that his *Sexual Behavior in the Human Female,* published in 1953, was even more controversial than the volume on males. The negative response came mostly from men, who were either shocked or threatened by Kinsey's figures on pre- and extramarital intercourse, and especially the greater-than-expected range of female sexual response. A New York rabbi claimed it was "a libel on all womankind." New York Congressman Louis B. Heller demanded complete censorship in a public letter to the postmaster general of the United States, followed by this statement: "He is hurling the insult of the century against our mothers, wives, daughters, and sisters under the pretext of making a great contribution to scientific research."[48] A special House Committee, charged with investigating projects funded by tax-exempt nonprofit foundations, was called in to put pressure on both Indiana University and the Rockefeller Foundation, the latter of which funded Kinsey's Institute for Sex Research. The investigation eventually led to the termination of his research funds in

1954. Kinsey died two years later, having never seen the remarkable change his project brought on science and society.[49]

What really got Kinsey in trouble was his acceptance of behavioral variation that excluded moral judgment. If moral species have no unchanging essence—no permanent and fixed typology by which to judge right and wrong behavior—then how can we derive an absolute ethical system? If morality is to be based on what people actually do, and what they do varies widely, then of what value is binary absolutism? Kinsey demonstrated that while "social forms, legal restrictions, and moral codes may be, as the social scientist would contend, the codification of human experience," they are, like all statistical and population generalizations, "of little significance when applied to particular individuals."[50] The problem is that laws are constructed around unambiguous yeses and noes, but human behavior is a continuum, expansive in variation and individuality. In many ways, laws tell us more about the lawmakers than they do about the lawbreakers, as Kinsey concluded:

> Prescriptions are merely public confessions of prescriptionists. . . . What is right for one individual may be wrong for the next; and what is sin and abomination to one may be a worthwhile part of the next individual's life. The range of individual variation, in any particular case, is usually much greater than is generally understood. Some of the structural characters in my insects vary as much as twelve hundred percent. In some of the morphologic and physiologic characters which are basic to the human behavior which I am studying, the variation is a good twelve thousand percent. And yet social forms and moral codes are prescribed as though all individuals were identical; and we pass judgments, make awards, and heap penalties without regard to the diverse difficulties involved when such different people face uniform demands.[51]

Provisional ethics accommodates the range of individual variation found in human populations and suggests that we should pass judgments, make awards, and heap penalties only with regard to our great diversity. Such accommodational flexibility leads irrevocably toward greater tolerance, and more tolerance leads inexorably toward more peaceful ways of interacting with people, whether they are inside or outside of our group.

From Enmity to Amity: How the World Works

In modern state societies, methodological individualism and individual tolerance can only take us so far in our long-range goal of reaching the highest levels of the Bio-Cultural Evolutionary Pyramid. Preserving the planet's ecosystem and biodiversity and maximizing within-group amity and minimizing between-group enmity also require social and political action. The goals are too far reaching and the time frames involved are too long range for how we were programmed by nature to think. We evolved in a Paleolithic environment in which our concern for the environment and biodiversity was restricted to a few tens of miles and hundreds of species over the course of only a few decades. A global ecosystem and deep time was beyond anyone's conception until the past half millennium, which is too short a time for evolution to create a global morality and deep-time ethic. Likewise, the number of people our ancestors encountered in their lifetime could be numbered in the hundreds, so there was no reason for evolution to have produced an ethnically diverse principle of tolerance. To save the planet and ourselves, we need a new morality that incorporates global biodiversity, human ethnicity, and deep time. Provisional ethics is one system of morality that attempts to do just that.

As a professional skeptic, I am often asked, incredulously, "Do you believe *anything*?" The question is absurd but understandable given the common misuse of the term as a synonym for "cynic" or "nonbeliever." In fact, skeptics believe all sorts of things, not the least of which is the power of science to understand the natural world. If, by fiat, I had to reduce the theory of scientific provisionalism to four tenets, they would be as follows (in other words, this is what I believe):

1. *Metaphysics*: Provisional Reality.
2. *Epistemology*: Provisional Naturalism.
3. *Ethics*: Provisional Morality.
4. *Politics*: Provisional Libertarianism.

Provisional Reality and *Provisional Naturalism*. I believe that reality exists over and above human and social constructions of that reality. Science as a method and naturalism as a philosophy together form the best tool we have for understanding that reality. Because science is cumulative—that is, it builds on itself in a progressive fashion—we can strive to achieve an ever-greater understanding of reality. Our knowledge of nature remains provisional because we can never know if we have final Truth. Because science is a human activity and nature is complex and dynamic, fuzzy logic and fractional probabilities best describe both nature and the estimations of our approximation toward understanding that nature. There is no such thing as the paranormal and the supernatural; there is only the normal and the natural and mysteries we have yet to explain. What separates science from all other human activities is its belief in the provisional nature of all conclusions. In science, knowledge is fluid and certainty fleeting. That is the heart of its limitation. It is also its greatest strength.

Provisional Morality. I believe that morality is the natural outcome of evolutionary and historical forces operating on both individuals and groups. The moral feelings of doing the right thing (such as virtuousness) or doing the wrong thing (such as guilt) were generated by nature as part of human evolution. Although cultures differ on what they define as right and wrong, the moral feelings of doing the right or wrong thing are universal to all humans. Human universals are pervasive and powerful, and include at their core the fact that we are, by nature, moral and immoral, good and evil, altruistic and selfish, cooperative and competitive, peaceful and bellicose, virtuous and nonvirtuous. Individuals and groups vary on the expression of such universal traits, but everyone has them. Most people most of the time in most circumstances are good and do the right thing for themselves and for others. But some people some of the time in some circumstances are bad and do the wrong thing for themselves and for others. As a consequence, moral principles are provisionally true, where they apply to most people, in most cultures, in most circumstances, most of the time. At some point in the last 10,000 years (around the time of writing and the shift from bands and tribes to chiefdoms and states), religions began to codify moral precepts into moral codes.

I believe that although we live in a determined universe and are governed by the laws of nature and forces of culture and history, because we can never know in its entirety the near-infinite causal net that determines our actions, we are free moral agents. And although there is no absolute and ultimate divinity to dole out rewards and punishments in some unspecified future, since moral principles are provisionally true for most people most of the time in most circumstances, provisional justice can be derived from individual responsibility and culpability through social and cultural beliefs, customs, mores, and laws that produce feelings of virtuousness and guilt and administer rewards and punishments. Since morality evolved as a trait of the species transcendent of any individual member of the species, moral provisionalism stands as a solid pillar between the permissiveness of moral relativism and the intolerance of moral absolutism.

I believe that we can discern the difference between right and wrong through three principles. (1) The ask first principle: to find out whether an action is right or wrong, ask first. (2) The happiness principle: it is a higher moral principle to always seek happiness with someone else's happiness in mind, and never seek happiness when it leads to someone else's unhappiness. (3) The liberty principle: it is a higher moral principle to always seek liberty with someone else's liberty in mind, and never seek liberty when it leads to someone else's loss of liberty. To implement social change, the moderation principle states that when innocent people die, extremism in the defense of anything is no virtue, and moderation in the protection of everything is no vice.

Provisional Libertarianism. I believe that humans are primarily driven to seek greater happiness, but the definition of such is personal and cannot be dictated and should not be controlled by any group. The free market is the best system yet devised for allowing the most individuals in the most places most of the time to achieve the most happiness. Individuals should take personal responsibility for their actions and buck up and quit whining when the slings and arrows of life take their toll. Libertarianism is provisional, however, because it is conditional and restricted. Before writing this book, I was an unabashed, unadulterated libertarian in favor of what is called *anarcho-capitalism,* a stateless society governed entirely by free markets and private contracts. I have since decided that such a society probably would not work, because the balance between the moral and immoral nature of humanity is too close. There are too many defectors and cheaters, too much greed and avarice. I could be wrong, but until the social experiment is run—an extensive free-market society is established and successfully operated for a century—I remain skeptical of extreme libertarianism. It sounds good in theory, but I am a scientist, not a philosopher; I prefer an empirical experiment to a thought experiment. We are dealing here with people, not atoms. Social experiments are always more complex than physical or biological experiments, where the unintended consequences of a minor change can cascade through the system to create major effects.

Where Goods Do Not Cross Frontiers, Armies Will

As discussed in chapter 3, one of the prime triggers of between-group violence is competition for scarce resources. There are rarely enough resources to support all individuals in all groups. Even if, at some given time, there were, such a condition could only be a temporary one because populations naturally tend to increase to the carrying capacity of the environment. Once that is exceeded, the demand for those resources will exceed the supply. Such was the condition throughout most of the Paleolithic for most peoples in most areas. The formula is simple: *population abundance plus resource scarcity equals war.* Thus, one way to decrease between-group violence is to increase the supply of resources to meet the demands of those in need of them. Nineteenth-century French economist Frederic Bastiat expressed this relationship thusly: "Where goods do not cross frontiers, armies will."[52]

The indigenous peoples of New Guinea, whom Jared Diamond described as living in an almost constant state of between-group violence, in many areas have found peace. How did this happen? In the 1960s, Western colonial governments initially imposed peace on them, then ensured the peace by providing goods and supplies that they needed as well as the technologies to enable them to continue producing more resources. In less than one generation these same New Guineans were operating computers, flying planes, and running their own small businesses. In subsequent ethnographic studies, anthropologists discovered that, in many ways, these New Guineans were much happier living under colonial rule because the endemic wars were taking such a devastating physical and psychological toll.[53] Where resources crossed New Guinea frontiers, New Guinea armies did not.

A similar case study can be found in the Yąnomamö, the so-called fierce people. There is good reason for the moniker because, as we saw in our extensive discussion of them in chapter 3, warfare has long been a part of Yąnomamö life. However, as missionaries in the area have discovered, the Yąnomamö do not actually like fighting. When the missionaries (and, subsequently, the Venezuelan government to which their protected territories belong) provided food and the tools for the production and procurement of food, Yąnomamö wars were significantly reduced. As Napoleon Chagnon discovered, however, even without outside intervention the Yąnomamö are sophisticated traders as well as warriors. The reason is that trade creates alliances. If, as it is said, "the enemy of my enemy is my friend," one of the primary means of protecting one's group is to form alliances with other groups. Trade between groups is a powerful social adhesive (as is intervillage feasting, which they also do). One village cannot go to another village and announce that they are worried about being conquered by a third, more powerful village, since this would reveal weakness. Instead, "they conceal and subsume the true motive for the alliance in the vehicles of trading and feasting, developing these institutions over months and even years. In this manner they retain an apparent modicum of sovereignty and pride, while simultaneously attaining the ultimate objectives: intervillage solidarity and military interdependence."

Chagnon found that his charges purposefully designed a division of labor within villages in order to generate trade between villages. "Each village has one or more special products that it provides to its allies. These include items such as dogs, hallucinogenic drugs (both cultivated and collected), arrow points, arrow shafts, bows, cotton yarn, cotton and vine hammocks, baskets of several varieties, clay pots, and, in the case of contacted villages, steel tools, fishhooks, fishline, and aluminum pots." Although, in principle, each Yąnomamö group could produce its own goods for survival, in fact, they don't; they set up a division of labor and system of trade. They do this, says Chagnon, not because they are nascent capitalists, but because they want to form political alliances with other groups, and trade is an effective means of so doing. "Without these frequent contacts with neighbors, alliances would be much slower in formation and would be even more unstable once formed. A prerequisite to stable alliance is repetitive visiting and feasting, and the trading mechanism serves to bring about these visits."[54] Where goods cross Yąnomamö frontiers, Yąnomamö armies do not.

My point is this: just as I argued that morality evolved long before religion, I am claiming that trade evolved long before the state. There is now archaeological evidence, for example, that over the past 200,000 years stone tools and other artifacts such as seashells, flint, mammoth ivory, and beads were the objects of trade among our hominid ancestors, because they are often found hundreds of miles from where they were manufactured.[55] Shepard Krech, in his debunking of the "ecological Indian" myth, shows that the reason Europeans were so readily able to trade with Native Americans (beads for pelts, for example) was that the Indians were already well accustomed to trading among themselves.[56] The psychology of trade probably has as much to do with forming alliances between individuals and groups as it does increasing the supply of resources, but the end result is the same: cooperation and reciprocal altruism that goes into making trade successful accentuates amity and attenuates enmity, leading (in the language of provisional ethics) to greater happiness and liberty for more people in more places more of the time.

The Biology of Cooperation and Trade

There is now scientific research to support the thesis that trade is good for both individuals and groups. Recall the Prisoner's Dilemma experiments discussed in chapter 2, in which it was found that a cooperative trusting strategy was shunned by players in a one-trial game, but embraced when they played multiple rounds, particularly when players could interact with each other to establish trust. The best strategy in iterated contests was tit for tat with no initial defection. In nearly all cases the most selfish thing to do—that is, the way to gain the most number of points (or money) in the long run—was to begin by trusting and cooperating, and then do whatever your partner does. The most successful tactic in an extensive Prisoner's Dilemma contest was a computer program entitled "Firm but Fair," which cooperates with cooperators, cooperates after a mutual defection, quits playing with constant defectors, and defects with partners who always cooperate (called suckers).[57] In a related experiment, nine subjects were each given five dollars. If five or more of the nine cooperated by donating their five dollars to a general pot, all nine would receive ten dollars. Although it pays to be a cooperator (you get ten instead of five dollars), it pays even more to be a defector (fifteen instead of five dollars), as long as at least five other people cooperate. The results were mixed, with many groups of nine subjects failing to achieve the critical mass of five cooperators, because there was no trust. Then the experimenters added a step:

members of some groups were given the opportunity to discuss their strategy options before playing. Those groups that interacted before playing averaged eight cooperators, and 100 percent of these groups earned cooperative bonuses. By sharp contrast, those groups that did not interact before playing earned bonuses only 60 percent of the time.[58] Finally, in research on social dilemmas, psychologist Robyn Dawes found that groups given the opportunity to communicate face-to-face were more likely to cooperate than those who were not. "It is not just the successful group that prevails," Dawes concluded, "but the individuals who have a propensity to form such groups."[59]

These results remind me of President Ronald Reagan's cold-war strategy: *trust with verification.* Trust is built over time and through interactions, and trade is an effective tool for establishing trust. (One of the first things to go when trust breaks down between two nations is trade. If a country is especially untrustworthy, the international community may even impose economic sanctions that prohibit any trade with them. The end result is often war.) The psychological impulse to form relationships and alliances is the deeper cause that lies beneath the moral sentiment of trust, and trade is an effective medium that allows people to create trusting relationships with and form attachments to other trustworthy people. And, recall, it is not enough to fake being a cooperator, because over time and with experience deceivers are usually flushed out. You actually have to believe you are a cooperator, and there is no surer way to believe you are a cooperator than to be one.

Our brains even evolved a mechanism to reinforce this process—cooperation leads to stimulation of the pleasure centers in the brain. Scientists at Emory University had thirty-six subjects play Prisoner's Dilemma while undergoing a functional magnetic resonance imaging (fMRI) brain scan. They found that the areas of the brains of cooperators that lit up were the same areas activated in response to such stimuli as desserts, money, cocaine, attractive faces, and other basic pleasures. Specifically, there were two broad areas dense in neurons that responded, both rich in dopamine (a neurochemical related to addictive behaviors): the anteroventral striatum in the middle of the brain (the "pleasure center," for which rats will endlessly press a bar to have it stimulated, even going without food) and the orbitofrontal cortex just above the eyes, related to impulse control and the processing of rewards. Tellingly, the cooperative subjects reported increased feelings of trust toward and camaraderie with their game partners.[60] In addition to dopamine, neuroscientists Steven Quartz and Terrence Sejnowski have documented the connection between oxytocin—a brain chemical produced during eating, breast-feeding, and sexual orgasms, believed to play a vital role in human bonding—and pro-social behaviors, such as cooperation and exchange.[61]

There is now, in fact, a banquet of data that has spawned a new field of research on the cognitive neuroscience of human social behavior, demonstrating that humans evolved powerful neurological mechanisms to reinforce cooperation, accentuate pro-social behavior, and bond nonrelated people through the process of social exchange.[62] Jorge Moll and his colleagues, for example, monitoring fMRI brain scans, found that moral emotions activate the amygdala, or the emotion module in the brain, as well as the orbital and medial prefrontal cortex, a higher level of cognitive processing in the brain, indicating that moral behaviors are as much related to moral emotions as they are to moral reasons (figure 33).[63] In a similar technique utilizing fMRI scans on subjects participating in two-person "trust and reciprocity" games, Kevin McCabe and his colleagues found that areas of the prefrontal cortex—known to mediate impulse control and the delay of immediate gratification—are activated in the brains of cooperators (but not defectors), suggesting that cooperation requires "attention to mutual gains with the inhibition of immediate reward gratification to allow cooperative decisions."[64] The importance of the prefrontal cortex in humans and the other great apes was explored by Katerina Semendeferi and her colleagues, who found that area 10 of the frontal lobe in particular is linked to such higher cognitive functions as the undertaking of initiatives and the planning of future actions, and that this area, while larger in apes than in monkeys, is in humans "larger relative to the rest of the brain than it is in the apes" and has "more space available for connections with other higher-order association areas." She concludes that "the neural substrates supporting cognitive functions associated with this part of the cortex enlarged and became specialized during hominid evolution."[65]

I claim that the reason for this cortical expansion is that humans evolved to became the preeminent social and moral primate. The brain imaging research of Uta and Chris Frith of University College London also supports this hypothesis, showing that in order to be a moral agent, one must be both self-aware and aware that others are self-aware,

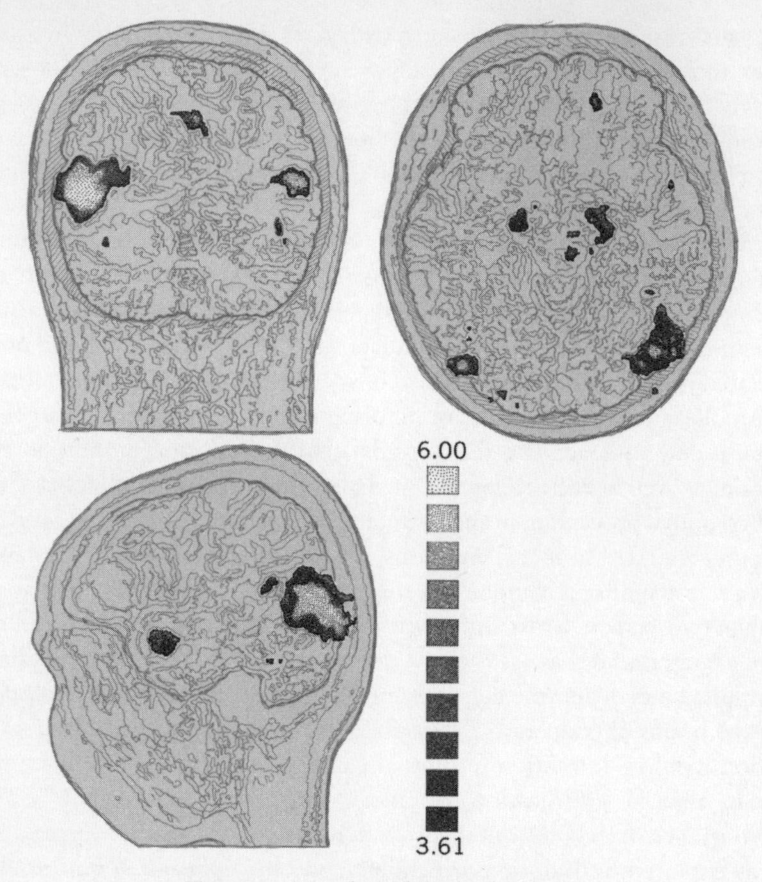

Figure 33. Moral Modules in the Brain

Jorge Moll and his colleagues employed functional magnetic resonance imaging (fMRI) to produce brain scans, and discovered that moral emotions activate, as seen in this figure, the prefrontal and temporal lobes (higher-level cognitive process in the brain) as well as the amygdala (the emotion module of the brain), indicating that moral behaviors are driven by both emotional and rational parts of the brain. (Rendered by Pat Linse, from Jorge Moll et al., "The Neural Correlates of Moral Sensitivity," in *The Journal of Neuroscience,* 2002, p. 2733)

functions that are located in two different areas of the brain. Self-awareness, at least in part, appears to be located in the medial prefrontal cortex, whereas representing others' actions and intentions appears to be centered in the temporal cortex. "We speculate that the precursors of mentalizing ability derive from a brain system that evolved for representing agents and actions, and the relationships between them."[66] It is those brain relationships that form the foundation of social relationships.

How does trust translate to trade? At the Center for Neuroeconomics Studies at Claremont Graduate University, Paul Zak has demonstrated the relationship between oxytocin, trust, and economic prosperity. He argues that economists have shown how trust is among the most powerful factors affecting economic growth, and that since trust is directly related to neurological chemicals such as oxytocin, it is vital for national prosperity that the country maximize social interactions among its members, as well as members of other countries. Free trade is one of the most effective means of socializing, as are education, increased civil liberties, freedom of the press, freedom of association (most notably by increasing telephones and roads), and even a cleaner environment (people in countries with polluted environments show higher levels of estrogen antagonists, thereby lowering their levels of oxytocin and thus their feelings of trust). Impoverished countries are poor, in part, because trust in the legal structures to protect business and personal investments is so low. "Differences in trust cause differences in living standards," Zak concludes. He has even computed that "a 15 percent increase in the proportion of people in a country who think others are trustworthy raises income per person by 1 percent per year for every year thereafter." For example, increasing levels of trust in the United States from its present 36 percent to 51 percent would raise the average income for every man, woman, and child in the country by $400 per year, or $30,000 over a lifetime.[67] It pays to trust (with verification, of course).

Although extrapolating directly from neurochemistry to national economies is surely oversimplifying matters, what all this research tells us is that on one level we cooperate for the same reason we copulate—because it feels good. On a deeper evolutionary level, the reason cooperating feels good is because it is good for us, individually and as a species. Thomas Jefferson realized this in 1814: "These good acts give pleasure, but how it happens that they give us pleasure? Because nature hath implanted in our breasts a love of others, a sense of duty to them, a moral instinct, in short, which prompts us irresistibly to feel and to succor their distresses."[68]

How do trust and trade reduce war and violence? In every case study of societies that made the transition from war to peace, there is a direct causal relationship between population size, ecological carrying capacity, and the availability and exchange of resources. As archaeologist Steven LeBlanc explains, "There is no change in the ability to shift to peacefulness as social complexity evolves. Rather, the shift occurs when the ecological relationships suddenly change, regardless of the type of social organization affected."[69] The primary engine driving the shift in these ecological relationships is trade. When populations grow beyond the carrying capacity of their environments, they are forced into competition, which leads to war, which leads to alliances, which leads to trade, which leads to peace. In other words, the solution to war—that is, to move a society from a warlike existence to a peacelike existence—is not to be found in a particular type of government or religion or ideology or worldview; it is in a particular type of social process called trade. The evolutionary origin of trade may have been political alliances, but one of the unintended consequences is that trade produces a division of labor that generates more goods for more people more of the time.

Moral Heroism: Skepticism as a Virtue

In the end, how shall we treat our fellow humans? Here we face ourselves in the penetrating mirror of humanity's 100,000-year journey, where the heroic in the human spirit is allowed to rise from the ashes of our primitive ancestry, imploring us to rise above the dark side of our nature. Religion has certainly inspired greatness out of ordinariness, and such heroics have been well documented throughout the ages, especially by the particular religions to whom the heroes professed worship. But religion has a built-in system of intolerance that logically follows from adherence to a fixed set of dogmas. I think we can do better.

I believe in the heroic nature of humanity and in the ability of human intelligence, reason, and creativity to triumph over problems and obstacles. There has been great progress in human history, as measured by greater amounts of liberty and prosperity for more people in more places more of the time. In some cases, religion, particularly Judaism and Christianity, has fostered freedom and free markets; where this influence was combined with the Enlightenment introduction of secular political and economic systems and the separation of church and state, more freedoms for more peoples were enjoyed over the past two centuries than accumulated over the previous two millennia. Patriarchal dominance, for example, is being systematically displaced by gender equality in societies where women are allowed to flourish. How did this happen? First, women decided that they were not going to put up with this arrangement any longer; second, women were given the freedom to act on this decision; and third, society realized that women also seek greater happiness and greater liberty, and that they can achieve them more readily without being in bondage to males. Increasing the happiness and liberty of half of society raises the overall happiness of the group. Libertarianism is the happiness principle and the liberty principle writ large.

Religious freedoms must always be protected, but the price for this security is the separation of religion from government. Historically, where church and state were wed, individual liberty suffered, including and especially religious liberty. Paradoxically—because many Christian conservatives today call for greater influence of their religion on politics—Christianity is at least partially responsible for this division between the sacred and the profane. Jesus himself admonished his followers (Matt. 22:21): "Render therefore unto Caesar the things which are Caesar's; and unto God the things that are God's." Historically, this resulted in two separate magisteria—spiritual and temporal, ecclesiastical and lay, religious and secular—each with its own laws, courts, and hierarchical authority.[70] (Historian Bernard Lewis, in fact, identifies the secularization of Western culture as one of the strongest reasons for its prosperity and progress in science, technology, and culture; and the lack of separation of church and state for "what went wrong" in Muslim countries that drove the Arab world from its medieval apex of human achievement to its status today as a cultural backwater.) Thus, in order to generate greater liberty for more people we must maintain the separation of church and state and foster the greater secularization of state society, by which I mean public morality should only be legislated by secular bodies (while private morality may be as religious as the individual prefers). The members of secular bodies may themselves be religious, but the body itself must remain religiously neutral. Because of the natural inclination to favor one's belief for preferential treatment even in secular systems, however, I personally prefer (although it should never be legislated) that public policy be governed by people with no religious preference at all. These are secularists.

Who are secularists? Secularists are nonbelievers, nontheists, atheists, agnostics, skeptics, free thinkers, humanists, and secular humanists. Unfortunately, many of these words carry pejorative baggage. Words matter and language counts. "Feminist" is a fine word that describes someone who believes in the need to secure the rights and opportunities for women equivalent to those provided for men. Unfortunately, thanks to conservatives like Rush Limbaugh, it has also come to be associated with sandal-wearing, tree-hugging, postmodern, deconstructionist, left-leaning liberals best scorned as "Femi-Nazis." Likewise, "atheist" is a descriptive term that simply means "without theism," and describes someone who does not believe in God(s). Unfortunately, thanks to religious fundamentalists, it has also come to be associated with sandal-wearing, tree-hugging, postmodern, deconstructionist, left-leaning liberals who are immoral, pinko communists hell-bent on corrupting the morals of America's youth. Speak the scorn into existence. A 1999 Gallup poll reflected this attitude. When asked, "If your party nominated a generally well-qualified person for president who happened to be an X would you vote for that person?" (with X representing Catholic, Jew, Baptist, Mormon, black, homosexual, woman, and atheist), while six of the eight received more than 90 percent approval, only 59 percent would vote for a homosexual and less than half, 49 percent, would vote for an atheist.

For the most part I avoid labels altogether and simply prefer to say what it is that I believe or do not believe. However, at some point labels are unavoidable (most likely due to the fact that the brain is wired to pigeonhole objects into linguistic categories), and thus one is forced to use identity language. Since the name of the magazine I cofounded is *Skeptic* and my monthly column in *Scientific American* is entitled "Skeptic," I usually just call myself a skeptic, from the Greek *skeptikos*, or "thoughtful." Etymologically, in fact, its Latin derivative is *scepticus*, for "inquiring" or "reflective." Further variations in the ancient Greek include "watchman" and "mark to aim at." Hence, skepticism is thoughtful and reflective inquiry. Skeptics are the watchmen of reasoning errors, aiming to expose bad ideas. Perhaps the closest fit for skeptic is "a seeker after truth; an inquirer who has not yet arrived at definite convictions." Skepticism is not "seek and ye shall find"—a classic case of what is called the confirmation bias in cognitive psychology—but "seek and keep an open mind." What does it mean to have an open mind? It is to find the essential balance between orthodoxy and heresy, between a total commitment to the status quo and the blind pursuit of new ideas, between being open-minded enough to accept radical new ideas and so open-minded that your brains fall out. The virtue of skepticism is about finding that balance.[71]

The nineteenth-century philosopher Robert G. Ingersoll, a secular moral hero if ever there was one, found additional freedoms in a naturalistic worldview, including freedom from

> the fear of eternal pain . . . from the winged monsters of the night . . . from devils, ghosts, and gods . . . no chains for my limbs—no lashes for my back—no fires for my flesh—no master's frown or threat—no following another's steps—no need to bow, or cringe, or crawl . . . I was free. I stood erect and fearlessly, joyously, faced all worlds. . . . And then my heart was filled with gratitude, with thankfulness, and went out in love to all the heroes, the thinkers who gave their lives for the liberty of hand and brain—for the freedom of labor and thought—to those who fell in the fierce fields of war, to those who died in dungeons bound with chains—to those who proudly mounted scaffold's stairs—to those whose bones were crushed, whose flesh was scarred and torn—to those by fire consumed—to all the wise, the good, the brave of every land, whose thoughts and deeds have given freedom to the sons of men. And then I vowed to grasp the torch that they had held, and hold it high, that light might conquer darkness still.[72]

The bright torch of science illuminates the darkness of humanity to reveal a human nature that is both moral and immoral, a product of our evolutionary heritage and our cultural history. We can construct a provisional ethical system that is neither dogmatically absolute nor irrationally relative, a more universal and tolerant morality that enhances the probability of the survival and well-being of all members of the species, and perhaps eventually of all species and even the biosphere, the only home we have ever known or will know until science leads us off the planet, out of the solar system, and to the stars. *Ad astra!*

APPENDIX I

THE DEVIL UNDER FORM OF BABOON: THE EVOLUTION OF EVOLUTIONARY ETHICS

> Our descent, then is the origin of our evil passions!!—The Devil under form of Baboon is our grandfather!
>
> —Charles Darwin, *M Notebook,* 1838

In a poetic form of authorial acknowledgment, Jonathan Swift famously gave the nod to those who came before him with this poem, a favorite among naturalists:

> *So, naturalists observe, a flea*
> *Hath smaller fleas that on him prey;*
> *And these have smaller still to bite 'em*
> *And so proceed* ad infinitum.
> *Thus every poet, in his kind,*
> *Is bit by him that comes behind.*

We are all bit by those behind us, but nowhere is this clearer than in the tradition of science because of its cumulative and progressive nature of building on those who came before. "If I have seen further it is by standing on ye shoulders of Giants," Sir Isaac Newton even more famously observed. The quote is itself part of a long historical tradition dating at least to the fourteenth-century scholar Bernard of Chartres, in reference to New Testament prophets standing on the shoulders of Old Testament prophets (as depicted in stone in one of the transepts of Chartres Cathedral): "We are like dwarfs seated on the shoulders of giants; we see more things than the ancients and things more distant.

But this is due neither to the sharpness of our own sight, nor to the greatness of our own stature, but because we are raised and borne aloft on that giant mass."[1]

The first half of this book, entitled "The Origins of Morality," was built on the theory of evolutionary ethics, a study that began in the late 1830s when, among his scattered thoughts on the implications for his budding evolutionary hypothesis, Charles Darwin penned this muse in his *M Notebook* (opened shortly after returning home from a five-year voyage around the world): "He who understands baboon would do more towards metaphysics than Locke." Biology, Darwin believed, not philosophy, is where we might find insights into our moral (not to mention our immoral) natures: "Our descent, then is the origin of our evil passions!!—The Devil under form of Baboon is our grandfather!"[2]

Darwin's founding of the study of evolutionary ethics belongs to a long tradition of ethical naturalism that dates back twenty-five centuries to Aristotle and eight centuries to Thomas Aquinas. Aquinas defined "natural law" as "that which nature has taught all animals"; thus, moral law must be rooted in the natural inclinations of the human animal. Because Darwin's biological explanation of morality was grounded in such natural inclinations, this has been called "Darwinian natural right."[3] Morality is subsumed under the umbrella of human nature, just one of a pantheon of thoughts and behaviors legitimately targeted as subjects of study within the natural sciences. Morality—as an expression of human thought and behavior related to the judgment and evaluation of one's own and others' thoughts and behaviors—can be explored and examined by psychologists, anthropologists, evolutionary psychologists, and other social scientists in the same manner that political beliefs, social attitudes, religious faith, and other human expressions of thought can be studied.[4]

Charles Darwin, then, was the first evolutionary psychologist and ethicist. In his 1871 book *The Descent of Man,* he made this logical inference from the data produced by zoology and anthropology: "The following proposition seems to me in a high degree probable—namely, that any animal whatever, endowed with well marked social instincts, the parental and filial affections being here included, would inevitably acquire a moral sense or conscience, as soon as its intellectual powers had become as well, or nearly as well developed, as in man." Even though it was Darwin, more than anyone else, who demonstrated that humans are animals too, it was the moral sense—of all our exalted characteristics—that separates us from all other animals: "I fully subscribe to the judgment of those writers who maintain that of all the differences between man and the lower animals, the moral sense or conscience is by far the most important."[5] The evolution of the moral sense was a step-by-step process, aided by the same emotions generated by religious rituals and expressions, that would result in "a highly complex sentiment, having its first origin in the social instinct, largely guided by the approbation of our fellow-men, ruled by reason, self-interest, and in later times by deep religious feelings, confirmed by instruction and habit, all combined, constitute our moral sense and conscience."[6]

Ever since Darwin the science of evolutionary ethics has waxed and waned, roughly passing through five stages: (1) *Origins*—from Darwin to the end of the First World War; (2) *Synthesis*—from the beginning of the modern synthesis of evolutionary theory in the 1920s through the mid-1970s; (3) *Controversy*—from the birth of sociobiology in 1975 to the early 1990s; (4) *Victory*—from the triumph of evolutionary psychology in the early 1990s to 2000; and (5) *Consolidation*—the incorporation of group selection and hierarchical evolutionary theory from 2000 to the present. While my historical summation of the first three stages is descriptive, the last two are prescriptive. That is, although historians of science will not find much to quibble with in my description of the field's origins, synthesis, and controversy, many scientists will disagree with my prescription of victory and consolidation.

Origins

In Charles Darwin's time there was no one more enthusiastic about applying natural selection than its codiscoverer Alfred Russel Wallace. Yet, in my biography of Wallace, *In Darwin's Shadow,* I argue that his purist mode of hyperselectionist thinking and his commitment to scientism led him, ironically, to conclude that natural selection cannot account for the human brain and morals. Like Darwin before him, Wallace minced no words about why humans and animals are different: "My view . . . was, and is, that there is a difference in kind, intellectually and morally, between man and other animals."[7] That difference, however, was not generated by natural selection, but was instead, Wallace concluded, the product of a higher intelligence because he could think of no possible reason nature would have selected for such a large and varied organ. He said as much in an article on "The Limits of Natural Selection as Applied to Man":

> In the brain of the lowest savages and, as far as we know, of the prehistoric races, we have an organ . . . little inferior in size and complexity to that of the highest types. . . . But the mental requirements of the lowest savages, such as the Australians or the Andaman Islanders, are very little above those of many animals. How then was an organ developed far beyond the needs of its possessor? Natural Selection could only have endowed the savage with a brain a little superior to that of an ape, whereas he actually possesses one but very little inferior to that of the average members of our learned societies.[8]

Therefore, Wallace concluded, "an Overruling Intelligence has watched over the action of those laws, so directing variations and so determining their accumulation, as finally to produce an organization sufficiently perfect to admit of, and even to aid in, the indefinite advancement of our mental and moral nature."[9] How did this evolution come about? Wallace argued that natural selection operated on the physical body of man long before a mind with consciousness existed. The races, represented by a "protoman," were fully developed physically before civilization began. Once the brain reached a certain level, however, natural selection would no longer operate on the body; man could now manipulate his environment. The creation of mind had lessened the effectiveness of natural selection (and therefore the process of evolution). Ironically, a propensity toward cooperation and mutual aid may have played a role in this attenuation: "In the rudest tribes the sick are assisted, at least with food; less robust health and vigour than the average does not entail death. The action of natural selection is therefore checked; the weaker, the dwarfish, those of less active limbs, or less piercing eyesight, do not suffer the extreme penalty which falls upon animals so defective."[10]

With this alteration of natural law, Wallace argued, came a shift from individual to group selection. While individuals would be protected by the group from the ravages of nature, groups themselves might continue evolving, especially those with high intelligence, foresight, sympathy, a sense of right, and self-restraint: "Tribes in which such mental and moral qualities were predominant would therefore have an advantage in the struggle for existence over other tribes in which they were less developed—would live and maintain their numbers, while the others would decrease and finally succumb."[11] Wallace argued that the harsher, more challenging climate of northern Europe had produced "a hardier, a more provident, and a more social race" than those from more southern climates. Indeed, he pointed out, European imperialism, particularly the British form, was causing whole races to disappear "from the inevitable effects of an unequal mental and physical struggle."[12] Ever the grand synthesizer, Wallace ends his argument with a flare of teleological purposefulness and an egalitarian hope for the future of humanity shaped via human-controlled group selection:

> If my conclusions are just, it must inevitably follow that the higher—the more intellectual and moral—must displace the lower and more degraded races; and the power of 'natural selection,' still acting on his mental organisation, must ever lead to the more perfect adaptation of man's higher faculties to the conditions of surrounding nature, and to the exigencies of the social state. While his external form will probably ever remain unchanged, except in the development of that perfect beauty which results from a healthy and well organised body, refined and ennobled by the highest intellectual faculties and sympathetic emotions, his mental constitution may continue to advance and improve, till the world is again inhabited by a single nearly homogeneous race, no individual of which will be inferior to the noblest specimens of existing humanity.[13]

One of Alfred Wallace's intellectual heroes was the philosopher, social scientist, and social Darwinist Herbert Spencer. So enamored of Spencer was Wallace that he named his firstborn son Herbert Spencer Wallace. Thus, we should not be surprised to know that their ideas on evolutionary theory, in particular evolutionary ethics, were well in accord. When Spencer read Wallace's 1864 paper "The Origin of the Races of Man" (quoted above), for example, Spencer immediately wrote Wallace and told him: "Its leading idea is, I think, undoubtedly true, and of much importance towards an interpretation of the facts. . . . I think it is quite clear, as you point out, that the small amounts of physical differences that have arisen between the various human races are due to the way in which mental modifications have served in place of physical ones."[14] In 1879 Wallace wrote to Spencer: "I doubt if evolution alone, even as you have exhibited its action, can account for the development of the advance and enthusiastic altruism that not only exists now, but apparently has always existed among men." But Spencer did not always go far enough for Wallace in speculating about the origins of morality: "If on this point I doubt, on another point I feel certain, and that is, not even your beautiful system of ethical science can act as a 'controlling agency' or in any way 'fill up the gap left by the disappearance of the code of supernatural ethics.' "[15]

Spencer was actually ahead of both Darwin and Wallace in attempting a scientific analysis of ethics (even if it was not a strictly evolutionary one) when he published, in 1851, *Social Statics; or The Conditions Essential to Human Happiness Specified, and the First of Them Developed.* This was not so much a descriptive theory on the origins of morality as it was a prescriptive theory on how morality should be applied to society in a rational and scientific manner. Spencer rejected utilitarian calculations of the greatest good for the greatest number. Instead, Spencer postulated, "the moral law of society, like its other laws, originates in some attribute of the human being." Although he called it science, Spencer was really doing philosophy, making his case through logical deduction rather than empirical facts, beginning with a divine origin of morality: "God wills man's happiness. Man's happiness can only be produced by the exercise of his faculties. Then God wills that he should exercise his faculties. But to exercise his faculties he must have liberty to do all that his faculties naturally impel him to do. Then God intends he should have that liberty. Therefore he has a right to that liberty." From this it follows that "All are bound to fulfill the Divine will by exercising them. All therefore must be free to do those things in which the exercise of them consists. That is, all must have rights to liberty of action." Of course, my freedom to swing my arm in any direction I choose ends at your nose. Spencer deduced this, of course, concluding, "Every man has freedom to do all that he wills, provided he infringes not the equal freedom of any other man."[16]

Spencer, as an editor of the magazine *The Economist,* was familiar with and accepted Adam Smith's theory that humans had the mental capacity for sympathy. Smith, before *The Wealth of Nations* made him the fountainhead of classical liberal free-market economics, was a professor of moral philosophy who argued that the foundation of morality was based on the ability we have to put our self in someone else's shoes. When you see someone grieving, you feel sympathy because you can project yourself into that situation and imagine how you would feel. An anticipation of self-grief generates genuine sympathy for the other person. In 1879 Spencer published *The Data of Ethics,* and in 1891 his *Principles of Ethics,* in which he abandoned supernatural intervention as a causal factor in the origins of morality and turned enthusiastically to Darwinian selection:

> We have to enter on the consideration of moral phenomena as phenomena of evolution; being forced to do this by finding that they form a part of the aggregate of phenomena which evolution has wrought out. If the entire visible universe has been evolved—if the solar system as a whole, the earth as a part of it, the life in general which the earth bears, as well as that of each individual organism—if the mental phenomena displayed by all creatures, up to the highest, in common with the phenomena presented by aggregates of these highest—if one and all conform to the laws of evolution; then the necessary implication is that those phenomena of conduct in these highest creatures with which Morality is concerned, also conform.[17]

While Spencer was arguably the most influential evolutionary ethicist outside of Darwin in the nineteenth century, he was not without strong critics, not the least of which was "Darwin's bulldog," Thomas Henry Huxley. Huxley was skeptical about how far evolutionary theory could be extended into the realm of ethics, but not for the same reason as Wallace, who questioned what selective advantage a system of ethics would have conferred on an individual or species. Instead, Huxley doubted that nature could be an ethical guide for us at all: "Cosmic evolution may teach us how the good and the evil tendencies of man may have come about; but, in itself, it is incompetent to furnish any better reason why what we call good is preferable to what we call evil than we had before." In one of the most powerful one-liners in the history of evolutionary thought (at least as it relates to ethics), Huxley definitely came down hard on the reality of what he saw as our brutal nature: "Let us understand, once for all, that the ethical process of society depends, not on imitating the cosmic process, still less in running away from it, but in combating it."[18]

After the initial excitement of evolutionary ethics wore off, its influence on both the public and academic philosophers and natural scientists faded into near oblivion. By the turn of the century the theory of evolution itself was experiencing something of a decline; scientists were openly expressing skepticism that natural selection could do all that Darwin said it could. Cambridge University philosopher G. E. Moore was especially contemptuous of evolutionary ethics, attacking it on the grounds that it violated the "naturalistic fallacy," mistakenly inferring the *ought* from the *is*, or prescribing the way things should be based on a description of the way things are. The "good," said Moore in his classic *Principia Ethica*, cannot be quantified like the utilitarians tried to do, nor could it be analyzed for its evolutionary adaptiveness. In fact, it cannot be defined by reference to some "other thing." Its existence had to be apprehended on its own without outside reference.[19]

Synthesis

By World War I, the study of evolutionary ethics was in serious decline. But as it moved into its second phase between the wars, Julian Huxley and C. H. Waddington revised it—in conjunction with the modern evolutionary synthesis. Julian Huxley reinvigorated evolutionary ethics by grafting it onto the larger intellectual and social movement known as humanism. Huxley, in fact, called himself a religious humanist, "but without belief in any personal God." Like his grandfather Thomas, Julian cared not at all for traditional religion and did not believe in God, but unlike the senior Huxley, Julian rejected Moore's charges, arguing that science can not only tell us the way things are, it can direct us toward the way things ought to be:

> In the broadest possible terms evolutionary ethics must be based on a combination of a few main principles: that it is right to realize ever new possibilities in evolution, notably those which are valued for their own sake; that it is right both to respect human individuality and to encourage its fullest development; that it is right to construct a mechanism for further social evolution which shall satisfy these prior conditions as fully, efficiently, and as rapidly as possible.[20]

Like Wallace, Julian Huxley's evolutionary ethics was based on a belief in the progressive nature of evolution, although he did not envision a socialist utopia as an ideal state. He shared with his grandfather an enthusiasm for science and evolutionary thinking, but his vision of ethics was far less combative against nature than Thomas Huxley's was. As Julian wrote:

> When we look at evolution as a whole, we find, among the many directions which it has taken, one which is characterized by introducing the evolving world-stuff to progressively higher levels of organization and so to new possibilities of being, action, and experience. This direction has culminated in the attainment of a state where the world-stuff (now moulded into human shape) finds that it experiences some of the new possibilities as having value in or for themselves; and further that among these it assigns higher and lower degrees of value, the higher values being those which are more intrinsically or more permanently satisfying, or involve a greater degree of perfection.[21]

The latter half of this second renaissance for evolutionary ethics saw it fade once again after the Second World War. This second waning was a result, in part, of an extreme antihereditarianism view in psychology and the social sciences—an understandable response to Nazi eugenics, ethnic cleansing, and especially the Holocaust. As a consequence, however, scientists steered clear of the study of the biological and evolutionary origins of morality, instead focusing on purely cultural explanations. It was a trend mirrored throughout the social sciences and humanities, as scientists and scholars began with a prima facie assumption that social and psychological behavior must primarily be influenced by the environment, not biology.

Controversy

Evolutionary ethics lay dormant for three decades until 1975 when Harvard evolutionary biologist Edward O. Wilson published his 700-page magnum opus, *Sociobiology: The New Synthesis*. Ironically, only the final chapter deals with humans ("Man: From Sociobiology to Sociology"), and only in one short section—barely two pages long—does the reader encounter ethics and its possible evolutionary origins. But what is said, when it is said, and who is doing the saying matters as much in science as it does in other human endeavors, and here is what

Wilson said: "Scientists and humanists should consider together the possibility that the time has come for ethics to be removed temporarily from the hands of the philosophers and biologicized."[22] Like Darwin's single line at the end of *The Origin of Species*—"light will be thrown on the origin of man and his history" (later editions added the modifier "much")—Wilson's one-liner fired a shot heard 'round the intellectual world. Academic philosophers were incensed that an outsider was encroaching on their turf and doing so in such cold scientific jargon: "Ethical philosophers intuit the deontological canons of morality by consulting the emotive centers of their own hypothalamic-limbic system," he wrote in *Sociobiology*, reducing thousands of years of philosophy to mere speculation on hormonally driven internal states. "Only by interpreting the activity of the emotive centers as a biological adaptation can the meaning of the canons be deciphered."[23] Wilson was vilified in the press, despite his disclaimer at the end of that brief section: "It should also be clear that no single set of moral standards can be applied to all human populations, let alone all sex-age classes within each population. To impose a uniform code is therefore to create complex, intractable moral dilemmas—these, of course, are the current condition of mankind."[24]

At stake in this battle—now known as the "evolution wars"—is nothing less than how human societies and families should be structured, how parents should raise children, how criminals should be handled, among other issues related to the nature of human nature. Also on the line, as the sociologist Ullica Segerstråle observed in her encyclopedic history of the evolution wars—*Defenders of the Truth*—is "the soul of science."[25]

The story of how an academic textbook by an entomologist could result in one of the most rancorous debates in all of science begins with the reactions to Wilson and his theory by his Harvard colleagues Stephen Jay Gould and Richard Lewontin. Their Sociobiology Study Group, along with the politically charged, left-leaning organization Science for the People, were involved in the now famous incident at the 1978 meeting of the American Association for the Advancement of Science in the Sheraton Park Hotel in Washington, D.C. When Wilson advanced to the podium, demonstrators chanted, "Racist Wilson you can't hide, we charge you with genocide!" Someone leaped up on the dais, grabbed a cup of ice water, and dumped it on Wilson's head, shouting, "Wilson, you are all wet!" This was too much even for Stephen Jay Gould, who admonished the demonstrators, telling them their actions were what Lenin had dismissively called "Infantile Leftism." The infamously bellicose anthropologist Napoleon Chagnon wasted no time in coming to Wilson's defense, grabbing one of the attackers and tossing him from the stage.

Why, Wilson wondered two decades later in his autobiographical book *Naturalist,* didn't Gould and Lewontin just come up to his office from theirs (one floor below in the same Harvard building) to discuss their concerns? Why attack him in the very public pages of the *New York Review of Books* when this all could have been handled in private? The reason is that science is not the private and always-rational enterprise it is often made out to be. Why, Gould and Lewontin could just as easily have asked, didn't Wilson come down one floor to their offices to discuss with them in private his ideas about applying principles of animal behavior to human societies? The answer is the same: if you want to get your theories out into the marketplace of ideas, you cannot sequester them in your office. You've got to make them public, and the more public the better. Hashing the debate out in public gives you the forum you would never get in private. (An analogy here will help. On March 14, 1994, I appeared on Phil Donahue's live national television show to debunk the Holocaust deniers.[26] The producers went to great lengths to keep me separated from them—different limos to the studio, different dressing rooms, different green rooms, different entrances to the set, and no talking during commercial breaks. Why? Because, I was told, they wanted the fresh drama of an initial encounter.) In the evolution wars and sociobiology debates, Gould and Lewontin had a scientific agenda that they wanted to air publicly—that adaptationist, gene-centered arguments in evolutionary theory can be carried too far, and that much in the history of life can be explained by nonadaptive processes and a multileveled analysis of genes, individuals, and groups. What better way to do it than to use Wilson as their foil? But who in the general public knows or cares about adaptations, exaptations, spandrels, contingencies, and other esoterica of evolutionary biology? What the public does understand quite well are Nazis, eugenics, race-purification programs, and other abuses of biology of the past century. Thus, sociobiology's critics reasoned, the best strategy is to begin with its ideological implications—particularly the racist overtones of genetic determinism—to capture an audience, then segue into the scientific arguments about the problems with hyperadaptationism. Gould said as much at a 1984 Harvard meeting Segerstråle attended: "We opened up the debate by taking a strong position. We took a definitive stand in order to open up the debate to scientific criticism. Until there is some legitimacy for expressing contrary opinions, scientists will shut up." From this (and numerous interviews with all parties involved), Segerstråle concludes: "What I take Gould to be saying here is that the controversy around Wilson's *Sociobiology* was, in fact, a vehicle for the real scientific controversy about adaptation! Far, then, from 'dragging politics into it,' or being 'dishonest' as [Ernst] Mayr accused Gould and Lewontin of being, their political involvement would have been instead a deliberate maneuver to gain a later hearing for their fundamentally scientific argument about adaptation. What Gould seems to have been saying here is that the scientific controversy about adaptation could not have been started without the political controversy about sociobiology."[27]

Before we accuse Gould and Lewontin of being overly Machiavellian in their political machinations, however, we should note that Wilson was not an innocent victim in this debate. It seems unlikely that a Harvard professor could author a book whose title defines a new science of applying biology to human social and moral behavior, in the middle of a decade that was defined by its ideological emphasis on egalitarian politics and cultural determinism, and not expect trouble. In point of fact, of course, all scientists have an agenda and the sooner we recognize that fact and come clean with our own, the better able the public will be to judge scientific theories. Certainly Gould and Lewontin went too far, as all social movements are wont to do. When I first met Ed Wilson, I was surprised at what a kind, generous, and soft-spoken man he is—anything but what I had expected from following the sociobiology debates. Then again, it would appear that Wilson knew exactly what he was doing all along. Throughout his long and illustrious career Wilson has brilliantly orchestrated a scientistic program of biologizing all of human behavior, from mate selection and maternal love to mass genocide and morality. No wonder the evolution wars have been so heated. As he has always done with such aplomb over the decades and through numerous scientific battles, Ed Wilson let his pen do the talking. He responded to his critics with a Pulitzer

Prize–winning answer in book form, *On Human Nature,* in which he succinctly threw down the gauntlet:

> Above all, for our own physical well-being if nothing else, ethical philosophy must not be left in the hands of the merely wise. Although human progress can be achieved by intuition and force of will, only hard-won empirical knowledge of our biological nature will allow us to make optimum choices among the competing criteria of progress.[28]

Victory

Ed Wilson's gauntlet was taken up by a cadre of scientists, philosophers, and scholars of many stripes, including anthropologist Donald Symons, psychologist Robert Axelrod, evolutionary biologist Robert Trivers, philosopher Michael Ruse, historian of science Robert Richards, biologist Richard Alexander, evolutionary theorist John Maynard Smith, sociologist James Q. Wilson, evolutionary biologist William D. Hamilton, primatologist Frans de Waal, and many others. Richard Dawkins's 1976 *The Selfish Gene* was especially influential in getting people to think about applying science and evolution to human behavior, including moral behavior. To the concept of genes as carriers of information, Dawkins added "memes"—cultural carriers of information that go beyond biology yet act much like genes in terms of propagation, selection, and mutation. He even treated religious ideas as virus memes that, like computer viruses, invade our mental software, destroying our programs for rational thought and behavior. One of the best books making the case for the evolutionary origins of moral behavior is Matt Ridley's 1997 *The Origins of Virtue: Human Instincts and the Evolution of Cooperation.* Robert Wright's 1994 *The Moral Animal* is an engaging history of evolutionary ethics; Paul Farber's 1994 *The Temptations of Evolutionary Ethics* is a scholarly history and critique; Philip Kitcher's 1995 *Vaulting Ambition* provides a strong critique; and Paul Thompson's 1995 *Issues in Evolutionary Ethics* is a useful collection of the most important works in the field, pro and con. At *Skeptic* magazine we devoted back-to-back special issues—with pro and con debates—on both evolutionary psychology and evolutionary ethics.[29]

As a final statement of victory, of sorts, in 1998 Ed Wilson published

Consilience: The Unity of Knowledge, a magisterial sweep of the history of knowledge from the Enlightenment to the present. Wilson devoted a full chapter to ethics, as discussed in chapter 1, updating his argument from two decades prior (in *Sociobiology* and *On Human Nature*) by reducing the debate about the origins of the moral sense to an either-or choice between transcendentalists and empiricists. Wilson, of course, is an empiricist for whom God is an unnecessary hypothesis. "In simplest terms, the options are as follows: I believe in the independence of moral values, whether from God or not, and I believe that moral values come from human beings alone, whether or not God exists." For Wilson, even Enlightenment atheist philosophers are still transcendentalists because "they tend to view natural law as a set of principles so powerful, whatever their origin, as to be self-evident to any rational person. In short, transcendental views are fundamentally the same whether God is invoked or not." For the empiricist, at least in a Wilsonian sense, "if we explore the biological roots of moral behavior, and explain their material origins and biases, we should be able to fashion a wise and enduring ethical consensus."[30] Wilson began his analysis by admonishing ethicists for not opening their arguments with a caveat such as: "this is my starting point, and it could be wrong." To his eternal (pardon the religious hyperbole) credit, Wilson had the intellectual integrity to end his own treatise on the material origins of ethics with this comment: "And yes—lest I forget—I may be wrong."[31]

Today, evolutionary psychology, and its subfield of evolutionary ethics, are budding sciences ripe with both testable hypotheses and not so testable just-so stories. The debate has proven to be a lively one, and its critics have many important points to make, with plenty of cautions and caveats to go around about inappropriately applying evolutionary theory to human thought and behavior, but in my opinion the theory of evolution has won the day in both psychology and ethics. The field moved into its fourth phase in the early 1990s when sociobiology—under a new covering cloth of evolutionary psychology—gained general acceptance among a sizable group of evolutionary biologists and theorists. But it is only fair that I acknowledge that not everyone will agree with me that the application of evolutionary theory to human psychology and morality achieved victory at any time. Critics of evolution in general, and evolutionary biologists skeptical of evolutionary psychology in particular, may take issue with me on this point.

Consolidation

The fifth phase in the history of evolutionary ethics involves a consolidation of many evolutionary concepts, most notably a pluralistic and hierarchical model of evolution that recognizes causal elements other than natural selection—such as group selection—that have operated in human evolution. As in my caveat above, however, there are those who would argue that the consolidation took place in the victory stage and that group selection and hierarchical theory are nothing more than a minor wrinkle in the overall fabric of life. Thus, what follows is as much prescriptive as it is descriptive.

GROUP SELECTION, HIERARCHICAL EVOLUTIONARY THEORY, AND THE EVOLUTION OF MORALITY

As discussed in chapter 2, group selection has been a controversial subject and remains a topic of hot debate among evolutionary theorists today. Evolutionary biologist David Sloan Wilson recounts the story of how group selection became anathema:

> George C. Williams is regarded as a hero by evolutionary biologists of the individualistic tradition. Specifically, he is the hero who severed the head of group selection and mounted it on a pole as an example of how not to think for future generations. As Williams tells the story (the last time I heard it was at the award ceremony for Sweden's Crafoord Prize, which Williams received in 1999 along with John Maynard Smith and Ernst Mayr), he was a young postdoctoral associate at the University of Chicago and attended a lecture by Alfred Emerson, a highly respected biologist who portrayed all of nature as like a big termite colony. Williams knew that the evolution of higher-level adaptation was not so simple. As he listened to Emerson he thought "if this is evolution, I want to do something else—like car insurance." Williams left the lecture muttering "Something must be done." That something was *Adaptation and Natural Selection,* first published in 1966 and still widely read. Williams was one of many evolutionists who reacted against the superorganismic perspective but he became the icon for its rejection. I wish I could report otherwise, but scientists need their heroes and heads on poles as much as any other human group.[32]

Wilson replied to Williams and other group selection critics in his own works, including *Unto Others: The Evolution and Psychology of*

Unselfish Behavior and *Darwin's Cathedral: Evolution, Religion, and the Nature of Society.*[33] In addition to support from Darwin himself, who applied group selection in a limited fashion, Wilson has support from two other evolutionary theorists: Ernst Mayr and Stephen Jay Gould.

Ernst Mayr as Group Selectionist. Ernst Mayr is unquestionably the dean of twentieth-century evolutionary theory. Throughout his long career he has promulgated a limited form of group selection—limited to the human social group. In 1959 he credited J. B. S. Haldane as the first to identify (in 1932) the "population-as-a-whole as a unit of selection" (as group selection was then called), and suggests that the rate of mutation, degree of outbreeding, distance of dispersal, aberrant sex ratios, and other mechanisms favoring either in-breeding or out-breeding are attributes of populations (but not individuals) that would be selected for or against. In 1982 Mayr identified V. C. Wynne-Edwards as the scientist who formalized the group selection process, but still limited it to the deme, or population level, with only a brief mention of social groups. In 1988 Mayr agreed with the critics of group selection when it comes to animal groups, but argued, "there is a great deal of evidence that human cultural groups, as wholes, can serve as the target of selection. Rather severe selection among such cultural groups has been going on throughout hominid history." In 1997 Mayr made a distinction between soft and hard group selection. Soft group selection "occurs whenever a particular group has more (or less) reproductive success than other groups simply because this success is due entirely to the mean selective value of the individuals of which the group is composed. Since every individual in sexually reproducing species belongs to a reproductive community, it follows that every case of individual selection is also a case of soft group selection, and nothing is gained by preferring the term soft group selection to the clearer traditional term individual selection." Soft group selection is just individual selection writ large. Hard group selection, by contrast, "occurs when the group as a whole has certain adaptive group characteristics that are not the simple sum of the fitness contributions of the individual members. The selective advantage of such a group is greater than the arithmetic mean of the selective values of the individual members. Such hard group selection occurs only when there is social facilitation among the members of the group or, in the case of the human species, the group has a culture which adds or detracts from the mean fitness value of the members of the cultural group." This is genuine group selection that differs qualitatively (not just quantitatively, as in soft group selection) from individual organismic selection, and is the subject of so much controversy.[34]

Finally, in 2000, in an extensive interview of Mayr for *Skeptic* magazine, Frank Sulloway and I queried him on group selection and the controversy it has generated. He surprised us when he said: "George Williams and Richard Dawkins have made a mistake, in my opinion, in completely rejecting group selection. But we have to be careful here to define what we mean by a group. There are different kinds of groups. There is one type of group that is a target of selection, and that is the social group. Hominid groups of hunter-gatherers were constantly competing with other hominid groups; some were superior and succeeded and others were not. It becomes quite clear that those groups who had highly cooperative and altruistic individuals were more successful than the ones torn apart by internal strife and egotism." So the social environment is as important as the physical environment. "The essential point is that if you are altruistic and make your group more successful, you thereby also increase the fitness of the altruistic individual (yourself)!" Critics, we noted, would argue that it is still the individual being selected for these characteristics, not the group. Mayr countered: "There is no question that the groups that were most successful had these individuals that were cooperative and altruistic, and those traits are genetic. But the group itself was the unit that was selected."[35]

Stephen Jay Gould as Group Selectionist. Just before his untimely death in May 2002, Stephen Jay Gould witnessed the publication of his magnum opus, *The Structure of Evolutionary Theory,* in which he defended group selection as a legitimate part of a hierarchical theory of selection that recognizes different levels at which the selection process occurs.[36] Gould begins by systematically building upon Darwin's cathedral, an apt metaphor as his tome begins with an architectural analysis of the Duomo (Cathedral) of Milan, showing how the original fourteenth-century foundational structure was appended over the centuries with spires and pinnacles, such that we can legitimately say a core structure remains intact while the finished building represents a far richer compendium of historical additions. Gould's mission is not to raze the Darwinian Gothic structure, or to tear down the neo-Darwinian Baroque facades, but to revise, refine, reinforce, and reconstruct those portions of Darwin's Duomo that have begun to crumble under the weathering effects of a century and a half of scientific research. The foundation of Darwin's Duomo rests upon three theoretical pillars that form the basis of a hierarchical theory of evolution: agency, efficacy, and scope.

Agency, or the level at which evolutionary change occurs. For Darwin, it is primarily individual organisms that are being selected for or against. Gould proposes a multitiered theory of evolution where change (and selection) occurs on six levels: gene, cell lineage, organism, deme, species, and clade. It is here where Gould defends his own version of group selection—which he calls species selection—without denying the power of Darwinian organismal selection. He does so with two caveats: "First, adjacent levels may interact in the full range of conceivable ways—in synergy, orthogonally, or in opposition. Second, the levels operate non-fractally, with fascinating and distinguishing differences in mode of functioning, and relative importance of components for each level. For example, the different mechanisms by which organisms and species maintain their equally strong individuality dictate that selection should dominate at the organismal level, while selection, drift, and drive should all play important and balanced roles at the species level."[37]

Efficacy, or the mechanism of evolutionary change. For Darwin it was natural selection (and its handmaiden, sexual selection, where females, for example, select for or against characteristics that they like or dislike by choosing certain males) that drives organisms to evolve. Gould does not deny the power of natural selection, but wishes to emphasize that in the three-billion-year history of the earth's rich panoply of life, there is more to the story. On top of the substratum of microevolution (short-term small changes) Gould adds macroevolution—long-term big changes caused by mass extinctions and other large-scale forces of change. To the bottom floor of adaptationism Gould attaches exaptationism—structures subsumed for later uses and whose original adaptive purposes are now lost to history.

Scope, or the range of effects wrought by natural selection. For

Darwin, gradual and systematic change extrapolated over geological expanses of time is all that is needed to account for life's diversity. For Gould, slow and steady sometimes wins the race, but life is also punctuated with catastrophic contingencies that fall in the realm of unique historical narratives rather than predictable natural laws. In Gould's view, history, not physics, should be evolutionary theory's model of science.

Revisions to these three branches of agency, efficacy, and scope (while the main Darwinian trunk retains its theoretical power), says Gould, produce a "distinct theoretical architecture, offering renewed pride in Darwin's vision and in the power of persistent critiques—a reconstitution and an improvement."[38] How can the paradigms of the original Darwinism, the neo-Darwinism of the synthesis, and Gouldian Darwinism coexist peacefully? We can ask the same question with regard to individual versus group selection, or organismal versus species selection as apparently competing paradigms. In science, doesn't one paradigm displace another in a way that makes them incompatible? No. Paradigms can build upon one another and cohabit the same scientific niche. Just as the Newtonian paradigm has been reconstituted to include the paradigms of relativity and quantum mechanics, the overarching Darwinian paradigm has been improved by, for example, the subsidiary punctuated equilibrium paradigm, which constitutes an improved reading of the herky-jerky fossil record whose numerous gaps so embarrassed Darwin. (The gaps, say Gould and his cotheorist Niles Eldredge, represent data of a speciation process that happens so rapidly that few "transitional" fossils are left in the historical record.) Darwinian gradualism and individual selectionism can be supplemented with Gouldian punctuationism and group selectionism. How?

Think of species not as billiard balls being knocked about the table of nature willy-nilly, but as polyhedrons, or multifaceted structures (picture an eight-sided die) that sit on a side until nudged by a potent force, and whose internal properties, Gould writes, "'push back' against external selection, thereby rendering evolution as a dialectic of inside and outside." Without discounting the outside, Gould wants us to look again inward (as so many evolutionary theorists did in Darwin's own day), where the restricting channels of both nature and history direct the selective forces in particular directions. Although individual selection is more potent and common than group selection, the latter may outdo the former under certain conditions, as is the case with the selection for human morality. "Organismic selection may trump species selection in principle when both processes operate at maximal efficiency, but if change associated with speciation operates as 'the only game in town,' then a weak force prevails while a potentially stronger force lies dormant. Nuclear bombs certainly make conventional firearms look risible as instruments of war, but if we choose not to employ the nukes, then bullets can be devastatingly effective."[39]

Ernst Mayr said we need to be precise in how we define a group. The same is true in defining what we mean by group selection, and how human morality may have evolved—at least in part—as a function of the group selection process. Darwin focused on tribes, and how they differed in terms of patriotism, fidelity, obedience, courage, and sympathy, and how these differences led to the selection of some tribes over others. Mayr talks about groups in terms of their social bonding and interactions, and how this might have led to certain groups being selected for survival over others. Gould's emphasis on species selection would seem beyond the scope of these more narrowly confined cohorts, but if we think of the rapid extinction of Neanderthals shortly after the arrival of anatomically modern Cro-Magnon humans—considered to be separate species by most paleoanthropologists—this too may be an example of group selection, where the group is the entire species. And the consequences of an entire hominid species disappearing forever off the face of the earth are profound.

APPENDIX II

MORAL AND RELIGIOUS UNIVERSALS AS A SUBSET OF HUMAN UNIVERSALS

The following 202 human universal traits are culled out of anthropologist Donald E. Brown's original list of 373 human universals, and are directly related to religious and moral behavior. They are presented as a demonstration of the universality of morality, and as further evidence of its evolutionary heritage. For these universals I have added parenthetical notes indicating the relation I believe each has to morality and religion. Brown's list is in: Donald E. Brown, *Human Universals,* New York: McGraw-Hill, 1991.

affection expressed and felt (necessary for altruism and cooperation to be reinforced)
age statuses (vital element in social hierarchy, dominance, respect for elder's wisdom)
anthropomorphization (basis of animism, anthropomorphic gods of Greece/Rome, attribution of human moral traits to the monotheistic God of Jews, Christians, and Muslims)
anticipation (vital for behaviors to have future consequences)
attachment (necessary for bonding, friendship, pro-social behavior)
belief in supernatural/religion (basis of the codification of morality)
beliefs about death (basis for belief in an afterlife)
beliefs about fortune and misfortune (superstition and religion)

binary cognitive distinctions (good and evil, moral and immoral)
biological mother and social mother normally the same person (elemental in inclusive fitness and kin altruism)
childhood fear of strangers (basis of xenophobia)
classification of behavioral propensities (basis of judging moral and immoral behaviors)
classification of inner states (basis of judging moral and immoral traits)
classification of kin (basis of inclusive fitness and kin altruism)
coalitions (foundation of social and group morality)
collective identities (basis of xenophobia, group selection)
conflict (creates moral problems)
conflict, consultation to deal with (resolution of moral problems)
conflict, means of dealing with (resolution of moral problems)
conflict, mediation of (foundation of much of moral behavior)
conjectural reasoning (necessary for moral judgment)
continua (ordering as cognitive pattern)
cooperation (part of altruism)
cooperative labor (part of kin, reciprocal, and blind altruism)
copulation normally conducted in privacy (moral presumption of exclusivity, unique relationship)
corporate (perpetual) statuses (moral ranking of groups)
coyness display (courtship, moral manipulation)
crying (sometimes expression of grief, moral pain)
customary greetings (part of conflict prevention and resolution)
dance (affiliated with many religious ceremonies)
death rituals (awareness of mortality behind many religious beliefs)
decision making (foundation of moral judgment and resolution of moral dilemmas)
decision making, collective (foundation of group moral judgment and resolution of group moral dilemmas)
differential valuations (necessary for moral judgment)
discrepancies between speech, thought, and action (moral intention v. behavior)
distinguishing right and wrong (foundation for all moral judgment and ethical systems)
divination (element in many religious ceremonies)
division of labor by age (form of status and hierarchy)
division of labor by sex (form of status and hierarchy)

dominance/submission (foundation of hierarchical social primate species)
dream interpretation (part of some religious shamanism)
economic inequalities, consciousness of (involved in status and hierarchy disputes)
emotions (necessary for moral sense)
empathy (necessary for moral sense)
entification (treating patterns and relations as things) (makes social moral relations and problems real)
envy (moral trait)
envy, symbolic means of coping with (dealing with moral trait)
ethnocentrism (xenophobia, group selection, war)
etiquette (enhances social relations)
facial communication (necessary for social relations)
facial expression of anger (communication of moral approval/disapproval)
facial expression of contempt (communication of moral approval/disapproval)
facial expression of disgust (communication of moral approval/disapproval)
facial expression of fear (communication of moral approval/disapproval)
facial expression of happiness (communication of moral approval/disapproval)
facial expression of sadness (communication of moral approval/disapproval)
facial expression of surprise (communication of moral approval/disapproval)
facial expressions, masking/modifying of (communication of moral approval/disapproval)
fairness (equity), concept of (foundation of social/moral justice)
family (or household) (the most basic social and moral unit)
father and mother, separate kin terms for (element in awareness of this social/moral unit)
fear of death (foundation of many religious customs and beliefs)
fears (generate much religious and moral behavior)
feasting (part of many religious rituals)
females do more direct childcare (division of labor in social hierarchical species)

figurative speech (symbolic communication necessary for moral reasoning)
folklore (part of gossip)
food sharing (form of cooperation and altruism)
future, attempts to predict (necessary for moral judgment)
generosity admired (reward for cooperative and altruistic behavior)
gestures (signs of recognition of others, conciliatory behavior)
gift giving (reward for cooperative and altruistic behavior)
good and bad distinguished (necessary for moral judgment)
gossip (necessary for monitoring social relations and to assess moral value)
government (social morality)
group living (social morality)
groups that are not based on family (necessary for higher moral reasoning and blind altruism)
healing the sick (or attempting to) (religious ritual)
hope (higher moral reasoning)
hospitality (enhances social relations)
identity, collective (necessary for group moral relations/xenophobia)
imagery (necessary for symbolic moral reasoning and judgment)
incest between mother and son unthinkable or tabooed (obvious evolutionary moral trait)
incest, prevention or avoidance (obvious evolutionary moral trait)
in-group, biases in favor of (necessary for group moral relations/xenophobia)
in-group distinguished from out-group(s) (necessary for group moral relations/xenophobia)
inheritance rules (reduces conflict within families and communities)
institutions (organized coactivities) (religion)
insulting (communication of moral approval/disapproval)
intention (part of moral reasoning and judgment)
interpolation (part of moral reasoning and judgment)
interpreting behavior (necessary for moral judgment)
judging others (foundation of moral approval/disapproval)
kin, close distinguished from distant (foundation of kin selection and kin altruism)
kin groups (foundation of kin selection/altruism and basic social group)
kin terms translatable by basic relations of procreation (foundation of kin selection and kin altruism)

kinship statuses (foundation of kin selection and kin altruism)
language employed to manipulate others (communication of moral approval/disapproval)
language employed to misinform or mislead (communication of moral approval/disapproval)
language not a simple reflection of reality (symbolic moral reasoning)
law (rights and obligations) (foundation of social harmony)
law (rules of membership) (foundation of social harmony)
likes and dislikes (foundation of moral judgment)
logical notion of "equivalent" (symbolic moral reasoning)
logical notion of "general/particular" (symbolic moral reasoning)
logical notion of "not" (symbolic moral reasoning)
logical notion of "opposite" (symbolic moral reasoning)
logical notion of "part/whole" (symbolic moral reasoning)
logical notions (symbolic moral reasoning)
magic (religion and superstitious behavior)
magic to increase life (religion and superstitious behavior)
magic to sustain life (religion and superstitious behavior)
magic to win love (moral manipulation)
making comparisons (necessary for moral judgments)
male and female and adult and child seen as having different natures (differences in moral behavior)
males dominate public/political realm (gender differences in moral behavior)
males engage in more coalitional violence (gender differences in moral behavior)
males more aggressive (gender differences in moral behavior)
males more prone to lethal violence (gender differences in moral behavior)
males more prone to theft (gender differences in moral behavior)
manipulate social relations (social moral control)
marriage (moral rules of foundational relationship)
mental maps (symbolic moral reasoning)
metaphor (symbolic moral reasoning)
mood- or consciousness-altering techniques and/or substances (religious ritual)
moral sentiments (the foundation of all morality)
moral sentiments, limited effective range of (parameters of moral foundation)

mother normally has consort during child-rearing years (gender differences in moral behavior)
mourning (expression of grief part of symbolic moral reasoning)
murder proscribed (moral judgment necessary in communities)
music related in part to religious activity (religious rituals)
myths (foundation of both religion and teaching of moral principles)
narrative (foundation of myths)
normal distinguished from abnormal states (necessary for moral judgment)
Oedipus complex (kin selection, moral assessment of appropriate/inappropriate relations)
oligarchy (de facto) (group social control)
past/present/future (necessary for symbolic moral reasoning)
person, concept of (foundation for moral judgment)
planning for future (foundation for moral judgment)
preference for own children and close kin (nepotism)
prestige inequalities (leads to moral conflict)
pride (a moral sense)
private inner life (foundation for moral reasoning and judgment)
promise (moral relations)
property (foundation of moral reasoning and judgment)
proverbs, sayings (teaching moral principles)
proverbs, sayings in mutually contradictory forms (teaching moral principles)
psychological defense mechanisms (psychology of moral sense and assessment)
rape (behavior leading to relational conflict)
rape proscribed (behavior morally judged)
reciprocal exchanges (of labor, goods, or services) (reciprocal altruism)
reciprocity, negative (revenge, retaliation) (reduces reciprocal altruism)
reciprocity, positive (enhances reciprocal altruism)
redress of wrongs (moral conflict resolution)
resistance to abuse of power, to dominance (moral conflict resolution)
rites of passage (religious ritual, foundation of moral assessment and judgment)
rituals (religion and morality)
sanctions (social moral control)
sanctions for crimes against the collectivity (social moral control)

sanctions include removal from the social unit (social moral control)
self distinguished from other self as neither wholly passive nor wholly autonomous
self is responsible (foundation of moral reasoning)
self-control (moral assessment and judgment)
self-image, awareness of (concern for what others think) (foundation of moral reasoning)
self-image, manipulation of (foundation of moral reasoning)
self-image, wanted to be positive (foundation of moral reasoning)
sex statuses (social hierarchy)
sexual attraction (foundation of major moral relations and tensions)
sexual attractiveness (foundation of major moral relations and tensions)
sexual jealousy (foundation of major moral relations and tensions)
sexual modesty (foundation of major moral relations and tensions)
sexual regulation (foundation of major moral relations and tensions)
sexual regulation includes incest prevention (foundation of major moral relations and tensions)
sexuality as focus of interest (foundation of major moral relations and tensions)
shame (moral sense)
snakes, wariness around (foundation of religious ritual and mythology)
social structure (foundation of morality in humans)
socialization (foundation of morality in humans)
socialization expected from senior kin (foundation of morality in humans)
statuses and roles (foundation of morality in humans)
statuses, ascribed and achieved (foundation of morality in humans)
statuses distinguished from individuals (foundation of morality in humans)
statuses on other than sex, age, or kinship bases (foundation of morality in humans)
stinginess, disapproval of (social and moral control mechanism)
succession (foundation of social hierarchy)
symbolic speech (foundation of moral reasoning and communication)
symbolism (foundation of moral reasoning)
tabooed foods (element in moral and social control)
tabooed utterances (communication of moral and social control)

taboos (moral and social control)
territoriality (social control)
time (religion, moral reasoning)
time, cyclicity of (religion, moral reasoning)
trade (part of social relations)
triangular awareness (assessing relationships among the self and two other people) (foundation of moral judgment)
true and false distinguished (necessary for moral assessment and judgment)
turn taking (conflict prevention)
violence, some forms of proscribed (moral and social control)
visiting (social relations)
weapons (conflict resolution)
worldview (foundation of all religion and morality)

NOTES

Prologue: One Long Argument

1. Cited in Timothy Ferris, *Red Limit* (New York: Random House, 1996).

2. I went so far as to elevate Darwin's observation to a dictum in my first column in my monthly series in *Scientific American,* April 2001, entitled "Colorful Pebbles and Darwin's Dictum."

3. Michael Shermer, *Why People Believe Weird Things: Pseudoscience, Superstitions, and Other Confusions of Our Time* (New York: W. H. Freeman, 1997; 2nd ed., New York: Henry Holt/Owl Books, 2002).

4. Michael Shermer, *How We Believe: Science, Skepticism, and the Search for God* (New York: W. H. Freeman, 1999; 2nd ed., New York: Henry Holt/Owl Books, 2003).

5. Thomas H. Huxley, *Evolution and Ethics* (New York: D. Appleton and Co., 1894), p. 238.

6. R. Sloan, E. Bagiella, and T. Powell, *The Lancet,* vol. 353 (February 20, 1999), pp. 664–67. See also: *Skeptic,* vol. 7, no. 1, p. 8. Article reviews recent studies on prayer and healing, noting a number of serious methodological flaws in the studies, including: (1) Lack of control of intervening variables (for example, most of these studies did not control for age, sex, education, ethnicity, socioeconomic status, marital status, and degree of religiosity or religious devotion, all of which can influence outcomes); (2) Failure to control for multiple comparisons (for example, in one study twenty-nine outcome variables were measured but only six were significantly altered by prayer, and in other studies different outcome variables were found to be significant, so there was no consistency across studies).

1: *Transcendent Morality*

1. Richard Dawkins, *River Out of Eden* (New York: Basic Books, 1995).

2. Charles Darwin, *The Origin of Species* (London: John Murray, 1859), p. 371.

3. See Paul Edwards, "Socrates," *Encyclopedia of Philosophy* (New York: Macmillan, 1967), 7:482.

4. Ibid.

5. E. O. Wilson, *Consilience: The Unity of Knowledge* (New York: Knopf, 1998), pp. 238–65. For a concise summary, see also E. O. Wilson, "The Biological Basis of Morality," *Atlantic Monthly,* vol. 281, no. 4 (1998), pp. 53–70.

6. This section was originally composed as an exercise in preparation for media interviews during the book tour planned for the publication of this book. These tours are arduous voyages across the cultural landscape of information dissemination. Early morning radio and television talk shows are followed by print interviews during lunch, afternoon signings at retail bookstores, a talk show or two tossed into the schedule in the late afternoon, followed by an evening lecture and signing, then a mad dash to the airport to get to the next city in order to do it all again the next day. It is déjà vu all over again until all cities come to look alike. Lights out is after midnight with a wake-up call before five the next morning. I have embarked on two such voyages, so I have learned to do what politicians on the campaign trail call "staying on message." What they mean is that no matter what question is asked or how many distractions are encountered, deliver the message you brought—quickly and succinctly. To an author in the modern world of sound-bite briefness, this entails distilling hundreds of pages of text into a handful of bulleted points and condensing thousands of hours of research into a five-minute summation. The exercise, however, is not a fruitless one because it forces the author to think through the theory carefully and sort out what is necessary and what is superfluous. In the end both author and reader profit.

7. Michael Novak, "Skeptical Inquirer," review of *How We Believe,* by Michael Shermer, *Washington Post,* February 13, 2000, p. 7.

2. *Why We Are Moral*

1. Barry H. Lopez, *Of Wolves and Men* (New York: Scribner's, 1978).

2. Gerald S. Wilkinson, "Reciprocal Food Sharing in the Vampire Bat," *Nature,* vol. 308 (1984), pp. 181–84. See also Lisa K. DeNault and Donald A. McFarlane, "Reciprocal Altruism Between Male Vampire Bats, *Desmodus rotundus,*" *Animal Behaviour,* vol. 49 (1995), pp. 855–56.

3. Frans B. de Waal, *Good Natured: The Origins of Right and Wrong in Humans and Other Animals* (Cambridge, Mass.: Harvard University Press, 1996), pp. 60–61; see also Frans B. de Waal, *Chimpanzee Politics: Power and Sex Among Apes* (Baltimore, Md.: Johns Hopkins University Press, 1989).

4. De Waal, *Good Natured,* p. 77.

5. In Michael Shermer, "The Pundit of Primate Politics: An Interview with Frans de Waal," *Skeptic,* vol. 8, no. 2 (2000), pp. 29–35.

6. Cited in de Waal, *Good Natured,* p. 42.

7. Cynthia Moss, *Elephant Memories: Thirteen Years in the Life of an Elephant Family* (New York: Fawcett Columbine, 1988).

8. Two outstanding sources of both include Christopher Boehm, *Hierarchy in the Forest: The Evolution of Egalitarian Behavior* (Cambridge, Mass.: Harvard University Press, 1999); Matt Ridley, *The Origins of Virtue: Human Instincts and the Evolution of Cooperation* (New York: Viking, 1997).

9. Jared Diamond, *Guns, Germs, and Steel: The Fates of Human Societies* (New York: W. W. Norton, 1997).

10. Jared Diamond, "The Religious Story," review of *Darwin's Cathedral,* by David Sloan Wilson, *New York Review of Books,* November 7, 2002.

11. Richard Leakey and Roger Lewin, *Origins Reconsidered: In Search of What Makes Us Human* (New York: Doubleday, 1992); Alison Jolly, *Lucy's Legacy: Sex and Intelligence in Human Evolution* (Cambridge, Mass.: Harvard University Press, 1999); Frans B. de Waal, ed., *Tree of Origin: What Primate Behavior Can Tell Us About Human Social Evolution* (Cambridge, Mass.: Harvard University Press, 2001); Steven A. LeBlanc, *Constant Battles: The Myth of the Peaceful, Noble Savage* (New York: St. Martin's Press, 2003).

12. Neil Roberts, *The Holocene: An Environmental History* (Oxford: Basil Blackwell, 1989).

13. For a more in-depth analysis of this relationship between politics and religion, see Jared Diamond, "The Religious Story," and Wilson, "The Biological Basis of Morality."

14. Rodney Stark and William Sims Bainbridge, *A Theory of Religion* (New Brunswick, N.J.: Rutgers University Press, 1987); Stark and Bainbridge, *Religion, Deviance, and Social Control* (New York: Routledge, 1997).

15. Diamond, "The Religious Story." See also the summary and analysis of hunter-gatherer social behavior and the propensity for violence against out-group members in all primate species, including and especially our hominid ancestors, in LeBlanc, *Constant Battles.*

16. For an evolutionary interpretation of the Old Testament and Talmud, see John Hartung, "Love Thy Neighbor: The Evolution of In-Group Morality," *Skeptic,* vol. 3, no. 4 (1995), pp. 86–99. Although I am using these examples to support a group selection model of evolution, Hartung rejects group selection; his interpretation is based strictly on individual, organismal selection.

17. Georges R. Tamarin, "The Influence of Ethnic and Religious Prejudice on Moral Judgment," *New Outlook,* vol. 9, no. 1 (1966), pp. 49–58, and Tamarin, *The Israeli Dilemma: Essays on a Warfare State* (Rotterdam, Netherlands: Rotterdam University Press, 1973).

18. It should be noted that many rabbis reject this in-group evolutionary interpretation of the Old Testament and the Talmud. They point to passages such as Deut. 10:19, in which God commands: "Love the sojourner therefore; for you were sojourners in the land of Egypt." And Exod. 22:21: "You shall not wrong a stranger or oppress him, for you were strangers in the land of Egypt." In *Skeptic* (vol. 4, no. 1, pp. 24–31), John Hartung and Rabbi Israel Chait, of Yeshiva B'Nei Torah in Far Rockaway, New York, exchanged views, with the latter defending Judaism as an inclusive and tolerant religion; the many in-group moral values found in the Old Testament, particularly in the Torah (the five books of Moses), are rare, taken out of context, or mistranslated from the original Hebrew, he argues. Neighbors, he says, also refers to non-Israelites. Hartung counters, however, that the original language of the Bible, even with these caveats, still makes a sharp distinction between those people who are in the group and those people who are out of the group. When the moral commandment is given to treat "strangers," "sojourners," and "sojourning strangers" with kindness, the references are to Israelites who were guests in another Israelite tribe, not non-Israelites who were in most circumstances to be dealt with rather harshly.

19. Aldous Huxley, *The Olive Tree* (New York: Harper & Bros., 1937).

20. Robert L. Bettinger, *Hunter-Gatherers: Archaeological and Evolutionary Theory* (New York: Plenum Press, 1991), p. 158.

21. Napoleon A. Chagnon. *Yąnomamö,* 4th ed. (New York: HBJ, 1992), pp. 80–86.

22. Robin Dunbar, *Grooming, Gossip and the Evolution of Language* (Cambridge, Mass.: Harvard University Press, 1996), pp. 61–79.

23. Dunbar, *Grooming, Gossip and the Evolution of Language,* p. 77.

24. Jerome H. Barkow, "Beneath New Culture is Old Psychology: Gossip and Social Stratification," in *The Adapted Mind,* ed. Jerome H. Barkow, Leda Cosmides, John Tooby (New York: Oxford University Press, 1992), pp. 627–28.

25. Personal correspondence, May 2000. There exists an impressive literature on the study of gossip. Here are a few references provided to me by Kari Konkola: Robin Dunbar, "The Chattering Classes (Have you heard the latest? Gossip, it seems, is what separates us from the animals. Far from being idle, it is vital for our safety—and even our survival)," *Times London Magazine,* February 5, 1994, pp. 28–29. M. Gluckman, "Gossip and Scandal," *Current Anthropology,* vol. 4 (1967), pp. 307–16. S. A. Hellweg, "Organizational Grapevines," in *Progress in Communication Sciences,* vol. 8, ed. Brenda Dervin and Melvin J. Voigt (Norwood, N.J.: Ablex, 1987). William A. Henry, III, "Pssst . . . Did You Hear About?" *Time,* vol. 135, no. 10 (March 5, 1990), p. 46. Blythe Holbrooke, *Gossip: How to Get It Before It Gets You and Other Suggestions for Social Survival* (New York: St. Martin's Press, 1983). Nicholas Lemann, "Gossip, The Inside Scoop: A History of American

Hearsay," *The New Republic* (November 5, 1990). Jack Levin and Arnold Arluke, *Gossip: The Inside Scoop* (New York: Plenum Press, 1987). S. E. Merry, "Rethinking Gossip and Scandal," in *Toward a General Theory of Social Control,* vol. 1: *Fundamentals,* ed. Donald Black (New York: Academic Press, 1984), pp. 271–302. L. Morrow, "The Morals of Gossip," *The Economist* (October 26, 1991), p. 64. Ralph L. Rosnow and G. A. Fine, *Rumor and Gossip: The Social Psychology of Hearsay* (New York: Elesevier Scientific, 1976). Tamotsu Shibutani, *Improvised News: A Sociological Study of Rumor* (Indianapolis: Bobbs-Merrill, 1966). Patricia Meyer Spacks, *Gossip* (Chicago: University of Chicago Press, 1986). Charles K. West, *The Social and Psychological Distortion of Information* (Chicago: Nelson-Hall, 1981).

26. I defend this statement in greater depth in my book *How We Believe* (see chapters 4 and 7 in particular).

27. E. O. Wilson, *On Human Nature* (Cambridge, Mass.: Harvard University Press, 1978).

28. Richard D. Alexander, *Darwinism and Human Affairs* (Seattle: University of Washington Press, 1979), and Alexander, *The Biology of Moral Systems* (New York: Aldine De Gruyter, 1987). F. Miele, "The (Im)moral Animal: A Quick and Dirty Guide to Evolutionary Psychology and the Nature of Human Nature," *Skeptic,* vol. 4, no. 1, (1996), pp. 42–49.

29. E. O. Wilson, *Biophilia* (Cambridge, Mass.: Harvard University Press, 1984).

30. Charles Darwin, *The Descent of Man, and Selection in Relation to Sex,* 2 vols. (London: John Murray, 1871), 1:163.

31. Ibid., p. 166. Although this is indeed what we would today call group selection, we must be careful not to hold up Darwin as an unalloyed champion of group selection. He turned to group selection only as a last resort after his attempts to explicate morality through individual selection had failed. Even as he apparently argued for group selection, Darwin proposed what we today would call *reciprocal altruism,* which is fully explained through individual selection: "As the reasoning powers and foresight of the members became improved, each man would soon learn from experience that if he aided his fellow-men, he would commonly receive aid in return. From this low motive he might acquire the habit of aiding his fellows" (*The Descent of Man,* 1:163).

If anyone had the motivation to employ Darwin in the service of defending group selection, it was Stephen Jay Gould, yet it was Gould who discovered Darwin's reluctance to be a group selectionist because "in permitting a true exception to organismal selection, Darwin's primary attitude exudes extreme reluctance—restriction to minimal groupiness, provision of other explanations in the ordinary organismal mode, limitation to a unique circumstance in a single species (human consciousness for the spread of an idea against the force of organismal selection), and placement within a more general argument for sexual selection, the strongest form of the orthodox mode." In an interesting deduction from this historical nugget, however, Gould argued that this bodes well for group selection: "The recognition that Darwin, despite such strong reluctance, could not avoid some role for species selection, builds a strong historical argument for the ineluctability of a hierarchical theory of selection." (Stephen Jay Gould, *The Structure of Evolutionary Theory* [Cambridge, Mass.: Harvard University Press, 2002], pp. 135–36.)

32. Michael T. Ghiselin, *The Economy of Nature and the Evolution of Sex* (Berkeley: University of California Press, 1974), p. 247.

33. Elliott Sober and David Sloan Wilson, *Unto Others: The Evolution and Psychology of Unselfish Behavior* (Cambridge, Mass.: Harvard University Press, 1998), pp. 27, 92.

34. W. D. Hamilton, "Innate Social Aptitudes of Man: An Approach From Evolutionary Genetics," in *Biosocial Anthropology,* ed. Robin Fox (New York: John Wiley and Sons, 1975), pp. 133–35.

35. David Sloan Wilson, "Nonzero and Nonsense: Group Selection, Nonzerosumness, and the Human Gaia Hypothesis," *Skeptic,* vol. 8, no. 1 (2000), p. 85. See also his elaboration on this model in David Sloan Wilson, *Darwin's Cathedral: Evolution, Religion, and the Nature of Society* (Chicago: University of Chicago Press, 2002).

For a critique of Wilson's theory, see Michael Ruse, "Can Selection Explain the Presbyterians?" *Science,* vol. 297 (August 30, 2002), p. 1479, in which he answers his title question thusly: "I want hard figures on birth patterns before and after Calvin, and I want to know who had kids and who did not. I want these figures correlated with religious practice and belief. Then and only then will I start to feel comfortable."

I asked MIT linguist and evolutionary psychologist Steven Pinker's opinion of group selection theory: "I must admit to being skeptical about D. S. Wilson's new group selection. For one thing, in many cases Wilson seems to bend himself into a pretzel to redescribe gene selection as a kind of group selection (as when he calls an ant colony a group, or a pair of reciprocal altruists as a group). In those cases he's not wrong, but he doesn't add anything; the predictions are pretty much the same, and he just seems on a crusade to revive the concept. In other cases (such as individual sacrifice to benefit the group) it just doesn't seem to fit terribly well with the psychology. If humans were group-selected, life would be very different—we wouldn't have emotions directed at other individuals like anger, guilt, and jealousy; we wouldn't have affairs, compete for status, hoard when we can get away with it, lie and cheat, etc. Sure, people risk their lives for the group in warfare, but generally when the risk is probabilistic (so they're not sure to die), when they are rewarded with status (or in the old days, women), or when they are suckered by promises of 72 virgins in an afterlife." (Personal correspondence, September 27, 2002.)

36. Wilson, *Darwin's Cathedral,* p. 1.

37. Ibid., p. 21.

38. Signe Howell, *Society and Cosmos: Chewong of Peninsular Malaya* (Singapore: Oxford University Press, 1984).

39. Paul Ekman, *Telling Lies: Clues to Deceit in the Marketplace, Marriage, and Politics* (New York: W. W. Norton, 1992); Ekman, *Emotions Revealed: Recognizing Faces and Feelings to Improve Communication and Emotional Life* (New York: Times Books, 2003).

40. There is a sizable body of literature on game theory and cooperation. Here are several excellent resources:

Robert Axelrod, *The Evolution of Cooperation* (New York: Basic Books, 1984). Robert Axelrod and W. D. Hamilton, "The Evolution of Cooperation," *Science,* vol. 211 (1981), pp. 1390–96. Robert H. Frank, *Passions Within Reason: The Strategic Role of the Emotions* (New York: W. W. Norton, 1988). Douglas R. Hofstadter, "Metamagical Themas: Computer Tournaments of the Prisoner's Dilemma Suggest How Cooperation Evolves," *Scientific American,* vol. 248, no. 5 (1983), pp. 16–26. John Keith Murnighan, *Bargaining Games* (New York: William Morrow and Co., 1992). Matt Ridley, *The Origins of Virtue* (New York: Viking, 1996). Michael Taylor, *The Possibility of Cooperation* (Cambridge: Cambridge University Press, 1987). Robert L. Trivers, "The Evolution of Reciprocal Altruism," *Quarterly Review of Biology,* vol. 46 (1971), pp. 35–57. John Von Neumann and Oskar Morgenstern, *Theory of Games and Economic Behavior* (Princeton: Princeton University Press, 1980). On the Internet: Google search "prisoner's dilemma" will lead to thousands of sites, computer simulations, chat rooms, discussions, bibliographies, and so on, such as http://pespmc1.vub.ac.be/PRISDIL.html.

41. David Hume, *An Enquiry Concerning the Principles of Morals* (1751) p. xi.

42. MIT *Encyclopedia of Cognitive Science* (http://cognet.mit.edu/MITECS/Entry/brown).

43. Donald E. Brown, *Human Universals* (New York: McGraw-Hill, 1991), p. 142.

44. Ibid., p. 143.

3. *Why We Are Immoral*

1. Aaron Sorkin, *A Few Good Men* (New York: Samuel French, June 1990).

2. See the *Internet Encyclopedia of Philosophy,* online at http://www.utm.edu/research/iep/e/epicur.htm, or the *Aphorisms of Epicurus,* online at http://www.ag.wastholm.net/author/Epicurus.

The defense of God's goodness in spite of the obvious existence of evil is called theodicy, a word coined by German mathematician Gottfried Leibniz. Theodicy became an especially contentious philosophical issue after a November 1755 earthquake leveled the city of Lisbon. In response, the atheist French philosopher Voltaire wrote a bitter poem inquiring of Christians, where was God in Lisbon? Jean-Jacques Rousseau penned a sharp rejoinder to Voltaire, arguing that so-called natural disasters like earthquakes were not acts of God, but natural acts that affected people in unnatural ways because we live in unnatural conditions, like crowded cities.

3. Alexander Pope, *An Essay on Man* (1733–34), ed. by Frank Brady (Indianapolis: Bobbs-Merrill, 1997). Also available online at http://www.library.utoronto.ca/utel/rp/poems/pope10.html.

4. Mary Neubauer, "For McCaugheys, a Day to Be Thankful in Many Ways," *Athens Banner-Herald,* November 28, 1997. Online at http://www.athensnewspapers.com/1997/112897/1128.a3septuplets.html.

5. See my essay "Only God Can Do That?" in Michael Shermer, *The Borderlands of Science* (New York: Oxford University Press, 1997), pp. 66–79.

6. Misty Bernall, *She Said Yes: The Unlikely Martyrdom of Cassie Bernall* (New York: Plough Publishing House, 1999). There is some dispute about whether Cassie Bernall actually said "yes" to the question about her belief. See http://www.geocities.com/me2kangaru/CassiesYes.html for a dissenting view.

7. *Random House College Dictionary* usages of evil include, as adjectives, "morally wrong; immoral; wicked: characterized or accompanied by misfortune or suffering; unfortunate"; and as nouns, in addition to those quoted, "that which is evil; evil quality; intention, or conduct; a disease, as king's evil; in an evil manner."

8. Daniel Jonah Goldhagen, *Hitler's Willing Executioners: Ordinary Germans and the Holocaust* (New York: Alfred A. Knopf, 1996), p. 14.

9. Max Frankel, "Willing Executioners?" *New York Times Book Review,* August 9, 1998.

10. Stanley Milgram, *Obedience to Authority: An Experimental View* (New York: Harper, 1969). See also Hans Askenasy, *Are We All Nazis?* (Secaucus, N.J.: Lyle Stuart, 1978).

11. Philip Zimbardo, "The Human Choice: Individuation, Reason, and Order Versus Deindividuation, Impulse, and Chaos," in *Nebraska Symposium on Motivation, 1969,* ed. W. J. Arnold and D. Levine (Lincoln: University of Nebraska Press, 1970). See also Philip Zimbardo, *The Psychology of Attitude Change and Social Influence* (New York: McGraw-Hill, 1991).

12. In Gustave M. Gilbert, *Nuremberg Diary* (New York: Farrar, Straus, and Co., 1947).

13. Richard Breitman, *The Architect of Genocide: Himmler and the Final Solution* (New York: Knopf, 1991), p. 116.

14. Carol Tavris and C. Wade, *Psychology in Perspective,* 2nd ed. (New York: Longman/Addison Wesley, 1995), p. 332.

15. Cited in Y. Büchler, "Document: A Preparatory Document for the Wannsee 'Conference,'" *Holocaust and Genocide Studies,* vol. 9, no. 1, (spring 1995), pp. 121–29.

16. I first coined the phrase "the evil of banality" in Michael Shermer and A. Grobman, *Denying History: Who Says the Holocaust Never Happened and Why Do They Say It?* (Berkeley: University of California Press, 1999).

17. Ron Rosenbaum, *Explaining Hitler: The Search for the Origins of His Evil* (New York: Random House, 1998), p. xxv.

18. Ibid., p. xxi.

19. Ibid., p. xii.

20. C. Lanzmann, "The Obscenity of Understanding: An Evening with Claude Lanzmann," *American Imago,* vol. 48, no. 4, (1991), pp. 473–95.

21. Roy F. Baumeister, *Evil: Inside Human Violence and Cruelty* (New York: W. H. Freeman, 1997).

22. Quoted in Michael Berenbaum, *The World Must Know: The History of the Holocaust as Told in the United States Holocaust Memorial Museum* (New York: Little, Brown, 1993), p. 204.

23. Baumeister, *Evil,* p. 14.

24. Ibid., pp. 73–74.

25. Ibid., pp. 95–96.

26. Bart Kosko, *Fuzzy Engineering* (Englewood Cliffs, N.J.: Prentice Hall, 1992).

27. Armchair philosophical ponderings like the blueness of the sky are not the only applications of fuzzy logic. There are now a host of fuzzy machines, such as a fuzzy washing machine that assigns a fractional number to the "dirtiness" of the water (say .8 dirty), then injects the appropriate amount of soap. Or a fuzzy video camera that automatically adjusts the shutter opening to the amount of light, or a "stabilization" device that adjusts to the motion of the video camera operator. Cruise missiles are fuzzy weapons that adjust their speed and trajectory according to the changing terrain that is compared to an onboard map. Japanese firms have rejected Aristotle and committed themselves to fuzzy logic, designing fuzzy fuel injectors, fuzzy film, fuzzy computers, fuzzy automatic transmissions, fuzzy dishwashers, fuzzy dryers, fuzzy copy machines, and even a fuzzy golf swing diagnostic system.

28. Bart Kosko, *Fuzzy Thinking: The New Science of Fuzzy Logic* (New York: Hyperion, 1993), p. 250.

29. Carol Tavris, "All Bad or All Good? Neither," *Los Angeles Times,* September 20, 1998, p. M5.

30. Lawrence Kohlberg, "Moral Stages and Moralization: The Cognitive-Developmental Approach," in *Moral Development and Behavior,* ed. T. Lickona (New York: Holt, Rinehart, and Winston, 1976), and Kohlberg, *Essays on Moral Development,* vol. 2: *The Psychology of Moral Development: The Nature and Validity of Moral Stages* (San Francisco: Harper and Row, 1984).

31. Kurt Bergling, *Moral Development: The Validity of Kohlberg's Theory* (Stockholm: Almqvist & Wiksell, 1981).

32. R. L. Gorsuch and S. McFarland, "Single vs. Multiple-item Scales for Measuring Religious Values," *Journal for the Scientific Study of Religion,* vol. 13 (1972), pp. 281–307. S. Selig and G. Teller, "The Moral Development of Children in Three Different School Settings," *Religious Education,* vol. 70 (1975), pp. 406–15.

33. L. Nucci and E. Turiel, "God's Word, Religious Rules, and Their Relation to Christian and Jewish Children's Concepts of Morality," *Child Development,* vol. 64 (1993), pp. 1475–91. D. Ernsberger and G. Manaster, "Moral Development, Intrinsic/Extrinsic Religious Orientation and Denominational Teachings," *Genetic Psychology Monographs,* vol. 104 (1981), pp. 23–41.

34. S. Sanderson, *Religion, Politics, and Morality: An Approach to Religion and Political Belief Systems and Their Relation Through Kohlberg's Cognitive-Developmental Theory of Moral Judgment* (Lincoln: University of Nebraska Dissertation Abstracts International 346259B, 1974).

35. Carol Gilligan, *In a Different Voice: Psychological Theory and Women's Development* (Cambridge, Mass.: Harvard University Press, 1982). For a literature review of Kohlberg's theory and his critics, see J. M. Darley and T. R. Shultz, "Moral Rules: Their Content and Acquisition," *Annual Review of Psychology,* vol. 41 (1990), pp. 525–56.

36. Tavris, "All Bad or All Good? Neither."

37. Told to *Skeptic* senior editor Frank Miele, who was in attendance. Personal correspondence with Frank Miele, November 20, 2000. My interview with Patrick Tierney was on Monday, November 20, 2000, for the science edition of NPR affiliate KPCC's *Airtalk.*

38. Patrick Tierney, *Darkness in El Dorado: How Scientists and Journalists Devastated the Amazon* (New York: W. W. Norton, 2000), p. 15. Kenneth Good, *Into the Heart: One Man's Pursuit of Love and Knowledge Among the Yanomami* (New York: HarperCollins, 1991). Jacques Lizot, *Tales of the Yanomami: Daily Life in the Venezuelan Forest* (New York: Cambridge University Press, 1985).

39. Tierney, *Darkness in El Dorado,* pp. 132–33.

40. Margot Roosevelt, "Yanomami: What Have We Done to Them?" *Time,* vol. 156, no. 14 (October 2, 2000), pp. 77–78.

41. Since evolutionary psychologist Steven Pinker published a letter in defense of Chagnon in the *New York Times Book Review* (in response to John Horgan's surprisingly

uncritical review of Tierney's book there), I queried him about some of the specific charges. "The idea that Chagnon caused the Yanomamö to fight is preposterous and contradicted by every account of the Yanomamö and other non-state societies. Tierney is a zealot and a character assassin, and all his serious claims crumble upon scrutiny." What about the charge of ethical breaches? "There are, of course, serious issues about ethics in ethnography, and I don't doubt that some of Chagnon's practices, especially in the 1960s, were questionable (as were the practices in most fields, such as my own—for example, the Milgram studies). But the idea that the problems of Native Americans are caused by anthropologists is crazy. In the issues that matter to us—skepticism, scientific objectivity, classic liberalism, etc.—Chagnon is on the right side." Personal correspondence with Steven Pinker, December 1, 2000.

42. Derek Freeman, "Paradigms in Collision: Margaret Mead's Mistake and What It Has Done to Anthropology," *Skeptic,* vol. 5, no. 3 (1997), pp. 66–73.

43. Ibid. Freeman added this comment about the politics of science: "About this extraordinary action, which is based on the notion that the scientific status of propositions can be settled politically, Sir Karl Popper wrote me: 'Many sociologists and almost all sociologists of science believe in a relativist theory of truth. That is, truth is what the experts believe, or what the majority of the participants in a culture believe. Holding a view like this your opponents could not admit you were right. How could you be, when all their colleagues thought like they did? In fact, they could prove that you were wrong simply by taking a vote at a meeting of experts. That settled it. And your facts? They meant nothing if sufficiently many experts ignored them, or distorted them, or misinterpreted them.'" Subsequently, Freeman received a personal letter from philosopher Peter Munz, who cut to the core of what lies behind the anthropology wars: "I was so excited, when the scales fell from my eyes that I felt I had to write to you and say what a splendid book. I hope its importance will not get buried in controversies about the emotional attitudes of young Samoans. Its importance reaches far beyond this particular problem. The debate, at heart, is about evolution."

44. Napoleon Chagnon, *Yąnomamö* (New York: Harcourt Brace College Publishers, 1992), pp. xii–xiii.

45. Ibid., p. 7.

46. Ibid., p. 10.

47. In 1995 Chagnon told *Scientific American* that because male aggression is esteemed in Yąnomamö culture, aggression as a human trait is highly malleable and culturally influenced, an observation that might have been made by Stephen Jay Gould, considered by most sociobiologists to be Satan incarnate. "Steve Gould and I probably agree on a lot of things," Chagnon surprisingly concluded. (John Horgan, "The New Social Darwinists," *Scientific American* [October 1995], pp. 150–57.)

48. Napoleon Chagnon, "The Myth of the Noble Savage: Lessons From the Yanomamö People of the Amazon," paper presented at the Skeptics Society Conference on Evolutionary Psychology and Humanistic Ethics, March 30, 1996.

49. Ibid. In light of his data on warriors who are rewarded with more wives, one questioner wondered what happens to the men who get no wives, and if this means that the Yąnomamö are polygamous. Chagnon explained that, indeed, some Yąnomamö men have no wives and that it is often they who are the causes of violence as they either resort to rape or stir up trouble with men who have more than one wife. But he added an important proviso that indicates, once again, Chagnon's sensitivity to the nuances and complexities within all cultures, and the danger of gross generalizations based on binary logic: "Anthropologists tend to pigeonhole societies as monogamous or polygamous or polyandrous, as if these are three different kinds of societies. In fact, you have to look at marriage as a life-historical process in all societies. There are, for example, cases of monogamy in Yąnomamö society. In fact, monogamy is the most common type of marriage. But there are also polyandrous families where one woman marries two men, who tend to be brothers. There are, in fact, examples of all three types of marriage arrangements in Yąnomamö culture."

50. Interview with Kenneth Good, December 5, 2000. Columbia Pictures bought the rights to produce a dramatic film based on the book, and Good even received a phone call from actor Richard Gere, who was interested in playing him. That deal has since fallen through and others have shown interest in a film deal, but nothing has come of it to date.

51. Good, *Into the Heart,* p. 115.

52. Ibid., p. 116.

53. Ibid.

54. Chagnon, *Yąnomamö,* p. 1.

55. Interview with Jared Diamond, November 27, 2000.

56. Shermer, *The Borderlands of Science,* pp. 241–61.

57. B. S. Low, "Behavioral Ecology of Conservation in Traditional Societies," *Human Nature,* vol. 7, no. 4 (1996), pp. 353–79. On the Standard Cross-Cultural Sample, see G. P. Murdock and D. White, "Standard Cross-Cultural Sample," *Ethnology,* vol. 8 (1969), pp. 329–69.

58. Robert Edgerton, *Sick Societies: Challenging the Myth of Primitive Harmony* (New York: Free Press, 1992).

59. Shepard Krech, III, *The Ecological Indian: Myth and History* (New York: W. W. Norton, 1999).

60. On the overhunting hypothesis and debate, see G. S. Krantz, "Human Activities and Megafaunal Extinctions," *American Scientist,* vol. 58 (1970), pp. 164–70; P. S. Martin and R. G. Klein, eds., *Quaternary Extinctions* (Tucson: University of Arizona Press, 1984); C. A. Reed, "Extinction of Mammalian Megafauna in the Old World Late Quaternary," *Bio-Science,* vol. 20 (1970), pp. 284–88.

The alternative explanation for the mass faunal extinction—that dramatic environmental changes at the end of the last ice age killed or weakened the herds—makes no sense in the larger context. The weather got warmer, not colder, and ice ages have come and gone before without triggering such mass die-offs. Why now? Overhunting remains the best explanation.

61. Jane Goodall, *The Chimpanzees of Gombe* (Cambridge, Mass.: Harvard University Press, 1986), p. 530.

62. Lawrence H. Keeley, *War Before Civilization: The Myth of the Peaceful Savage* (New York: Oxford University Press, 1996). See also Arther Ferrill, *The Origins of War: From the Stone Age to Alexander the Great* (London: Thames and Hudson, 1988).

63. Keeley, *War Before Civilization,* pp. 64, 19, 50.

64. LeBlanc, *Constant Battles.* The list of peaceful societies is not long: Copper Eskimo, Ingalik Eskimo, the Gebusi of lowland New Guinea, the African !Kung bushmen, the Mbuti Pygmies of Central Africa, the Semang of peninsular Malaysia, the South American Siriono of Amazonia, the Yahgan of Tierra del Fuego, the Warrau of the Orinoco Delta of eastern Venezuela, and the Aborigines who lived along the west coast of Tasmania. However, LeBlanc notes that "some of these same 'peaceful' societies have extremely high homicide rates. Among the Copper Eskimo and the New Guinea Gebusi, for example, a third of all adult deaths were from homicide. . . . Which killing is considered a homicide and which killing is an act of warfare? Such questions and answers become somewhat fuzzy. So some of this so-called peacefulness is more dependent on the definition of homicide and warfare than on reality" (p. 202).

65. Ibid., p. 125.

66. Steven A. LeBlanc, *Prehistoric Warfare in the American Southwest* (Salt Lake City: University of Utah Press, 1999), p. 124.

67. LeBlanc, *Constant Battles,* pp. 224–28.

68. R. Cassels, "Faunal Extinction and Prehistoric Man in New Zealand and the Pacific Islands," in *Quaternary Extinctions,* ed. P. S. Martin and R. G. Klein (Tucson: University of Arizona Press, 1984), pp. 741–67. Alfred W. Crosby, *Germs, Seeds, and Animals: Studies in Ecological History* (London: M. E. Sharpe, 1994). Jared Diamond, *The Third Chimpanzee* (New York: HarperCollins, 1992).

69. See http://www.classicnote.com/ClassicNotes/Titles/grapeswrath/ for an excellent synopsis of the book with extensive commentary.

70. Image from http://www.planetwaves.net/pogo.html.

71. Aleksander Solzhenitsyn, *The Gulag Archipelago, 1918–1956: An Experiment in Literacy Investigation,* 3 vols. (New York: Harper & Row, 1974–78).

72. Aeschylus, *Prometheus Bound,* c. 430 B.C.E. Online at http://classics.mit.edu/Aeschylus/prometheus.html.

4. *Master of My Fate*

1. See http://www.law.umkc.edu/faculty/projects/ftrials/hinckley/hinckleytrial.html for a remarkably thorough analysis of the Hinckley case, including court documents from the trial, testimonies and depositions by Hinckley and the psychiatrists for both the defense and prosecution, and the judge's decision.

2. *Tormenta: The Execution of Robert Francois Damiens*, 1757. Available online at http://www.perno.com/european/docs/tormenta.htm.

3. C. S. Lewis, *A Grief Observed* (New York: Macmillan, 1963). See also the moving film *Shadowlands*, starring Anthony Hopkins as C. S. Lewis and Debra Winger as his wife, Joy.

4. John Milton, *Paradise Lost*, in *The Great Books of the Western World* (Chicago: University of Chicago Press, 1952).

5. René Descartes, *Principles of Philosophy* (trans. J. Veitch), Part I (London: Dent, 1649), p. 41.

6. C. S. Lewis, *Beyond Personality* (New York: Macmillan, 1945).

7. The complete story can be found online at http://www.ksu.edu/english/baker/english320/Maugham-AS.htm.

8. Pierre Simon Laplace, *A Philosophical Essay on Probabilities* (1814) (New York: Dover, 1951).

9. Pope, *An Essay on Man*.

10. See http://www.law.umkc.edu/faculty/projects/ftrials/hinckley/hinckleytrial.html. All quotes and facts discussed within the section of this chapter on the Hinckley case are from this Web page. Additional citations are included in note 11.

11. For additional information on the Hinckley case, see Lincoln Caplan, *The Insanity Defense and Trial of John W. Hinckley, Jr.* (New York: Dell Publishing Company, 1984); James W. Clarke, *On Being Mad or Merely Angry: John W. Hinckley, Jr., and Other Dangerous People* (Princeton, N.J.: Princeton University Press, 1990); Jack and Jo Ann Hinckley, *Breaking Points* (Grand Rapids, Mich.: Chosen Books, c. 1985); Peter Low, *The Trial of John W. Hinckley, Jr.* (New York: Foundation Press, 1985); Professional Educational Group, *Classics of the Courtroom: Vincent Fuller's Summation in United States v. John Hinckley* (1990); Rita J. Simon and David E. Aaronson, *The Insanity Defense: A Critical Assessment of Law and Policy in the Post-Hinckley Era* (New York: Praeger Publishers, 1988); Henry J. Steadman, *Before and After Hinckley: Evaluating Insanity Defense Reform* (New York: Guilford Press, 1993).

12. See Martin Gardner's excellent discussion of this issue in my interview of him. Michael Shermer, "The Annotated Gardner: An Interview with Martin Gardner—Founder of the Modern Skeptical Movement," *Skeptic*, vol. 4, no. 1 (1997), pp. 56–60.

13. Martin Gardner, *The Whys of a Philosophical Scrivener* (New York: William Morrow, 1983), pp. 272–75.

14. Ibid., p. 115.

15. Owen Flanagan, *The Problem of the Soul: Two Visions of Mind and How to Reconcile Them* (New York: Basic Books, 2002), pp. 126–27.

16. Murray Gell-Mann, *The Quark and the Jaguar* (New York: W. H. Freeman, 1994). The strongest case for the indeterminism argument was made by physicist Roger Penrose, *The Emperor's New Mind: Concerning Computers, Minds, and the Law of Physics* (London: Penguin, 1991). See also Roger Penrose, *Shadows of the Mind: A Search for the Missing Science of Consciousness* (Oxford: Oxford University Press, 1996).

17. Daniel C. Dennett, *Elbow Room: The Varieties of Free Will Worth Wanting* (Cambridge, Mass.: MIT Press, 1984).

18. Roy F. Baumeister and Sara R. Wotman, *Breaking Hearts: The Two Sides of Unrequited Love* (New York: Guilford Press, 1992).

19. G. Kreiman, I. Fried, and C. Koch, "Single Neuron Correlates of Subjective Vision in the Human Medial Temporal Lobe," *Proceedings of the National Academy of Sciences*, no. 99 (2002), pp. 8378–83.

20. Michael Shermer, "Demon-Haunted Brain," *Scientific American* (March 2003), p. 32.

21. Michael A. Persinger, *Neuropsychological Bases of God Beliefs* (New York: Praeger, 1987), and Persinger, "Paranormal and Religious Beliefs May Be Mediated Differently by Subcortical and Cortical Phenomenological Processes of the Temporal (Limbic) Lobes," *Perceptual and Motor Skills,* vol. 76 (1993), pp. 247–51.

22. Olaf S. Blanke, T. Ortigue, T. Landis, and M. Seeck, "Neuropsychology: Stimulating Illusory Own-body Perceptions," *Nature,* vol. 419 (September 19, 2002), pp. 269–70.

23. Andrew Newberg, Eugene D'Aquili, and Vince Rause, *Why God Won't Go Away* (New York: Ballantine Books, 2001).

24. Peggy La Cerra and Roger Bingham, *The Origin of Minds: Evolution, Uniqueness, and the New Science of the Self* (New York: Harmony Books, 2003).

25. Ibid., pp. 224–26.

26. Steven Pinker, *The Blank Slate: The Modern Denial of Human Nature* (New York: Viking, 2002), p. 175.

27. Per-Olof Astrand and Kaare Rodahl, *Textbook of Work Physiology* (New York: McGraw Hill, 1986).

28. Only half in jest I sometimes wonder if there isn't a metagene gene—a gene that causes people to think that everything is in our genes. Here's an evolutionary just-so story that critics of evolutionary psychology could have a field day with: people tend to view behavior as genetically caused because back in the Paleolithic era, those individuals who were more inclined to view behavior as genetically determined won more copulations and thus passed on their metagene genes through more offspring. Of course, Paleolithic cave persons knew nothing about genes, so we might postulate that they tended to view the actions of others as either largely capricious or largely determined. The latter would be high in metagene genes, and they, of course, would be better adapted and more successful because believing one lives in a deterministic world better allows one to determine cause and effect relationships, and that is what leads to enhanced survival and the propagation of one's genes, including one's metagenes.

29. Matt Ridley, *Genome: The Autobiography of a Species in 23 Chapters* (New York: HarperCollins, 2001).

30. Daniel C. Dennett, *Freedom Evolves* (New York: Viking. 2003).

31. Ibid., p. 238.

32. Ibid., p. 251.

33. Michael Shermer, "The Chaos of History: On a Chaotic Model That Represents the Role of Contingency and Necessity in Historical Sequences," *Nonlinear Science Today,* vol. 2, no. 4 (1993), pp. 1–13; Shermer, "Exorcising LaPlace's Demon: Chaos and Antichaos, History and Metahistory," *History and Theory,* vol. 34, no. 1 (1995), pp. 59–83; Shermer, "The Crooked Timber of History," *Complexity,* vol. 2, no. 6 (1997), pp. 23–29.

34. Frank J. Tipler, *The Physics of Immortality* (New York: Doubleday, 1994).

35. William Ernest Henley, "Invictus," in *Modern British Poetry,* ed. Louis Untermeyer (New York: Harcourt Brace, 1920). Also available online at http://www.bartleby.com/103/7.html.

5. Can We Be Good Without God?

1. Quoted in an Associated Press release, 1999. Available online at http://zanazl.tripod.com/Columbine/Articles/KipKinkel.html.

2. Janelle Brown, "Doom, Quake and Mass Murder: Gamers Search Their Souls After Discovering the Littleton Killers Were Part of Their Clan," *Salon*.com (April 23, 1999). Online at http://www.salon.com/tech/feature/1999/04/23/gamers/.

3. Julian Whitaker, "Health & Healing," 1999. Online at http://www.drwhitaker.com/wit_abouthh.php.

4. Quoted in B. A. Robinson, "Why Did the Columbine Shooting Happen?" Ontario Consultants on Religious Tolerance, 1999; updated December 3, 2001. Online at http://www.religioustolerance.org/sch_vio1.htm.

5. Quoted in an Associated Press release, 1999. Available online at http://zanazl.tripod.com/Columbine/Articles/KipKinkel.html.

6. See L. Chibbaro, Jr., "Young Gays Traumatized by Shooting," *Washington Blade,* May 7, 1999.

7. Quoted in Robinson, "Why Did the Columbine Shooting Happen?"

8. Ibid.

9. Ibid.

10. Quoted in Jacob Weisberg, "What Do You Mean by 'Violence'?" *Slate*.com (May 15, 1999). Online at http://slate.msn.com/default.aspx?id=28168.

11. The letter/speech is online at http://majoritywhip.house.gov/news.asp?formmode=SingleRelease&gcid=123.

12. Quoted in a review of the PBS series *Evolution,* in which much was made of blaming the theory for human tragedies like Columbine, by Julie Salamon, "A Stark Explanation for Mankind from an Unlikely Rebel," *New York Times,* September 24, 2001.

13. Quoted in Andrea Szalanski, "Columbine Report to Vindicate Nonbelievers," *Secular Humanist Bulletin,* vol. 16, no 2 (2000). Online at http://www.secularhumanism.org/library/shb/szalanski_16_2.htm.

14. Wendy Murray Zoba, "Church, State, and Columbine," Excerpted from *Day of Reckoning* (Brazos Press), *Christianity Today,* vol. 45, no. 5 (April 2, 2001), p. 54. Online at http://www.christianitytoday.com/ct/2001/005/3.54.html.

15. L. Stammer, "Anglican Leader Visits L.A.," *Los Angeles Times,* May 25, 1996, pp. B1–3.

16. Fyodor Dostoyevsky, *The Brothers Karamazov* (1880) (Chicago: University of Chicago Press, 1952), p. 132.

17. Ibid.

18. Ibid.

19. In John Hick, *The Existence of God* (New York: Collier Books, 1964).

20. J. Wiscombe, "'I Don't Do Therapy.' Dr. Laura Schlessinger, the Country's Top Female Radio Personality, Calls Herself a Prophet," *Los Angeles Times Magazine,* January 18, 1998, p. 11.

21. Laura Schlessinger, editorial, *The Calgary Sun,* September 9, 1997, p. 22.

22. Laura Schlessinger, *How Could You Do That?!: The Abdication of Character, Courage, and Conscience* (New York: HarperCollins, 1996), p. 9.

23. Adolf Hitler, *Mein Kampf* (1925) (New York: Houghton Mifflin Co., 1943), pp. 267–68.

24. Martin Broszat, *The Hitler State: The Foundation and Development of the Internal Structure of the Third Reich* (New York: Longman, 1981).

25. Louis L. Snyder, ed., *Hitler's Third Reich: A Documentary History* (Chicago: Nelson Hall, 1981), p. 167.

26. Ibid., p. 168.

27. D. B. Barrett, G. T. Kurian, and T. M. Johnson, eds., *World Christian Encyclopedia: A Comparative Survey of Churches and Religions in the Modern World,* 2 vols. (New York: Oxford University Press, 2001).

6. *How We Are Moral*

1. Immanuel Kant, *Fundamental Principles of The Metaphysic of Morals* (1785), translated by T. K. Abbott (Chicago: University of Chicago Press, 1952).

2. Francis A. Schaeffer, *How Then Should We Live?: The Rise and Decline of Western Thought and Culture* (New York: Fleming H. Revell Co., 1976).

3. Barrett, Kurian, and Johnson, *World Christian Encyclopedia.* (Barrett is head of the Global Evangelization Movement, making one wonder if all this data is being collected to calibrate how long it will take to reduce this rich religious diversity to one cosmo-macro-mega Christian religion.)

4. Robert Goodwin Olson, *An Introduction to Existentialism* (New York: Dover Publishers, 1962). Robert C. Solomon, ed., *Existentialism* (New York: Modern Library, Random House, 1974). See also *Existentialism: A Primer* online at www.tameri.com/csw/exist/.

5. Jeremy Bentham, *The Principles of Morals and Legislation* (1789) (New York: Macmillan, 1948), p. 80.

6. Ibid., p. 30.

7. Ibid.

8. Stephen Jay Gould, *Hen's Teeth and Horse's Toes* (New York: W. W. Norton, 1983), p. 25.

9. Bart Kosko, *Fuzzy Thinking: The New Science of Fuzzy Logic* (New York: Hyperion, 1993), p. 250.

10. Adam Smith, *An Inquiry into the Nature and Causes of the Wealth of Nations* (1776) (New York: Modern Library, Random House, 1965).

11. For example, Marilyn vos Savant was bombarded with angry letters when she revealed the correct solution in her weekly column in *Parade* magazine. Marilyn vos Savant, "Ask Marilyn," *Parade,* September 9, 1990, February 17, 1991, and July 7, 1991. You can actually play the three-door game online at http://utstat.toronto.edu/david/MH.html#1. And on other Web sites you can find computer programs that have run hundreds of thousands of simulations of the game, proving that in the long run it is better to switch doors. See also Leonard Gillman, "The Car and the Goats," *American Mathematical Monthly,* vol. 99, no. 1 (January 1992), pp. 3–7.

12. Thomas Gilovich, Robert Vallone, and Amos Tversky, "The Hot Hand in Basketball: On the Misperception of Random Sequences," *Cognitive Psychology,* vol. 17 (1985), pp. 295–314. Since we are dealing with professional basketball players instead of coins, adjustments have to be made. If a player's shooting percentage is 60 percent, for example, we would expect, by chance, that he will sink six baskets in a row once for every twenty sequences of six shots attempted.

13. Barry Glassner, *The Culture of Fear: Why Americans Are Afraid of the Wrong Things* (New York: Basic Books, 1999).

14. Ibid.

15. Ibid., p. 123.

16. Ibid., p. 167.

17. Ibid.

18. Stardate: 1672.1 Earthdate: October 6, 1966. *Star Trek,* episode 5, "The Enemy Within," produced by Gene Roddenberry, directed by Leo Penn, written by Richard Matheson.

19. First presented as a principle in Michael Shermer, "The Captain Kirk Principle," *Scientific American* (December 2002), p. 32.

20. David G. Myers, *Intuition: Its Powers and Perils* (New Haven, Conn.: Yale University Press, 2002), p. 1.

21. Nalini Ambady and Robert Rosenthal, "Thin Slices of Expressive Behavior as Predictors of Interpersonal Consequences: A Meta-Analysis," *Psychological Bulletin,* vol. 3 (1992), pp. 256–74, and Ambady and Rosenthal, "Half a Minute: Predicting Teacher Evaluations from Thin Slices of Nonverbal Behavior and Physical Attractiveness," *Journal of Personality and Social Psychology,* vol. 64, pp. 431–41.

22. J. A. Krosnick et al., "Subliminal Conditioning of Attitudes," *Personality and Social Psychology Bulletin,* vol. 18 (1992), pp. 152–62.

23. Myers, *Intuition,* p. 119.

24. Ibid., pp. 36–37.

25. Ibid., pp. 174–76.

26. Ibid., pp. 44–45.
27. Ibid., pp. 46–47.
28. Ibid., pp. 46–49. See also Paul Ekman, *Emotions Revealed: Recognizing Faces and Feelings to Improve Communication and Emotional Life* (New York: Times Books, 2003).
29. Jonathan Haidt, "The Emotional Dog and Its Rational Tail: A Social Intuitionist Approach to Moral Judgment," *Psychological Review,* vol. 108 (2001), pp. 814–34; Haidt, "The Positive Emotion of Elevation," *Prevention and Treatment 3,* article 3 (2000); Haidt, "The Moral Emotions," in *Handbook of Affective Sciences,* ed. R. J. Davidson, K. Scherer, and H. H. Goldschmidt (Oxford: Oxford University Press, 2003).
30. J. Greene et al., "An fMRI Investigation of Emotional Engagement in Moral Judgment," *Science,* vol. 293 (2001), pp. 2105–08.
31. Steven N. Brenner and Earl A. Molander, "Is the Ethics of Business Changing?" *Harvard Business Review* (January-February 1977), pp. 57–71.
32. P. A. M. Van Lange, T. W. Taris, and R. Vonk, "Dilemmas of Academic Practice: Perceptions of Superiority Among Social Psychologists," *European Journal of Social Psychology,* vol. 27 (1997), pp. 675–85.
33. "Oprah: A Heavenly Body? Survey Finds Talk-Show Host a Celestial Shoo-in," *U.S. News and World Report* (March 31, 1997), p. 18.
34. J. A. White and S. Plous, "Self-Enhancement and Social Responsibility: On Caring More, but Doing Less, Than Others," *Journal of Applied Social Psychology,* vol. 25 (1995), pp. 1297–1318.
35. J. Kruger, "Personal Beliefs and Cultural Stereotypes About Racial Characteristics," *Journal of Personality and Social Psychology,* vol. 71 (1996), pp. 536–48.
36. Brad J. Sagarin, Kelton V. L. Rhoads, and Robert B. Cialdini, "Deceiver's Distrust: Denigration as a Consequence of Undiscovered Deception," *Personality and Social Psychology Bulletin,* vol. 24 (1998), pp. 1167–76.

7. *How We Are Immoral*

1. David Buss, *The Dangerous Passion: Why Jealousy Is as Necessary as Love and Sex* (New York: Free Press, 2002).
2. J. Stuart Snelson, *Win-Win Theory for Win-Win Success: Science, Theory, and Strategy of Win-Win Success for Family and Business, Community and Nation* (forthcoming).
3. Gottfried Leibniz, *Discourse on Metaphysics* (1826), translated by G. R. Montgomery (Buffalo, N.Y.: Prometheus Books, 1992).
4. Jonathan D. Glater, "Adultery May Be a Sin, but It's a Crime No More," *New York Times,* April 17, 2003, p. A12.
5. Buss, *The Dangerous Passion,* p. 103.
6. D. H. Lawrence, *Pornography and So On* (London: Faber and Faber, 1936). Stewart statement rendered in case judgment *Jacobellis v. Ohio,* 1964. 378 U.S. 184.
7. Anaïs Nin, *Little Birds* (New York: Harcourt, 1963), p. 123.
8. D. Henson and H. Rubin, "Voluntary Control of Eroticism," *Journal of Applied Behavior Analysis,* vol. 4 (1971), pp. 37–44. K. Kelley and D. Byrne, "Assessment of Sexual Responding: Arousal, Affect, and Behavior," in *Social Psychophysiology,* ed. John Cacioppo and Richard Petty (New York: Guilford, 1983). D. Przbyla and D. Byrne, "The Mediating Role of Cognitive Process in Self Reported Sexual Arousal," *Journal of Research in Personality,* vol. 18 (1984), pp. 54–63. M. Zuckerman, "Physiological Measures of Sexual Arousal in the Human," in *Technical Reports on the Commission on Obscenity and Pornography,* vol. 1. (Washington, D.C.: U.S. Government Printing Office, 1971).
9. Catharine MacKinnon, "Pornography: A Feminist Perspective," position paper presented to the Minneapolis City Council, 1983. Andrea Dworkin, *Letters From a War Zone* (New York: Dutton, 1988).

10. Berl Kutchinsky, "The Effect of Easy Availability of Pornography on the Incidence of Sex Crimes: The Danish Experience," in *Pornography and Censorship,* ed. David Copp and Susan Wendell (Amherst, N.Y.: Prometheus Books, 1983), p. 307.

11. Fred R. Berger, "Pornography, Sex, and Censorship," in *Pornography and Sexual Deviance,* ed. Michael Joseph Goldstein and Harold S. Kant (Berkeley: University of California Press, 1973).

12. Luis T. Garcia, "Exposure to Pornography and Attitudes About Women and Rape; A Correlational Study," *Journal of Sex Research,* vol. 22 (1986), pp. 378–85. Cynthia S. Gentry, "Pornography and Rape: An Empirical Analysis," *Deviant Behavior,* vol. 12 (1991), pp. 277–88. Berl Kutchinsky, "Pornography and Rape: Theory and Practice?" *International Journal of Law and Psychiatry,* vol. 14 (1991), pp. 47–64.

13. Paul Abramson and Haruo Hayashi, "Pornography in Japan: Cross-Cultural and Theoretical Considerations," in *Pornography and Sexual Aggression,* ed. Neil Malamuth and Edward Donnerstein (Orlando: Academic Press, 1984). Berl Kutchinsky, "Pornography and its Effects in Denmark and the United States: A Rejoinder and Beyond," *Comparative Social Research: An Annual,* vol. 8 (1985), pp. 301–30.

14. Edward Donnerstein and Leonard Berkowitz, "Victims' Reactions in Aggressive Erotic Films as a Factor in Violence Against Women," in Malamuth and Donnerstein, *Pornography and Sexual Aggression.*

15. Edward Donnerstein, "Erotica and Human Aggression," in *Aggression: Theoretical and Empirical Review,* vol. 2, ed. Russell Geen and Edward Donnerstein (New York: Academic Press, 1983).

16. W. A. Fisher and D. Byrne, "Individual Differences in Affective, Evaluative and Behavioral Responses to an Erotic Film," *Journal of Applied Social Psychology* (August 1978), pp. 355–65.

17. Neil Malamuth and James Check, "The Effects of Aggressive Pornography on Beliefs of Rape Myths: Individual Differences," *Journal of Research in Personality,* vol. 19 (1985), pp. 299–320.

18. Goldstein and Kant, eds., *Pornography and Sexual Deviance.*

19. Naomi Wolf, "Our Bodies, Our Souls," *The New Republic* (October 16, 1995). In this article for *The New Republic,* feminist author Naomi Wolf shocked the pro-choice movement by claiming that the fetus at all stages is a human individual and therefore abortion is immoral (although she still supports free choice). In Wolf's 6,700-word essay, however, there is not a single scientific fact presented in support of her claim for fetal human individuality. Instead, we get emotional references to "lapel pins with the little feet," "framed sonogram photos," and "detailed drawings of the fetus" from the popular pregnancy book *What to Expect When You're Expecting.* With similar shortcomings, in a 1995 PBS *Firing Line* debate, Arianna Huffington claimed that scientists have proven that life begins at conception, yet no facts were offered in support of this claim.

20. Roy Rivenburg, a writer for the *Los Angeles Times,* succinctly summarized the pro-choice and pro-life positions: Roy Rivenburg, "A Decision Between a Woman and God," *Los Angeles Times,* May 24, 1996.

21. The Supreme Court's decision is posted all over the Internet. A Google search generates 141,000 hits; see, for example, http://members.aol.com/abtrbng/410b1.htm or www2.law.cornell.edu.

22. See, for example, the *Amici Curiae Brief in Support of Appellees,* William L. Webster et al., *Appellants v. Reproductive Health Services et al., Appellees,* 1988.

23. B. L. Koops, L. J. Morgan, and F. C. Battaglia, "Neonatal Mortality Risk in Relation to Birth Weight and Gestational Age: Update," *Journal of Pediatrics,* vol. 101 (1982), pp. 969–77. Milner and Beard, "Limit of Fetal Viability," *Lancet,* vol. 1, 1984, p. 1079. Pleasure, Dhand, and Kaur, "What Is the Lower Limit of Viability?" *American Journal of Diseases of Children,* vol. 138 (1984), p. 783.

24. I. R. Beddis et al., "New Technique for Servo-Control of Arterial Oxygen Tension in Preterm Infants," *Archives of Disease in Childhood,* vol. 54 (1979), pp. 278–80.

25. Michael Flower, "Neuromaturation and the Moral Status of Human Fetal Life," in *Abortion Rights and Fetal Personhood,* ed. Edd Doerr and James W. Prescott (Long Beach, Calif.: Centerline Press, 1989). M. E. Molliver, I. Kostovic, and H. Van Der Loos, "The Development of Synapses in Cerebral Cortex of the Human Fetus," *Brain Research,* vol. 50 (1973), pp. 403–7. D. P. Purpura, "Morphogenesis of Visual Cortex in the Preterm Infant," in *Growth and Development of the Brain,* ed. Mary A. B. Brazier (New York: Raven Press, 1975).

26. For a good general discussion of the terms of this debate, see Carl Sagan and Ann Druyan, *Shadows of Forgotten Ancestors* (New York: Random House, 1992).

27. Claudia Kalb, "Treating the Tiniest Patients," *Newsweek* (June 9, 2003), pp. 48–51.

28. Debra Rosenberg, "The War Over Fetal Rights," *Newsweek* (June 9, 2003), pp. 40–47.

29. John-Thor Dahlburg, "Firm Says It Created First Human Clone," *Los Angeles Times,* December 28, 2002, p. A13.

30. Quoted in Jeffrey Kluger, "Will We Follow the Sheep?" *Time* (March 10, 1997), pp. 67–70, 72.

31. Quoted in *Cloning Human Beings: Report and Recommendations* (Rockville, Md.: National Bioethics Advisory Commission, 1997).

32. See Nancy L. Segal, *Entwined Lives: Twins and What They Tell Us About Human Behavior* (New York: Dutton, 1999). Segal convincingly shows that genes influence our behavior and personality in innumerable ways that cannot be ignored. Comparing identical twins reared apart with identical twins reared together, fraternal (nonidentical) twins reared together, siblings reared together, and pseudotwins (genetically different adopted children) reared together, identical twins reared apart are more alike on almost all measures than the comparison groups, including a number of striking similarities between identicals reared apart—from the sublime, such as Harold Shapiro (head of the National Bioethics Advisory Commission charged by President Clinton to pass a moral ruling on cloning) and his twin both growing up to become university presidents, to the ridiculous, such as a preference for a rare Swedish toothpaste called Vademecum. Despite these similarities, identical twins can be surprisingly different. Even for such characteristics as height and weight, which have a heritability of over 90 percent, it is striking how different many of these identical twins turned out to be (photographs appear throughout Segal's book). And when we consider traits that show heritabilities in the 50 percent range, it is clear how much environment counts. Segal and her colleagues who study twins have revealed that heredity counts a great deal more than it was recently fashionable to believe, and their science is solid. See also Michael Shermer, "I, Clone," *Scientific American,* vol. 288, no. 4 (April 2003).

33. Quoted in Dahlburg, "Firm Says It Created First Human Clone."

34. Quoted in Kluger, "Will We Follow the Sheep?," p. 72.

35. Ibid., 71.

36. Quoted in M. L. Rantala and Arthur J. Milgram, eds., *Cloning: For and Against* (Chicago: Open Court, 1999), p. 157. This theme has been circulating for decades, from Ted Howard and Jeremy Rifkin's 1977 *Who Should Play God?: The Artificial Creation of Life and What it Means for the Future of the Human Race* (New York: Delacorte Press) to Ted Peters's 1997 *Playing God?: Genetic Determinism and Human Freedom* (New York: Routledge), to a flurry of godly warnings following the brouhaha over Dolly the cloned sheep, such as this one from Kenneth Woodward, opining in *Newsweek:* "Perhaps the message of Dolly is that society should reconsider its casual ethical slide toward assuming mastery over human life. Do we really want to play God?" (Kenneth L. Woodward, "Today the Sheep . . . Tomorrow the Shepherd?" *Newsweek* [March 10, 1997]).

37. Quoted in Dahlburg, "Firm Says It Created First Human Clone."

38. Associated Press, "Clerics Denounce Cloned Baby Claim," *Los Angeles Times,* December 29, 2002, p. A15.

39. Isaac Asimov, *I, Robot* (New York: Random House, 1950). In *Robots and Empire,* Asimov presented the "Zeroth Law" as a prequel to the three laws of robotics: "0. A robot may not injure humanity or, through inaction, allow humanity to come to harm." Asimov

explained: "Unlike the Three Laws, however, the Zeroth Law is not a fundamental part of positronic robotic engineering, is not part of all positronic robots, and, in fact, requires a very sophisticated robot to even accept it." Asimov claimed that the Three Laws originated on December 23, 1940, from a conversation he had with the science fiction publisher John W. Campbell. The Three Laws did not appear in Asimov's first two robot stories, "Robbie" and "Reason," but the First Law was stated in Asimov's third robot story "Liar!" The first story to explicitly state the Three Laws was "Runaround," which appeared in the March 1942 issue of *Astounding Science Fiction.* They were finally codified in *I, Robot* in 1950.

40. These barriers, and others, are outlined in Steven M. Wise, *Science and the Case for Animal Rights* (Boston: Perseus Books, 2002).

41. Verlyn Klinkenborg, "Cow Parts," *Discover* (August 2001), pp. 53–62.

42. Stephen Jay Gould, *The Mismeasure of Man* (New York: W. W. Norton, 1981).

43. Carol Tavris, *The Mismeasure of Woman* (New York: Simon and Schuster, 1992).

44. David Brion Davis, "The Enduring Legacy of the South's Civil War Victory," *New York Times,* August 26, 2001, section 4, p. 1. See also Davis, *Slavery and Human Progress* (New York: Oxford University Press, 1984).

45. Ernst Mayr, *The Growth of Biological Thought* (Cambridge, Mass.: Harvard University Press, 1982), p. 273. Mayr notes that the "'actual vs. potential' distinction is unnecessary since 'reproductively isolated' refers to the possession of isolating mechanisms, and it is irrelevant for species status whether or not they are challenged at a given moment." Mayr offers these "more descriptive" definitions: "A species is a reproductive community of populations (reproductively isolated from others) that occupies a specific niche in nature." And: "Species are the real units of evolution, as the temporary incarnation of harmonious, well-integrated gene complexities" (Ernst Mayr, *Animal Species and Evolution* [Cambridge: Harvard University Press, 1963]).

46. Vercors (Jean Bruller), *You Shall Know Them,* translated by R. Barisse (New York: Pocket Books, 1955).

47. Richard G. Klein, *The Human Career: Human Biological and Cultural Origins* (Chicago: University of Chicago Press, 1999).

48. Wise, *Science and the Case for Animal Rights.* See also Steven M. Wise, *Rattling the Cage: Toward Legal Rights for Animals* (Boston, Mass.: Perseus, 2000). Before reading Wise's book, I was unconvinced by the arguments of animal rights' activists who, it seemed to me, did not appear to understand Aristotle's moral guideline of "all things in moderation." By setting a goal of achieving all rights for all mammals right now, they have, de facto, procured no rights for any animals ever. That's not strictly correct, of course. There have been many legal victories, particularly with regard to protecting animals from cruelty. But for most of us in the sciences, the animal rights movement has been too political, too extreme, and too ignorant of science. Wise's book does not suffer from these shortcomings.

49. See, for example, Jeffrey Moussaieff Masson and Susan McCarthy, *When Elephants Weep: The Emotional Lives of Animals* (New York: Delacorte Press, 1995).

50. Christopher Boehm, *Hierarchy in the Forest: The Evolution of Egalitarian Behavior* (Cambridge, Mass.: Harvard University Press). Marian Stamp Dawkins, *Through Our Eyes Only: The Search for Animal Consciousness* (New York: W. H. Freeman, 1993). Daniel C. Dennett, *Kinds of Minds: Toward an Understanding of Consciousness* (New York: Basic Books, 1996). Donald R. Griffin, *Animal Minds: Beyond Cognition to Consciousness* (Chicago: University of Chicago Press, 2001). Marc Hauser, *Wild Minds: What Animals Really Think* (New York: Henry Holt, 2000). Sue Taylor Parker and M. L. McKinney, eds., *Self-Awareness in Animals and Humans: Developmental Perspectives* (Cambridge: Cambridge University Press, 1994). Irene Pepperberg, *The Alex Studies: Cognitive and Communicative Abilities of Parrots* (Cambridge, Mass.: Harvard University Press, 1999). Richard D. Ryder, *Animal Revolution: Changing Attitudes Toward Speciesism* (London: Basil Blackwell, 1989). Richard Sorabji, *Animal Minds and Human Morals: The Origin of the Western Debate* (Ithaca, N.Y.: Cornell University Press, 1993).

51. John D. Bonvillian and Francine G. P. Patterson, "Sign Language Acquisition and the Development of Meaning in a Lowland Gorilla," in C. Mandell and A. McCabe, eds.,

The Problem of Meaning: Behavioral and Cognitive Perspectives (Amsterdam: Elsevier, 1997). Francine G. P. Patterson and Wendy Gordon, "The Case for the Personhood of Gorillas," in *The Great Ape Project: Equality Beyond Humanity,* ed. Paola Cavalieri and Peter Singer (New York: St. Martin's Press, 1993). Francine G. P. Patterson and Eugene Linden, *The Education of Koko* (New York: Holt, Rinehart, and Winston, 1981). Richard Byrne, *The Thinking Ape: Evolutionary Origins of Intelligence* (Oxford: Oxford University Press, 1995). Sue Taylor Parker, Robert W. Mitchell, and H. Lyn Miles, eds., *The Mentalities of Gorillas and Orangutans* (Cambridge: Cambridge University Press, 1999).

52. Birute M. F. Galdikas, *Reflections of Eden: My Years with the Orangutans of Borneo* (Boston: Little, Brown, 1995). H. Lyn Miles, "Simon Says: The Development of Imitation in an Encultured Orangutan," in *Reaching Into Thought: The Minds of the Great Apes,* ed. Anne E. Russon et al. (Cambridge: Cambridge University Press, 1996), and Miles, "ME CHANTEK: The Development of Self-Awareness in a Signing Orangutan," in *Self-Awareness in Animals and Humans: Developmental Perspectives,* ed. Sue Taylor Parker et al. (Cambridge: Cambridge University Press, 1994). Lesley J. Rogers, *Minds of Their Own: Thinking and Awareness in Animals* (Boulder, Colo.: Westview Press, 1998).

53. Diana Reiss and Lori Marino, "Mirror Self-Recognition in the Bottlenose Dolphin: A Case of Cognitive Convergence," *Proceedings of the National Academy of Sciences,* May 8, 2001, pp. 5937–42. Marc Bekoff, ed., *The Smile of a Dolphin: Remarkable Accounts of Animal Emotions* (New York: Crown Books, 2000). Karen Pryor and Kenneth S. Norris, eds., *Dolphin Societies: Discoveries and Puzzles* (Chicago: University of Chicago Press, 2000).

54. Louis M. Herman and Palmer Morrel-Samuels, "Knowledge Acquisition and Asymmetry Between Language Comprehension and Production: Dolphins and Apes as General Models for Animals," in *Interpretation and Explanation in the Study of Animal Behavior,* eds. Marc Bekoff and Dale Jamieson (Boulder, Colo.: Westview Press, 1990). Lori Marino, "A Comparison of Encephalization Between Odontocete Cetaceans and Anthropoid Primates," *Brain, Behavior, and Evolution,* vol. 51 (1988), p. 230. Sam H. Ridgway, "Physiological Observations on Dolphin Brains," in *Dolphin Cognition and Behavior: A Comparative Approach,* ed. Ronald J. Schusterman et al. (New Jersey: Lawrence Erlbaum, 1986), pp. 32–33.

8. Rise Above

1. See David Gerrold, *The World of "Star Trek"* (New York: Ballantine Books, 1979); and Stephen E. Whitfield and Gene Roddenberry, *The Making of "Star Trek"* (New York: Ballantine Books, 1968).

2. Diamond, *The Third Chimpanzee.*

3. Michael Ghiglieri, *The Dark Side of Man: Tracing the Origins of Male Violence* (Reading, Mass.: Perseus Books, 1999).

4. Brian Hare, Michelle Brown, Christina Williamson, and Michael Tomasello, "The Domestication of Social Cognition in Dogs," *Science,* vol. 298 (November 22, 2002), pp. 1634–36. Jennifer A. Leonard et al., "Ancient DNA Evidence for Old World Origin of New World Dogs," *Science,* vol. 298 (November 22, 2002), pp. 1613–15. Peter Savolainen et al., "Genetic Evidence for an East Asian Origin of Domestic Dogs," *Science,* vol. 298 (November 22, 2002), pp. 1610–12.

5. Lyudmila N. Trut, "Domestication of the Fox: Roots and Effects," *Scientifur,* vol. 19 (1995), pp. 11–18. Lyudmila N. Trut, "Early Canid Domestication: The Farm-Fox Experiment," *American Scientist* (March/April 1999).

6. Richard Wrangham and Dale Peterson, *Demonic Males: Apes and the Origins of Human Violence* (New York: Houghton Mifflin, 1996). The area 13 work is described in: Katerina Semendeferi et al., "Limbic Frontal Cortex in Hominoids: A Comparative Study of Area 13," *American Journal of Physical Anthropology,* vol. 106 (1998), pp. 129–55. In this paper the authors discuss how area 13 was shown in rhesus monkeys to be involved in the disinhibition of emotional responses.

7. Robert Sapolsky explained that the reason for his skepticism is threefold:

1. The levels of serotonin in synapses, or in whole brain regions may be quite different in closely related species, simply reflecting different nuts and bolts aspects of the synthesis and breakdown of serotonin, rather than something like, "Species X, which is more aggressive, has less serotonin than Species Y." One example, new world monkeys, have vastly higher levels of corticosteroids than do old world monkeys—about an order of magnitude difference. Scientists have concocted all sorts of elaborate stories about how life is so much more stressful for a new world monkey . . . until it was discovered that the order of magnitude increase in corticosteroid levels is accompanied by an order of magnitude decrease in sensitivity of corticosteroid receptors.

2. Body size. There is a strange but convincing literature showing that body size is a confound in serotonin studies. What's that about? Serotonin is taken from a lumbar tap. The taller the person, the further serotonin has to travel from the brain to the bottom of the spine, thus being diluted. So low serotonin (taken from the spine) could just reflect body size. Correct for that and some of the serotonin/behavior findings disappear. So, there's big body size differences in different primate species, making direct cross-species comparisons difficult.

3. Finally, the social meaning of aggression can be dramatically different in different species. Is aggression for real or symbolic? Is it something that is needed in a bottom up or a top down hierarchical system? Does aggression increase the likelihood of dominance (most old world monkeys), or decrease it (male vervet monkeys)? And so on. So amid there not being any data on direct comparisons of serotonin among the apes, including us, I don't think such a comparison would be terribly meaningful. (Personal correspondence, June 5, 2003.)

Richard Wrangham responded to this critique as follows: "Sapolsky's caution is fair and the issue is moot because of the shortage of data. But I would bet that he's wrong. That is, I predict that there will prove to be a recognizable and consistent species correlation with impulsive aggression, based on what has happened to serotonin levels in domesticated species compared to wild (good data from foxes, mink and rats). The only way to get serotonin from bonobos and chimps, in practice, is to get it from autopsies, and I've asked Joe Erwin about doing so and he has agreed, pending my providing something in writing which I have yet to do. So that's where I'm at . . . ." (Personal correspondence, June 5, 2003.) Paul J. Zak, "Trust," *Journal of Financial Transformation* (CAPCO Institute) 7, 18–24.

8. Richard Wrangham and David Pilbeam, "Apes as Time Machines," in *African Apes*, vol. 1: *All Apes Great and Small*, ed. Birute M. F. Galdikas et al. (New York: Kluwer Academic/Plenum Publishers, 2002), pp. 5–18. Richard Wrangham, "Is Military Incompetence Adaptive?" *Evolution and Human Behavior*, vol. 20 (1999), pp. 3–17. See also Craig B. Stanford, *The Hunting Apes: Meat Eating and the Origins of Human Behavior* (Princeton, N.J.: Princeton University Press, 1999).

9. My shift from a religious to a scientific way of thinking took several years. If there was a moment that could be called defining, it came on the day I removed the silver ichthus from around my neck. Recall the now-embarrassing fashions of the 1970s that included—in addition to bell-bottom pants and puffy shirts—gold and silver necklaces. The ichthus is the famous Jesus "fish" embossed with Greek letters roughly translated as "Jesus Christ, Son of God, Savior" that has since found itself embroiled in a bumper-sticker war with Darwin "fish" of various species—with and without feet, with a wrench, mounting a Christian fish, a Christian fish devouring a Darwin fish, and so forth.

10. See http://www.holysmoke.org/skhok/aa011.htm.

11. Reported in the September/October 1997 issue of *Freedom Writer*, published by the Institute for First Amendment Studies; survey conducted by the Gliss Institute of the University of Ohio (1,200 contacted, 600 responses, representing a cross section of between 200,000 and 400,000 Americans). Additional findings included: 86 percent believe in a literal interpretation of the Bible and affirm that Satan is real and Jesus is the only way to salvation; 60 percent believe the world will end in Armageddon; 95 percent favored the outlawing of

abortion; 93 percent support school vouchers; and 69 percent agreed that "environmental protection laws have gone too far and should be reversed." Of those who called themselves "conservative Christians," 71 percent identified themselves as evangelical Protestant.

12. Clay F. Naff, commentary on Metanexus Discussion Web page, 2002. FUTURES@LISTSERV.METANEXUS.NET or claynaff@yahoo.com.

13. Ibid.

14. Salman Rushdie, "Religion, As Ever, Is the Poison in India's Blood," *The Guardian,* March 9, 2002. Online at http://books.guardian.co.uk/departments/politicsphilosophyandsociety/story/0,6000,664342,00.html.

15. Quoted in Nicholas D. Kristof, "All-American Osamas," *New York Times,* June 7, 2002, p. A27.

16. Ibid.

17. Kristof, "All-American Osamas." Michael Isikoff, "Flushed From the Woods," *Newsweek* (June 9, 2003), p. 35.

18. A. N. Franzblau, "Religious Belief and Character Among Jewish Adolescents," *Teachers College Contributions to Education,* no. 634 (1934).

19. Murray G. Ross, *Religious Beliefs of Youth* (New York: Association Press, 1950).

20. Travis Hirschi and Rodney Stark, "Hellfire and Delinquency," *Social Problems,* vol. 17 (1969), pp. 202–13.

21. R. E. Smith, G. Wheeler, and E. Diener, "Faith Without Works: Jesus People, Resistance to Temptation, and Altruism," *Journal of Applied Social Psychology,* vol. 5 (1975), pp. 320–30.

22. David M. Wulff, *Psychology of Religion: Classic and Contemporary Views* (New York: Wiley, 1991), pp. 219–20.

23. George Barna, *Index of Leading Spiritual Indicators,* 1996 and 2001. Survey data available online at www.barna.org in the Research Archives files. Barna defines a "born again" Christian as one who answers yes to two questions: "Have you ever made a personal commitment to Jesus Christ that is still important in your life today?" and "When I die, I will go to Heaven because I have confessed my sins and have accepted Jesus Christ as my savior." As for divorce rates among religious denominations, Barna found, "surprisingly," that "the Christian denomination whose adherents have the highest likelihood of getting divorced are Baptists." Twenty-nine percent of Baptists have been to divorce court, although nondenominational Christians (small sects and independents) show an even higher rate of 34 percent. Catholics and Lutherans have the lowest percentage of divorces at 21 percent. Mainline Protestants "experience divorce on par with the national average (25 percent)." Mormons—"renowned for their emphasis upon strong families"—come in at an indistinguishable divorce rate of 24 percent.

24. J. A. Adande, "The World According to Reverend Reggie," *Los Angeles Times,* April 4, 1998, p. 3.

25. Quoted in Carl Sagan, *Demon-Haunted World: Science as a Candle in the Dark* (New York: Random House, 1995), p. 430.

26. Edwin S. Gaustad, Philip L. Barlow, and Richard Dishno, eds., *New Historical Atlas of Religion in America* (New York: Oxford University Press, 2001).

27. John Rawls, *A Theory of Justice* (Cambridge: Belknap/Harvard University Press, 1971), p. 56.

28. Robert Nozick, *Anarchy, State, and Utopia* (New York: Basic Books, 1974), p. ix.

29. Ayn Rand, "Introducing Objectivism," *Los Angeles Times,* June 17, 1962.

30. Ayn Rand, *Atlas Shrugged* (New York: Random House, 1957).

31. Nathaniel Branden, *Judgment Day: My Years With Ayn Rand* (Boston: Houghton Mifflin, 1989), pp. 255–256. With later irony Branden added that since he was declared by Rand to be her "intellectual heir" and "an ideal exponent of her philosophy, he is to be accorded only marginally less reverence than Ayn Rand herself."

32. Barbara Branden, *The Passion of Ayn Rand* (New York: Doubleday, 1986), p. 227.

33. Milton Friedman, "Say 'No' to Intolerance," *Liberty* (July 18, 1991).

34. Murray N. Rothbard, *The Sociology of the Ayn Rand Cult* (monograph) (Port Townsend, Wash.: Liberty Publishing, 1987).

35. Ayn Rand, "How Does One Lead a Rational Life in an Irrational Society?" in *The Virtue of Selfishness* (New York: New American Library, 1964), p. 91. For Rand's pronouncements on hundreds of subjects, see Harry Binswanger, *The Ayn Rand Lexicon: Objectivism From A to Z* (New York: New American Library, 1986).

36. See *The Interpreter's Bible,* vol. 7, pp. 324–26, for a lengthy discussion of this issue.

37. Ayn Rand, from John Galt's speech in *Atlas Shrugged,* reprinted in *For the New Intellectual* (New York: New American Library, 1961), p. 216.

38. Cornelia V. Christenson, *Kinsey: A Biography* (Indianapolis: Indiana University Press, 1971), pp. 126–27.

39. Ibid., p. 4.

40. Alfred C. Kinsey, Wardell Baxter Pomeroy, and Clyde E. Martin, *Sexual Behavior in the Human Male* (Philadelphia: W. B. Saunders, 1948), p. 20.

41. Ibid., pp. 638–47.

42. Ibid., p. 639.

43. Quoted in Christenson, *Kinsey,* p. 5.

44. Kinsey, Pomeroy, and Martin, *Sexual Behavior in the Human Male,* p. 678.

45. Quoted in Christenson, *Kinsey,* pp. 8–9.

46. Ibid., p. 223.

47. Ibid., p. 213.

48. Ibid., pp. 163–66.

49. I am aware of the controversy surrounding Kinsey's data, and the accusations against him regarding the validity of his data collection techniques and the frequencies of certain sexual behaviors he reported in his publications (which critics claim are greatly exaggerated). My concern here, however, is not with Kinsey's data per se, but with his methodology, and how methodological individualism leads to a greater understanding of the diversity of human behavior. Kinsey's use of enormous sample sizes for both wasps and humans is sound, regardless of whether specific findings turn out to be corroborated or not.

50. Quoted in Christenson, *Kinsey,* p. 6.

51. Ibid., pp. 6–7.

52. The quote, and slight variations on it, is always attributed to Bastiat, although I have been unable to track down the source. Bastiat certainly could have (or would have) said it, as he was a champion of free trade. See, for example, his most classic work "The Law" in *Selected Essays on Political Economy,* ed. George B. de Huszar (Irvington-on-Hudson, N.Y.: Economic Education, 1995).

53. Robert B. Edgerton, *Sick Societies: Challenging the Myth of Primitive Harmony* (New York: Free Press, 1992). For an excellent treatise on the evolutionary history of violence and aggression see: Michael P. Ghiglieri, *The Dark Side of Man.*

54. Napoleon Chagnon, *Yąnomamö,* p. 162. See also: Ronald M. Berndt, "The Walmadjeri and Gugadja" in *Hunters and Gatherers Today,* ed. M. G. Bicchieri (Prospect Heights, Ill.: Waveland Press, Inc., 1988). In his study of two Australian aboriginal tribes, the Walmadjeri and Gugadja, Berndt says of trade: "Trade goods are passed, so to speak, from one interactory zone to the next. When large ceremonies and rituals are held, some of the participants come from places a great distance apart; they provide, therefore, an ideal opportunity for bartering. Trade takes place within the context of ritual and often is not seen as being something separate" (p. 188).

55. Clive Gamble, *Timewalkers: The Prehistory of Global Colonization* (London: Alan Sutton, 1993). T. Douglas Price and James A. Brown, eds., *Prehistoric Hunter-Gatherers: The Emergence of Cultural Complexity* (San Diego, Calif.: Academic Press, 1985).

56. Shepard Krech, III, *The Ecological Indian: Myth and History* (New York: W. W. Norton, 1999), p. 152.

57. V. C. L. Hutson and G. T. Vickers, "The Spatial Struggle of Tit-for-Tat and Defect," *Philosophical Transactions of the Royal Society of London B,* vol. 348 (1995), pp. 393–404. Kenneth Binmore, *Game Theory and the Social Contract. Volume 1: Playing Fair* (Cambridge, Mass.: MIT Press, 1994).

58. Bruce Bower, "Getting Out from Number One: Selfishness May Not Dominate Human Behavior," *Science News,* vol. 137, no. 17 (1990), pp. 266–67.

59. Robyn M. Dawes, Alphons van de Kragt, and John M. Orbell, "Cooperation for the Benefit of Us—Not Me, or My Conscience," in *Beyond Self-Interest,* ed. Jane Mansbridge (Chicago: University of Chicago Press, 1990), pp. 97–110.

60. James K. Rilling et al., "A Neural Basis for Social Cooperation," *Neuron,* vol. 35 (July 18, 2002), pp. 395–404. See also: Kimberly A. Wade-Benzoni et al., "Cognitions and Behavior in Asymmetric Social Dilemmas: A Comparison of Two Cultures," *Journal of Applied Psychology,* vol. 87, no. 1 (2002), pp. 87–95. Robyn M. Dawes and D. M. Messick, "Social Dilemmas," *International Journal of Psychology,* vol. 35, no. 2, pp. 111–16.

61. Steven R. Quartz and Terrence J. Sejnowski, *Liars, Lovers, and Heroes: What the New Brain Science Reveals about How We Become Who We Are* (New York: William Morrow, 2002).

62. Ralph Adolphs, "Cognitive Neuroscience of Human Social Behavior," *Nature Reviews Neuroscience,* vol. 4 (March 2003), pp. 165–70. R. J. Dolan, "Emotion, Cognition, and Behavior," *Science,* vol. 298 (November 8, 2002), pp. 1191–94.

63. Jorge R. Moll et al., "The Neural Correlates of Moral Sensitivity: A Functional Magnetic Resonance Imaging Investigation of Basic and Moral Emotions," *Journal of Neuroscience,* vol. 22, no. 7 (April 1, 2002), pp. 2730–36.

64. Kevin McCabe et. al., "A Functional Imaging Study of Cooperation in Two-Person Reciprocal Exchange," *Proceedings of the National Academy of Science,* vol. 98, no. 20 (September 25, 2001), pp. 11832–35.

65. Katerina Semendeferi et al., "Prefrontal Cortex in Humans and Apes: A Comparative Study of Area 10," *American Journal of Physical Anthropology,* vol. 114 (2001), pp. 224–41.

66. Uta Frith and Chris Frith, "The Biological Basis of Social Interaction," *Current Directions in Psychological Science,* vol. 10, no. 5 (October 2001), pp. 151–55.

67. Paul J. Zak, "Trust," *Journal of Financial Transformation* (CAPCO Institute), vol. 7 (2002), pp. 18–24. For general discussions on cooperation and trust see also: Sarah Blaffer Hrdy, *Mother Nature: A History of Mothers, Infants, and Natural Selection* (New York: Pantheon, 1999). Shelley E. Taylor, *The Tending Instinct: How Nurturing Is Essential to Who We Are and How We Live* (New York: Times Books, 2002).

68. Noble E. Cunningham, Jr., *In Pursuit of Reason: The Life of Thomas Jefferson* (Baton Rouge, La.: Louisiana State University Press, 1987).

69. LeBlanc, *Constant Battles,* p. 207.

70. Bernard Lewis, *What Went Wrong?: Western Impact and Middle Eastern Response* (New York: Oxford University Press, 2002), pp. 96–116.

71. At the April 2003 annual conference of the Atheist Alliance International, at which I spoke and discussed the labeling problem, a new label was proposed by Paul Geisert and Mynga Futrell of Sacramento, California, who note that, by analogy, homosexuals used to suffer a similar labeling problem when they were called homos, queers, fruits, fags, and fairies. Their solution was to change the label to a more neutral term—*gay.* Over the past couple of decades, gays have won significant liberties for themselves, starting with gay pride and gay marches that have led to gay rights. Analogously, instead of calling ourselves nonbelievers, nontheists, atheists, and the like, it was suggested that we call ourselves *Brights.* "A Bright is a person whose worldview is naturalistic—free of supernatural and mystical elements. Brights base their ethics and actions on a naturalistic worldview." "Bright" is a good word, meaning "cheerful and lively," "showing an ability to think, learn, or respond quickly," and "reflecting or giving off strong light." Thus, it is a positive word, and myself, the evolutionary biologist Richard Dawkins, the philosopher Daniel Dennett, and the magician and paranormal investigator James Randi all signed up to be Brights (the Brights is not an organization; it is a constituency which, if it grows large enough, may one day be capable of wielding political influence; for more information go to www.the-brights.net). Unfortunately, the brand name Bright was never market-tested on those who might want to use it. When I announced to the 25,000 readers of our electronic e-skeptic newsletter that I was a Bright, I received hundreds of e-mails, roughly 95 percent of which were emphatically negative about the term and indicated that in no uncertain terms would they call themselves Brights. The primary reason given was

that the word sounds elitist, especially since the natural antonym is "Dims." Subsequently, I organized a focus group of a dozen people unaffiliated with skeptics or atheists in which ten out of the twelve rejected the term outright as being too snobby and off-puttting. Nevertheless, the Brights Web page continues to log new members, already numbering in the thousands in seventy-five countries, so it remains to be seen whether the Brights as a new meme will take hold. For more information write Paul Geisert and Mynga Futrell at P.O. Box 163418, Sacramento, CA 95816, e-mail: TheBrightsNet@aol.com.

72. Robert Ingersoll, *Ingersoll's Greatest Lectures* (New York: The Freethought Press Association, 1944).

Appendix I

1. Cited in Robert K. Merton, *On the Shoulders of Giants* (New York: Harcourt Brace, 1965). See also Richard Hardison, *Upon the Shoulders of Giants* (Baltimore, Md.: University Press of America, 1988). Astronomer and historian of science John Gribbon is convinced that Newton intended the comment to be read sarcastically in an attempt to attenuate any credit being given to one of his arch rivals, Robert Hooke. See John Gribbon, "On the Shoulders of Midgets?" *Skeptic,* vol. 10, no. 1 (2003), pp. 36–39.

2. Quoted in Paul H. Barrett, ed., *Metaphysics, Materialism, and the Evolution of Mind: Early Writings of Charles Darwin* (Chicago: University of Chicago Press, 1974), pp. 57, 63.

3. This term was coined by Larry Arnhart in his book of the same title. As Arnhart explained: "When morality is thus understood as part of human nature, moral conduct does not require religious belief or any belief in transcendent moral norms. Morality is important to us because as social animals who try to act in the light of past experience and future expectations, we need norms of right and wrong conduct to secure the social cooperation required to satisfy our natural desires" (Larry Arnhart, *Darwinian Natural Right: The Biological Ethics of Human Nature* [Albany: State University of New York Press, 1998], p. 34).

4. See, for example, my analysis of religious beliefs and attitudes in Shermer, *How We Believe.*

5. Darwin. *The Descent of Man,* pp. 71–72.

6. Ibid., pp. 165–66.

7. Alfred R. Wallace, "Evolution and Character," *Fortnightly Review,* vol. 83 (1908), pp. 1–24.

8. Alfred R. Wallace, "The Limits of Natural Selection as Applied to Man," in *Contributions to the Theory of Natural Selection* (London: Macmillan, 1870), pp. 391–392.

9. Ibid.

10. Alfred R. Wallace, "The Origin of Human Races and the Antiquity of Man Deduced from the Theory of 'Natural Selection,'" *J. ASL,* vol. 2 (1864), p. 173.

11. Ibid., p. 174.

12. Ibid., pp. 177–78.

13. Ibid., pp. 185–86.

14. Herbert Spencer, *The Principles of Ethics,* vol. 1 (Indianapolis: Liberty Classics, 1893), p. 31.

15. David Duncan, ed., *Life and Letters of Herbert Spencer,* 2 vols. (New York: D. Appleton, 1968).

16. Herbert Spencer, *Social Statics; or The Conditions Essential to Human Happiness Specified, and the First of Them Developed* (London: John Chapman, 1851), pp. 29, 93, 121.

17. Herbert Spencer, *The Principles of Ethics,* vol. 2 (New York: D. Appleton, 1891), p. x.

18. Huxley, *Evolution and Ethics.*

19. G. E. Moore, *Principia Ethica* (Cambridge: Cambridge University Press, 1912), p. 38.

20. Thomas H. Huxley and Julian S. Huxley, *Touchstone for Ethics 1893–1943* (New York: Harper and Brothers, 1947), p. 136.

21. Ibid., p. 137.

22. E. O. Wilson, *Sociobiology: The New Synthesis* (Cambridge, Mass.: Harvard University Press, 1975), p. 562.

23. Ibid., p. 563.

24. Ibid., p. 564.

25. Ullica Segerstråle, *Defenders of the Truth: The Battle for Science in the Sociobiology Debate and Beyond* (New York: Oxford University Press, 2000).

26. Recounted in detail in Shermer and Grobman, *Denying History.*

27. Segerstråle, *Defenders of the Truth.*

28. Wilson, *On Human Nature,* p. 7.

29. Richard D. Alexander, *Darwinism and Human Affairs* (Seattle: University of Washington Press, 1979), and Alexander, *The Biology of Moral Systems* (New York: Aldine De Gruyter, 1987). Axelrod, *The Evolution of Co-operation.* Dawkins, *The Selfish Gene.* De Waal, *Good Natured.* Paul Lawrence Farber, *The Temptations of Evolutionary Ethics* (Berkeley: University of California Press, 1994). W. D. Hamilton, *Narrow Roads of Gene Land,* vol. 1: *Evolution of Social Behaviour* (New York: W. H. Freeman, 1996). Philip Kitcher, *Vaulting Ambition: Sociobiology and the Quest for Human Nature* (Cambridge: MIT Press, 1985). John Maynard Smith, *Did Darwin Get it Right?* (New York: Chapman and Hall, 1992). Robert J. Richards, *Darwin and the Emergence of Evolutionary Theories of Mind and Behavior* (Chicago: University of Chicago Press, 1987). Michael Ruse, *Taking Darwin Seriously: A Naturalistic Approach to Philosophy* (Oxford: Basil Blackwell, 1986). Donald Symons, *The Evolution of Human Sexuality* (New York: Oxford University Press, 1979). Paul Thompson, *Issues in Evolutionary Ethics* (Albany, N.Y.: SUNY Press, 1995). Robert Trivers, *Social Evolution* (Menlo Park, Calif.: Benjamin/ Cummings, 1985). James Q. Wilson, *The Moral Sense* (New York: Free Press, 1993). Robert Wright, *The Moral Animal* (New York: Random House, 1994).

30. Wilson, *Consilience,* pp. 238–65. For a concise summary, see also Wilson, "The Biological Basis of Morality," pp. 53–70.

31. Ibid., p. 241.

32. Wilson, "Nonzero and Nonsense," pp. 84–89.

33. Wilson, *Darwin's Cathedral.* Sober and Wilson, *Unto Others.*

34. Ernst Mayr, "Where Are We?" *Cold Spring Harbor Symposium on Quantitative Biology,* vol. 24 (1959), pp. 409–40. Reprinted in Ernst Mayr, *Evolution and Diversity of Life* (Cambridge, Mass.: Harvard University Press, 1976); Ernst Mayr, *The Growth of Biological Thought,* p. 595; Mayr, *Toward a New Philosophy of Biology* (Cambridge, Mass.: Harvard University Press, 1988), p. 79; Mayr, *This Is Biology* (Cambridge, Mass.: Harvard University Press, 1997), p. 202.

35. Michael Shermer and Frank Sulloway, "The Grand Old Man of Evolution. An Interview with Evolutionary Biologist Ernst Mayr," *Skeptic,* vol. 8, no. 1 (2000), pp. 76–83.

36. Stephen Jay Gould, *The Structure of Evolutionary Theory* (Cambridge, Mass.: Harvard University Press, 2002).

37. Ibid., p. 73.

38. Some of those critiques, however, have been aimed not at Darwin's Duomo, but at Gould's Pinnacles. To his credit, Gould unhesitatingly allows his critics to speak, but the price they pay is facing the buzz saw of his rhetorical brilliance and literary erudition, as in this maximally insulting cut of one critic—the philosopher Dan Dennett who penned a fifty-page critique of Gould—when he quotes Schiller: "Mit Dummheit kämpfen die Götter selbst vergebens"—"even the gods cannot fight with stupidity" (Gould, *The Structure of Evolutionary Theory,* p. 1009). One persistent misunderstanding about Gould's remodeling of Darwin's Duomo stems from what I call the "paradigm paradox" (Michael Shermer, *The Borderlands of Science: Where Sense Meets Nonsense* [New York: Oxford University Press, 2001], p. 98).

39. Gould, *The Structure of Evolutionary Theory,* pp. 71, 652.

BIBLIOGRAPHY

Abramson, Paul, and Haruo Hayashi. "Pornography in Japan: Cross-Cultural and Theoretical Considerations." In *Pornography and Sexual Aggression,* edited by Neil Malamuth and Edward Donnerstein. Orlando: Academic Press, 1984.

Adande, J. A. "The World According to Reverend Reggie." *Los Angeles Times,* April 4, 1998, p. 3.

Adolphs, Ralph. "Cognitive Neuroscience of Human Social Behavior." *Nature Reviews Neuroscience,* vol. 4 (March 2003), pp. 165–70.

Alexander, Richard D. *The Biology of Moral Systems.* New York: Aldine De Gruyter, 1987.

———. *Darwinism and Human Affairs.* Seattle: University of Washington Press, 1979.

Ambady, Nalini, and Robert Rosenthal. "Half a Minute: Predicting Teacher Evaluations from Thin Slices of Nonverbal Behavior and Physical Attractiveness." *Journal of Personality and Social Psychology,* vol. 64 (1993), pp. 431–41.

———. "Thin Slices of Expressive Behavior as Predictors of Interpersonal Consequences: A Meta-Analysis." *Psychological Bulletin,* vol. 3 (1992), pp. 256–74.

Arnhart, Larry. *Darwinian Natural Right: The Biological Ethics of Human Nature.* Albany: State University of New York Press, 1998.

Asimov, Isaac. *I, Robot.* New York: Random House, 1950.

Askenasy, Hans. *Are We All Nazis?* Secaucus, N.J.: Lyle Stuart, 1978.

Astrand, Per-Olof, and Kaare Rodahl. *Textbook of Work Physiology.* New York: McGraw Hill, 1986.

Axelrod, Robert. *The Evolution of Cooperation.* New York: Basic Books, 1984.

Axelrod, Robert, and W. D. Hamilton. "The Evolution of Cooperation." *Science,* vol. 211 (1981), pp. 1390–6.

Barkow, Jerome H. "Beneath New Culture Is Old Psychology: Gossip and Social Stratification." In *The Adapted Mind,* edited by Jerome H. Barkow, Leda Cosmides, John Tooby. New York: Oxford University Press, 1992.

Barna, George. *Index of Leading Spiritual Indicators,* 1996 and 2001. Survey data available online at www.barna.org.

Barrett, D. B., G. T. Kurian, and T. M. Johnson, eds. *World Christian Encyclopedia: A Comparative Survey of Churches and Religions in the Modern World.* 2 vols. New York: Oxford University Press, 2001.

Barrett, Paul H., ed. *Metaphysics, Materialism, and the Evolution of Mind: Early Writings of Charles Darwin.* Chicago: University of Chicago Press, 1974.

Bastiat, Frederic. "The Law." In *Selected Essays on Political Economy,* edited by George B. de Huszar. Irvington-on-Hudson, N.Y.: Foundation for Economic Education, 1995.

Baumeister, Roy F. *Evil: Inside Human Violence and Cruelty.* New York: W. H. Freeman, 1997.

Baumeister, Roy F., and Sara R. Wotman. *Breaking Hearts: The Two Sides of Unrequited Love.* New York: Guilford Press, 1992.

Beddis, I. R. et al. "New Technique for Servo-Control of Arterial Oxygen Tension in Preterm Infants." *Archives of Disease in Childhood,* vol. 54 (1979), pp. 278–80.

Bekoff, Marc, ed. *The Smile of a Dolphin: Remarkable Accounts of Animal Emotions.* New York: Crown Books, 2000.

Bentham, Jeremy. *The Principles of Morals and Legislation* (1789). New York: Macmillan, 1948.

Berenbaum, Michael. *The World Must Know: The History of the Holocaust as Told in the United States Holocaust Memorial Museum.* New York: Little, Brown, 1993.

Berger, Fred R. "Pornography, Sex, and Censorship." In *Pornography and Sexual Deviance,* edited by Michael Joseph Goldstein and Harold S. Kant. Berkeley: University of California Press, 1973.

Bergling, Kurt. *Moral Development: The Validity of Kohlberg's Theory.* Stockholm: Almqvist & Wiksell, 1981.

Bernall, Misty. *She Said Yes: The Unlikely Martyrdom of Cassie Bernall.* New York: Plough Publishing House, 1999.

Berndt, Ronald M. "The Walmadjeri and Gugadja." In *Hunters and Gatherers Today,* edited by M. G. Bicchieri. Prospect Heights, Ill.: Waveland Press, Inc., 1988.

Bettinger, Robert L. *Hunter-Gatherers: Archaeological and Evolutionary Theory.* New York: Plenum Press, 1991.

Binmore, Kenneth. *Game Theory and the Social Contract. Vol. 1: Playing Fair.* Cambridge, Mass.: MIT Press, 1994.

Blanke, Olaf S., T. Ortigue, T. Landis, and M. Seeck. "Neuropsychology: Stimulating Illusory Own-body Perceptions." *Nature,* vol. 419 (September 19, 2002), pp. 269–70.

Boehm, Christopher. *Hierarchy in the Forest: The Evolution of Egalitarian Behavior.* Cambridge, Mass.: Harvard University Press, 1999.

Bonvillian, John D., and Francine G. P. Patterson. "Sign Language Acquisition and the Development of Meaning in a Lowland Gorilla." In *The Problem of Meaning: Behavioral and Cognitive Perspectives,* edited by C. Mandell and A. McCabe. Amsterdam: Elsevier, 1997.

Bower, Bruce. "Getting Out From Number One: Selfishness May Not Dominate Human Behavior." *Science News,* vol. 137, no. 17 (1990), pp. 266–67.

Branden, Barbara. *The Passion of Ayn Rand.* New York: Doubleday, 1986, p. 227.

Branden, Nathaniel. *Judgment Day: My Years With Ayn Rand.* Boston: Houghton Mifflin, 1989.

Breitman, Richard. *The Architect of Genocide: Himmler and the Final Solution.* New York: Knopf, 1991.

Brenner, Steven N., and Earl A. Molander. "Is the Ethics of Business Changing?" *Harvard Business Review* (January–February 1977), pp. 57–71.

Broszat, Martin. *The Hitler State: The Foundation and Development of the Internal Structure of the Third Reich.* New York: Longman, 1981.

Brown, Donald E. *Human Universals.* New York: McGraw-Hill, 1991.

Brown, Janelle. "Doom, Quake and Mass Murder: Gamers Search Their Souls After Discovering the Littleton Killers Were Part of Their Clan." *Salon*.com (April 23, 1999).

Büchler, Y. "Document: A Preparatory Document for the Wannsee 'Conference.'" *Holocaust and Genocide Studies,* vol. 9, no. 1 (spring 1995), pp. 121–29.

Buss, David. *The Dangerous Passion: Why Jealousy Is as Necessary as Love and Sex.* New York: Free Press, 2002.

Byrne, Richard. *The Thinking Ape: Evolutionary Origins of Intelligence.* Oxford: Oxford University Press, 1995.

Caplan, Lincoln. *The Insanity Defense and Trial of John W. Hinckley, Jr.* New York: Dell Publishing Company, 1984.

Cassels, R. "Faunal Extinction and Prehistoric Man in New Zealand and the Pacific Islands." In *Quaternary Extinctions,* edited by P. S. Martin and R. G. Klein, pp. 741–67. Tucson: University of Arizona Press, 1984.

Chagnon, Napoleon. "The Myth of the Noble Savage: Lessons From the Yanomamö People of the Amazon." Paper presented at the Skeptics Society Conference on Evolutionary Psychology and Humanistic Ethics, March 30, 1996.

———. *Yąnomamö.* 4th ed. New York: HBJ, 1992.

Chibbaro, L., Jr. "Young Gays Traumatized by Shooting." *Washington Blade,* May 7, 1999.

Christenson, Cornelia V. *Kinsey: A Biography.* Indianapolis: Indiana University Press, 1971.

Clarke, James W. *On Being Mad or Merely Angry: John W. Hinckley, Jr. and Other Dangerous People.* Princeton, N.J.: Princeton University Press, 1990.

Crosby, Alfred W. *Germs, Seeds, and Animals: Studies in Ecological History.* London: M. E. Sharpe, 1994.

Cunningham, Jr., Noble E. *In Pursuit of Reason: The Life of Thomas Jefferson.* Baton Rouge, La.: Louisiana State University Press, 1987.

Dahlburg, John-Thor. "Firm Says It Created First Human Clone." *Los Angeles Times,* December 28, 2002, p. A13.

Darley, J. M., and T. R. Shultz. "Moral Rules: Their Content and Acquisition." *Annual Review of Psychology,* vol. 41 (1990), pp. 525–56.

Darwin, Charles. *The Descent of Man, and Selection in Relation to Sex.* 2 vols. London: John Murray, 1871.

———. *The Origin of Species.* London: John Murray, 1859.

Davis, David Brion. "The Enduring Legacy of the South's Civil War Victory." *New York Times,* August 26, 2001, section 4, p. 1.

———. *Slavery and Human Progress.* New York: Oxford University Press, 1984.

Dawes, Robyn M., and D. M. Messick. "Social Dilemmas." *International Journal of Psychology,* vol. 35, no. 2, pp. 111–16.

Dawes, Robyn M., Alphons van de Kragt, and John M. Orbell. "Cooperation for the Benefit of Us—Not Me, or My Conscience." In *Beyond Self-Interest,* edited by Jane Mansbridge. Chicago: University of Chicago Press, 1990, pp. 97–110.

Dawkins, Marian Stamp. *Through Our Eyes Only: The Search for Animal Consciousness.* New York: W. H. Freeman, 1993.

Dawkins, Richard. *The Selfish Gene.* New York: Oxford University Press, 1976.

DeNault, Lisa K., and Donald A. McFarlane. "Reciprocal Altruism Between Male Vampire Bats, *Desmodus rotundus.*" *Animal Behaviour,* vol. 49 (1995), pp. 855–56.

Dennett, Daniel C. *Freedom Evolves.* New York: Viking. 2003.

———. *Kinds of Minds: Toward an Understanding of Consciousness.* New York: Basic Books, 1996.

———. *Elbow Room: The Varieties of Free Will Worth Wanting.* Cambridge, Mass.: MIT Press, 1984.

Descartes, René. *Principles of Philosophy.* Part I, 1649.

de Waal, Frans B. *Good Natured: The Origins of Right and Wrong in Humans and Other Animals.* Cambridge, Mass.: Harvard University Press, 1996.

———. *Chimpanzee Politics: Power and Sex Among Apes.* Baltimore, Md.: Johns Hopkins University Press, 1989.

———, ed. *Tree of Origin: What Primate Behavior Can Tell Us About Human Social Evolution.* Cambridge, Mass.: Harvard University Press, 2001.

Diamond, Jared. "The Religious Story." Review of *Darwin's Cathedral,* by David Sloan Wilson. *New York Review of Books,* November 7, 2002.

———. *Guns, Germs, and Steel: The Fates of Human Societies.* New York: W. W. Norton, 1997.

———. *The Third Chimpanzee.* New York: HarperCollins, 1992.

Dolan, R. J. "Emotion, Cognition, and Behavior." *Science,* vol. 298 (November 8, 2002), pp. 1991–94.

Donnerstein, Edward. "Erotica and Human Aggression." In *Aggression: Theoretical and Empirical Review.* vol. 2, edited by Russell Geen and Edward Donnerstein. New York: Academic Press, 1983.

Donnerstein, Edward, and Leonard Berkowitz. "Victims' Reactions in Aggressive Erotic Films as a Factor in Violence Against Women." In *Pornography and Sexual Aggression,* edited by Malamuth and Donnerstein.

Dostoyevsky, Fyodor. *The Brothers Karamazov.* Chicago: University of Chicago Press, 1880.

Dunbar, Robin. *Grooming, Gossip and the Evolution of Language*. Cambridge, Mass.: Harvard University Press, 1996.

———. "The Chattering Classes (Have you heard the latest? Gossip, it seems, is what separates us from the animals. Far from being idle, it is vital for our safety—and even our survival)." *Times London Magazine,* February 5, 1994, pp. 28–29.

Duncan, David, ed. *Life and Letters of Herbert Spencer.* 2 vols. New York: D. Appleton, 1968.

Dworkin, Andrea. *Letters From a War Zone*. New York: Dutton, 1988.

Edgerton, Robert. *Sick Societies: Challenging the Myth of Primitive Harmony.* New York: Free Press, 1992.

Edwards, Paul. "Socrates." *Encyclopedia of Philosophy,* vol. 7. New York: Macmillan. 1967.

Ekman, Paul. *Emotions Revealed: Recognizing Faces and Feelings to Improve Communication and Emotional Life*. New York: Times Books, 2003.

———. *Telling Lies: Clues to Deceit in the Marketplace, Marriage, and Politics*. New York: W. W. Norton, 1992.

Ernsberger, D., and G. Manaster. "Moral Development, Intrinsic/Extrinsic Religious Orientation and Denominational Teachings." *Genetic Psychology Monographs,* vol. 104 (1981), pp. 23–41.

Farber, Paul Lawrence. *The Temptations of Evolutionary Ethics*. Berkeley: University of California Press, 1994.

Ferrill, Arther. *The Origins of War: From the Stone Age to Alexander the Great*. London: Thames and Hudson, 1988.

Ferris, Timothy. *Red Limit*. New York: Random House, 1996.

Fisher, W. A., and D. Byrne. "Individual Differences in Affective, Evaluative and Behavioral Responses to an Erotic Film." *Journal of Applied Social Psychology,* August 1978, pp. 355–365.

Flanagan, Owen. *The Problem of the Soul: Two Visions of Mind and How to Reconcile Them*. New York: Basic Books, 2002.

Flower, Michael. "Neuromaturation and the Moral Status of Human Fetal Life." In *Abortion Rights and Fetal Personhood,* edited by Edd Doerr and James W. Prescott. Long Beach, Calif.: Centerline Press, 1989.

Frank, Robert H. *Passions Within Reason: The Strategic Role of the Emotions*. New York: W. W. Norton, 1988.

Frankel, Max. "Willing Executioners?" *New York Times Book Review,* August 9, 1998.

Franzblau, A. N. "Religious Belief and Character Among Jewish Adolescents." *Teachers College Contributions to Education,* no. 634, 1934.

Freeman, Derek. "Paradigms in Collision: Margaret Mead's Mistake and What It Has Done to Anthropology." *Skeptic,* vol. 5, no. 3 (1997), pp. 66–73.

Friedman, Milton. "Say 'No' to Intolerance." *Liberty* (July 18, 1991).

Frith, Uta, and Chris Firth. "The Biological Basis of Social Interaction." *Current Directions in Psychological Science,* vol. 10, no. 5 (October 2001), pp. 151–55.

Galdikas, Birute M. F. *Reflections of Eden: My Years with the Orangutans of Borneo*. Boston: Little, Brown, 1995.

Gamble, Clive. *Timewalkers: The Prehistory of Global Colonization*. London: Alan Sutton, 1993.

Garcia, Luis T. "Exposure to Pornography and Attitudes About Women and Rape; A Correlational Study." *Journal of Sex Research,* vol. 22 (1986), pp. 378–85.

Gardner, Martin. *The Whys of a Philosophical Scrivener.* New York: William Morrow, 1983.
Gaustad, Edwin S., Philip L. Barlow, and Richard Dishno, eds. *New Historical Atlas of Religion in America.* New York: Oxford University Press, 2001.
Gell-Mann, Murray. *The Quark and the Jaguar.* New York: W. H. Freeman, 1994.
Gerrold, David. *The World of "Star Trek."* New York: Ballantine Books, 1979.
Gentry, Cynthia S. "Pornography and Rape: An Empirical Analysis." *Deviant Behavior,* vol. 12 (1991), pp. 277–288.
Ghiglieri, Michael. *The Dark Side of Man: Tracing the Origins of Male Violence.* Reading, Mass.: Perseus Books, 1999.
Ghiselin, Michael T. *The Economy of Nature and the Evolution of Sex.* Berkeley: University of California Press, 1974.
Gilbert, Gustave M. *Nuremberg Diary.* New York: Farrar, Straus, and Co., 1947.
Gilligan, Carol. *In a Different Voice: Psychological Theory and Women's Development.* Cambridge, Mass.: Harvard University Press, 1982.
Gillman, Leonard. "The Car and the Goats." *American Mathematical Monthly,* vol. 99, no. 1 (January 3–7, 1992), pp. 3–7.
Gilovich, Thomas, Robert Vallone, and Amos Tversky. "The Hot Hand in Basketball: On the Misperception of Random Sequences." *Cognitive Psychology,* vol. 17 (1985), pp. 295–314.
Glassner, Barry. *The Culture of Fear: Why Americans Are Afraid of the Wrong Things.* New York: Basic Books, 1999.
Glater, Jonathan D. "Adultery May Be a Sin, but It's a Crime No More." *New York Times,* April 17, 2003, p. A12.
Gluckman, M. "Gossip and Scandal," *Current Anthropology,* vol. 4 (1967), pp. 307–316.
Goldhagen, Daniel Jonah. *Hitler's Willing Executioners: Ordinary Germans and the Holocaust.* New York: Alfred A. Knopf, 1996.
Goldstein, Michael Joseph, and Harold S. Kant, eds. *Pornography and Sexual Deviance.* Berkeley: University of California Press, 1973.
Good, Kenneth. *Into the Heart: One Man's Pursuit of Love and Knowledge Among the Yanomami.* New York: HarperCollins, 1991.
Goodall, Jane. *The Chimpanzees of Gombe.* Cambridge, Mass.: Harvard University Press, 1986.
Gorsuch, R. L., and S. McFarland. "Single vs. Multiple-item Scales for Measuring Religious Values." *Journal for the Scientific Study of Religion,* vol. 13 (1972), pp. 281–307.
Gould, Stephen Jay. *The Structure of Evolutionary Theory.* Cambridge, Mass.: Harvard University Press, 2002.
———. *Hen's Teeth and Horse's Toes.* New York: W. W. Norton, 1983.
———. *The Mismeasure of Man.* New York: W. W. Norton, 1981.
Greene, J. et al. "An fMRI Investigation of Emotional Engagement in Moral Judgment." *Science,* vol. 293 (2001), pp. 2105–08.
Gribbon, John. "On the Shoulders of Midgets?" *Skeptic,* vol. 10, no. 1 (2003).
Griffin, Donald R. *Animal Minds: Beyond Cognition to Consciousness.* Chicago: University of Chicago Press, 2001.
Haidt, Jonathan. "The Moral Emotions." In *Handbook of Affective Sciences,* edited by R. J. Davidson, K. Scherer, and H. H. Goldschmidt. Oxford: Oxford University Press, 2003.
———. "The Emotional Dog and Its Rational Tail: A Social Intuitionist Approach to Moral Judgment." *Psychological Review,* vol. 108 (2001), pp. 814–34.
———. "The Positive Emotion of Elevation." *Prevention and Treatment* 3, article 3, 2000.
Hamilton, W. D. *Narrow Roads of Gene Land,* vol. 1: *Evolution of Social Behaviour.* New York: W. H. Freeman, 1996.
———. "Innate Social Aptitudes of Man: An Approach From Evolutionary Genetics." In *Biosocial Anthropology,* edited by Robin Fox, pp. 133–35. New York: John Wiley and Sons, 1975.
Hardison, Richard. *Upon the Shoulders of Giants.* Baltimore, Md.: University Press of America, 1988.

Hare, Brian, Michelle Brown, Christina Williamson, and Michael Tomasello. "The Domestication of Social Cognition in Dogs." *Science,* vol. 298 (November 22, 2002), pp. 1634–36.

Hartung, John. "Love Thy Neighbor: The Evolution of In-Group Morality." *Skeptic,* vol. 3, no. 4 (1995), pp. 86–99.

Hauser, Marc. *Wild Minds: What Animals Really Think.* New York: Henry Holt, 2000.

Hayman, Ronald. *Sartre: A Biography.* New York: Simon and Schuster, 1987.

Hellweg, S. A. "Organizational Grapevines." In *Progress in Communication Sciences.* vol. 8, edited by Brenda Dervin and Melvin J. Voigt. Norwood, N.J.: Ablex, 1987.

Henley, William Ernest. "Invictus." In *Modern British Poetry,* edited by Louis Untermeyer. New York: Harcourt Brace, 1920.

Henry, William A., III. "Pssst . . . Did You Hear About?" *Time,* vol. 135, no. 10 (March 5, 1990), p. 46.

Henson, D., and H. Rubin. "Voluntary Control of Eroticism." *Journal of Applied Behavior Analysis,* vol. 4 (1971), pp. 37–44.

Herman, Louis M., and Palmer Morrel-Samuels. "Knowledge Acquisition and Asymmetry Between Language Comprehension and Production: Dolphins and Apes as General Models for Animals." In *Interpretation and Explanation in the Study of Animal Behavior,* edited by Marc Bekoff and Dale Jamieson. Boulder, Colo.: Westview Press, 1990.

Hick, John. *The Existence of God.* New York: Collier Books, 1964.

Hinckley, Jack and Jo Ann. *Breaking Points.* Grand Rapids, Mich.: Chosen Books, c. 1985.

Hirschi, Travis, and Rodney Stark. "Hellfire and Delinquency." *Social Problems,* vol. 17 (1969), pp. 202–13.

Hitler, Adolf. *Mein Kampf* (1925). New York: Houghton Mifflin Co., 1943.

Hofstadter, Douglas R. "Metamagical Themas: Computer Tournaments of the Prisoner's Dilemma Suggest How Cooperation Evolves." *Scientific American,* vol. 248, no. 5 (1983), pp. 16–26.

Holbrooke, Blythe. *Gossip: How to Get It Before It Gets You and Other Suggestions for Social Survival.* New York: St. Martin's Press, 1983.

Horgan, John. "The New Social Darwinists." *Scientific American* (October 1995), pp. 150–57.

Howard, Ted, and Jeremy Rifkin. *Who Should Play God?: The Artificial Creation of Life and What it Means for the Future of the Human Race.* New York: Delacorte Press, 1977.

Howell, Signe. *Society and Cosmos: Chewong of Peninsular Malaya.* Singapore: Oxford University Press, 1984.

Hrdy, Sarah Blaffer. *Mother Nature: A History of Mothers, Infants, and Natural Selection.* New York: Pantheon, 1999.

Hume, David. *An Enquiry Concerning the Principles of Morals,* 1751.

Hutson, V. C. L., and G. T. Vickers. "The Spatial Struggle of Tit-for-Tat and Defect." *Philosophical Transactions of the Royal Society of London B,* vol. 348 (1995), pp. 393–404.

Huxley, Aldous. *The Olive Tree.* New York: Harper & Bros., 1937.

Huxley, Thomas H. *Evolution and Ethics.* New York: D. Appleton and Co., 1894.

Huxley, Thomas H., and Julian S. Huxley. *Touchstone for Ethics 1893–1943.* New York: Harper and Brothers, 1947.

Ingersoll, Robert. *Ingersoll's Greatest Lectures.* New York: The Freethought Press Association, 1944.

Isikoff, Michael. "Flushed From the Woods." *Newsweek* (June 9, 2003), p. 35.

Jolly, Alison. *Lucy's Legacy: Sex and Intelligence in Human Evolution.* Cambridge, Mass.: Harvard University Press, 1999.

Kalb, Claudia. "Treating the Tiniest Patients." *Newsweek* (June 9, 2003), pp. 48–51.

Kant, Immanuel. *Fundamental Principles of The Metaphysic of Morals* (1785). Translated by T. K. Abbott. Chicago: University of Chicago Press, 1952.

Keeley, Lawrence H. *War Before Civilization: The Myth of the Peaceful Savage.* New York: Oxford University Press, 1996.

Kelley, K., and D. Byrne. "Assessment of Sexual Responding: Arousal, Affect, and Behavior." In *Social Psychophysiology,* edited by John Cacioppo and Richard Petty. New York: Guilford, 1983.

Kinsey, Alfred C., Wardell Baxter Pomeroy, and Clyde E. Martin. *Sexual Behavior in the Human Male.* Philadelphia, Pa.: W. B. Saunders, 1948.

Kitcher, Philip. *Vaulting Ambition: Sociobiology and the Quest for Human Nature.* Cambridge, Mass.: MIT Press, 1985.

Klein, Richard G. *The Dawn of Human Culture.* New York: John Wiley, 2002.

———. *The Human Career: Human Biological and Cultural Origins.* Chicago: University of Chicago Press, 1999.

Klinkenborg, Verlyn. "Cow Parts." *Discover* (August 2001).

Kluger, Jeffrey. "Will We Follow the Sheep?" *Time* (March 10, 1997), pp. 67–70, 72.

Kohlberg, Lawrence. *Essays on Moral Development,* vol. 2: *The Psychology of Moral Development: The Nature and Validity of Moral Stages.* San Francisco: Harper and Row, 1984.

———. "Moral Stages and Moralization: The Cognitive-Developmental Approach." In *Moral Development and Behavior,* edited by T. Lickona. New York: Holt, Rinehart, and Winston, 1976.

Koops, B. L., L. J. Morgan, and F. C. Battaglia. "Neonatal Mortality Risk in Relation to Birth Weight and Gestational Age: Update." *Journal of Pediatrics,* vol. 101 (1982), pp. 969–77.

Kosko, Bart. *Fuzzy Thinking: The New Science of Fuzzy Logic.* New York: Hyperion, 1993.

———. *Fuzzy Engineering.* Englewood Cliffs, N.J.: Prentice Hall, 1992.

Krantz, G. S. "Human Activities and Megafaunal Extinctions," *American Scientist,* vol. 58 (1970), pp. 164–70.

Krech, Shepard, III. *The Ecological Indian: Myth and History.* New York: W. W. Norton, 1999.

Kreiman, G., I. Fried, and C. Koch. "Single Neuron Correlates of Subjective Vision in the Human Medial Temporal Lobe." *Proceedings of the National Academy of Sciences,* no. 99 (2002), pp. 8378–83.

Kristof, Nicholas D. "All-American Osamas." *The New York Times,* June 7, 2002, p. A27.

Krosnick, J. A. et al. "Subliminal Conditioning of Attitudes." *Personality and Social Psychology Bulletin,* vol. 18 (1992), pp. 152–62.

Kruger, J. "Personal Beliefs and Cultural Stereotypes About Racial Characteristics." *Journal of Personality and Social Psychology,* vol. 71 (1996), pp. 536–48.

Kutchinsky, Berl. "Pornography and Rape: Theory and Practice?" *International Journal of Law and Psychiatry,* vol. 14 (1991), pp. 47–64.

———. "Pornography and Its Effects in Denmark and the United States: A Rejoinder and Beyond." *Comparative Social Research: An Annual,* vol. 8 (1985), pp. 301–30.

———. "The Effect of Easy Availability of Pornography on the Incidence of Sex Crimes: The Danish Experience." In *Pornography and Censorship,* edited by David Copp and Susan Wendell, p. 307. Amherst, N.Y.: Prometheus Books, 1983.

La Cerra, Peggy, and Roger Bingham. *The Origin of Minds: Evolution, Uniqueness, and the New Science of the Self.* New York: Harmony Books, 2003.

Lanzmann, C. "The Obscenity of Understanding: An Evening with Claude Lanzmann." *American Imago,* vol. 48, no. 4 (1991), pp. 473–95.

Laplace, Pierre Simon. *A Philosophical Essay on Probabilities* (1814). New York: Dover, 1951.

Lawrence, D. H. *Pornography and So On.* London: Faber and Faber, 1936.

Leakey, Richard, and Roger Lewin. *Origins Reconsidered: In Search of What Makes Us Human.* New York: Doubleday, 1992.

LeBlanc, Steven A. *Constant Battles: The Myth of the Peaceful, Noble Savage.* New York: St. Martin's Press, 2003.

———. *Prehistoric Warfare in the American Southwest.* Salt Lake City: University of Utah Press, 1999.

Leibniz, Gottfried. *Discourse on Metaphysics* (1826). Translated by G. R. Montgomery. Buffalo, N.Y.: Prometheus Books, 1992.

Lemann, Nicholas. "Gossip, The Inside Scoop: A History of American Hearsay." *The New Republic* (November 5, 1990).

Leonard, Jennifer A. et al. "Ancient DNA Evidence for Old World Origin of New World Dogs." *Science*, vol. 298 (November 22, 2002), pp. 1613–15.

Levin, Jack, and Arnold Arluke. *Gossip: The Inside Scoop*. New York: Plenum Press, 1987.

Lewis, Bernard. *What Went Wrong? Western Impact and Middle Eastern Response*. New York: Oxford University Press, 2002.

Lewis, C. S. *A Grief Observed*. New York: Macmillan, 1963.

———. *Beyond Personality*. New York: Macmillan, 1945.

Lizot, Jacques. *Tales of the Yanomami: Daily Life in the Venezuelan Forest*. New York: Cambridge University Press, 1985.

Lopez, Barry H. *Of Wolves and Men*. New York: Scribner's, 1978.

Low, B. S. "Behavioral Ecology of Conservation in Traditional Societies." *Human Nature*, vol. 7, no. 4 (1996), pp. 353–79.

Low, Peter W., John Calvin Jeffries, Jr., and Richard J. Bonnie. *The Trial of John W. Hinckley, Jr.: A Case Study in the Insanity Defense*. Mineola, N.Y.: Foundation Press, 1986.

MacKinnon, Catharine. "Pornography: A Feminist Perspective." Position paper presented to the Minneapolis City Council, 1983.

Malamuth, Neil, and James Check. "The Effects of Aggressive Pornography on Beliefs of Rape Myths: Individual Differences." *Journal of Research in Personality*, vol. 19 (1985), pp. 299–320.

Marino, Lori. "A Comparison of Encephalization Between Odontocete Cetaceans and Anthropoid Primates." *Brain, Behavior, and Evolution*, vol. 51 (1988), p. 230.

Martin, P. S., and R. G. Klein, eds. *Quaternary Extinctions*. Tucson: University of Arizona Press, 1984.

Masson, Jeffrey Moussaieff, and Susan McCarthy. *When Elephants Weep: The Emotional Lives of Animals*. New York: Delacorte Press, 1995.

Mayr, Ernst. *This Is Biology*. Cambridge, Mass.: Harvard University Press, 1997.

———. *Toward a New Philosophy of Biology*. Cambridge, Mass.: Harvard University Press, 1988.

———. *The Growth of Biological Thought*. Cambridge, Mass.: Harvard University Press, 1982.

———. "Where Are We?" *Cold Spring Harbor Symposium on Quantitative Biology*, vol. 24 (1959), pp. 409–40. Reprinted in Ernst Mayr. *Evolution and Diversity of Life*. Cambridge, Mass.: Harvard University Press, 1976.

———. *Animal Species and Evolution*. Cambridge, Mass.: Harvard University Press, 1963.

McCabe, Kevin et al. "A Functional Imaging Study of Cooperation in Two-Person Reciprocal Exchange." *Proceedings of the National Academy of Science*, vol. 98, no. 20 (September 25, 2001), pp. 11832–35.

McElhaney, James W., ed. *Vincent Fuller's Summation in U.S. v. Hinckley (1981)*, Classics of the Courtroom, Trial Transcript Series, vol. 12. Minnetonka, Minn.: Professional Education Group, 1988.

Merry, S. E. "Rethinking Gossip and Scandal." In *Toward a General Theory of Social Control*, vol. 1: *Fundamentals*, edited by Donald Black, pp. 271–302. New York: Academic Press, 1984.

Merton, Robert K. *On the Shoulders of Giants*. New York: Harcourt Brace, 1965.

Miele, F. "The (Im)moral Animal: A Quick and Dirty Guide to Evolutionary Psychology and the Nature of Human Nature." *Skeptic*, vol. 4, no. 1 (1996), pp. 42–49.

Miles, H. Lyn. "Simon Says: The Development of Imitation in an Encultured Orangutan." In *Reaching Into Thought: The Minds of the Great Apes*, edited by Anne E. Russon et al. Cambridge: Cambridge University Press, 1996.

———. "ME CHANTEK: The Development of Self-Awareness in a Signing Orangutan." In *Self-Awareness in Animals and Humans: Developmental Perspectives*, edited by Sue Taylor Parker et al. Cambridge: Cambridge University Press, 1994.

Milgram, Stanley. *Obedience to Authority: An Experimental View.* New York: Harper, 1969.

Milner P., and R. W. Beard. "Limit of Fetal Viability." *Lancet,* vol. 1 (1984), p. 1079.

Milton, John. *Paradise Lost.* In *The Great Books of the Western World.* Chicago: University of Chicago Press, 1952.

Moll, Jorge R. et al. "The Neural Correlates of Moral Sensitivity: A Functional Magnetic Resonance Imaging Investigation of Basic and Moral Emotions." *Journal of Neuroscience,* vol. 22, no. 7 (April 1, 2002), pp. 2730–36.

Molliver, M. E., I. Kostovic, and H. Van Der Loos. "The Development of Synapses in Cerebral Cortex of the Human Fetus." *Brain Research,* vol. 50 (1973), pp. 403–07.

Moore, G. E. *Principia Ethica.* Cambridge: Cambridge University Press, 1912.

Morrow, L. "The Morals of Gossip." *The Economist* (October 26, 1991), p. 64.

Moss, Cynthia. *Elephant Memories: Thirteen Years in the Life of an Elephant Family.* New York: Fawcett Columbine, 1988.

Murdock, G. P., and D. White. "Standard Cross-Cultural Sample." *Ethnology,* vol. 8 (1969), pp. 329–69.

Murnighan, John Keith. *Bargaining Games.* New York: William Morrow and Co., 1992.

Myers, David G. *Intuition: Its Powers and Perils.* New Haven, Conn.: Yale University Press, 2002.

National Bioethics Advisory Commission. *Cloning Human Beings: Report and Recommendations.* Rockville, Md.: National Bioethics Advisory Commission, 1997.

Neubauer, Mary. "For McCaugheys, a Day to be Thankful in Many Ways." *Athens Banner-Herald,* November 28, 1997.

Newberg, Andrew, Eugene D'Aquili, and Vince Rause. *Why God Won't Go Away.* New York: Ballantine Books, 2001.

Nin, Anaïs. *Little Birds.* New York: Harcourt, 1963.

Novak, Michael. "Skeptical Inquirer." Review of *How We Believe,* by Michael Shermer. *Washington Post,* February 13, 2000, p. 7.

Nozick, Robert. *Anarchy, State, and Utopia.* New York: Basic Books, 1974.

Nucci, L., and E. Turiel. "God's Word, Religious Rules, and Their Relation to Christian and Jewish Children's Concepts of Morality." *Child Development,* vol. 64 (1993), pp. 1475–91.

Olson, Robert Goodwin. *An Introduction to Existentialism.* New York: Dover Publishers, 1962.

Parker, Sue Taylor, Robert W. Mitchell, and H. Lyn Miles, eds. *The Mentalities of Gorillas and Orangutans.* Cambridge: Cambridge University Press, 1999.

———. *Self-Awareness in Animals and Humans: Developmental Perspectives.* Cambridge: Cambridge University Press, 1994.

Patterson, Francine G. P., and Wendy Gordon. "The Case for the Personhood of Gorillas." In *The Great Ape Project: Equality Beyond Humanity,* edited by Paola Cavalieri and Peter Singer. New York: St. Martin's Press, 1993.

Patterson, Francine G. P., and Eugene Linden. *The Education of Koko.* New York: Holt, Rinehart, and Winston, 1981.

Penrose, Roger. *Shadows of the Mind: A Search for the Missing Science of Consciousness.* Oxford: Oxford University Press, 1996.

———. *The Emperor's New Mind: Concerning Computers, Minds, and the Law of Physics.* London: Penguin, 1991.

Pepperberg, Irene. *The Alex Studies: Cognitive and Communicative Abilities of Parrots.* Cambridge, Mass.: Harvard University Press, 1999.

Persinger, Michael A. "Paranormal and Religious Beliefs May Be Mediated Differently by Subcortical and Cortical Phenomenological Processes of the Temporal (Limbic) Lobes." *Perceptual and Motor Skills,* vol. 76 (1993), pp. 247–51.

———. *Neuropsychological Bases of God Beliefs.* New York: Praeger, 1987.

Peters, Ted. *Playing God?: Genetic Determinism and Human Freedom* (New York: Routledge, 1997.

Pinker, Steven. *The Blank Slate: The Modern Denial of Human Nature.* New York: Viking, 2002.

Pleasure, Dhand, and S. Kaur. "What Is the Lower Limit of Viability?" *American Journal of Diseases of Children,* vol. 138 (1984), p. 783.

Pope, Alexander. *An Essay on Man* (1733–34). Edited by Frank Brady. Indianapolis: Bobbs-Merrill, 1997.

Price, T. Douglas, and James A. Brown, eds. *Prehistoric Hunter-Gatherers: The Emergence of Cultural Complexity.* Academic Press, 1985.

Pryor, Karen, and Kenneth S. Norris, eds. *Dolphin Societies: Discoveries and Puzzles.* Chicago: University of Chicago Press, 2000.

Przbyla, D., and D. Byrne. "The Mediating Role of Cognitive Process in Self Reported Sexual Arousal." *Journal of Research in Personality,* vol. 18 (1984), pp. 54–63.

Purpura, D. P., "Morphogenesis of Visual Cortex in the Preterm Infant." In *Growth and Development of the Brain,* edited by Mary A. B. Brazier. New York: Raven Press, 1975.

Quartz, Steven R., and Terrence J. Sejnowski. *Liars, Lovers, and Heroes: What the New Brain Science Reveals About How We Become Who We Are.* New York: William Morrow, 2002.

Rand, Ayn. "How Does One Lead a Rational Life in an Irrational Society?" In *The Virtue of Selfishness.* New York: New American Library, 1964.

———. "Introducing Objectivism." *Los Angeles Times,* June 17, 1962.

———. *Atlas Shrugged.* New York: Random House, 1957.

Rantala, M. L., and Arthur J. Milgram, eds. *Cloning: For and Against.* Chicago: Open Court, 1999.

Rawls, John. *A Theory of Justice.* Cambridge, Mass.: Belknap/Harvard University Press, 1971.

Reed, C. A. "Extinction of Mammalian Megafauna in the Old World Late Quaternary," *BioScience,* vol. 20 (1970), pp. 284–88.

Reiss, Diana, and Lori Marino. "Mirror Self-Recognition in the Bottlenose Dolphin: A Case of Cognitive Convergence." *Proceedings of the National Academy of Sciences* (May 8, 2001), pp. 5937–42.

Richards, Robert J. *Darwin and the Emergence of Evolutionary Theories of Mind and Behavior.* Chicago: University of Chicago Press, 1987.

Ridgway, Sam H. "Physiological Observations on Dolphin Brains." In *Dolphin Cognition and Behavior: A Comparative Approach,* edited by Ronald J. Schusterman et al., pp. 32–33.New Jersey: Lawrence Erlbaum, 1986.

Ridley, Matt. *Genome: The Autobiography of a Species in 23 Chapters.* New York: HarperCollins, 2001.

———. *The Origins of Virtue: Human Instincts and the Evolution of Cooperation.* New York: Viking, 1997.

Rilling, James K. et al. "A Neural Basis for Social Cooperation." *Neuron,* vol. 35 (July 18, 2002), pp. 395–404.

Rivenburg, Roy. "A Decision Between a Woman and God." *Los Angeles Times,* May 24, 1996.

Roberts, Neil. *The Holocene: An Environmental History.* Oxford: Basil Blackwell, 1989.

Robinson, B. A. "Why Did the Columbine Shooting Happen?" Ontario Consultants on Religious Tolerance, 1999; updated December 3, 2001. Online at http://www.religioustolerance.org/sch_vio1.htm.

Rogers, Lesley J. *Minds of Their Own: Thinking and Awareness in Animals.* Boulder, Colo.: Westview Press, 1998.

Roosevelt, Margot. "Yanomami: What Have We Done to Them?" *Time,* vol. 156, no. 14 (October 2, 2000), pp. 77–78.

Rosenbaum, Ron. *Explaining Hitler: The Search for the Origins of His Evil.* New York: Random House, 1998.

Rosenberg, Debra. "The War Over Fetal Rights." *Newsweek* (June 9, 2003), pp. 40–47.

Rosnow, Ralph L., and G. A. Fine. *Rumor and Gossip: The Social Psychology of Hearsay.* New York: Elesevier Scientific, 1976.

Ross, Murray G. *Religious Beliefs of Youth.* New York: Association Press, 1950.

Rothbard, Murray N. *The Sociology of the Ayn Rand Cult.* Monograph. Port Townsend, Wash.: Liberty Publishing, 1987.

Ruse, Michael. "Can Selection Explain the Presbyterians?" *Science,* vol. 297 (August 30, 2002), p. 1479.
———. *Taking Darwin Seriously: A Naturalistic Approach to Philosophy.* Oxford: Basil Blackwell, 1986.
Rushdie, Salman. "Religion, As Ever, Is the Poison in India's Blood." *The Guardian* (March 9, 2002).
Ryder, Richard D. *Animal Revolution: Changing Attitudes Toward Speciesism.* London: Basil Blackwell, 1989.
Sagan, Carl. *Demon-Haunted World: Science as a Candle in the Dark.* New York: Random House, 1995.
Sagan, Carl, and Ann Druyan. *Shadows of Forgotten Ancestors.* New York: Random House, 1992.
Sagarin, Brad J., Kelton V. L. Rhoads, and Robert B. Cialdini. "Deceiver's Distrust: Denigration as a Consequence of Undiscovered Deception." *Personality and Social Psychology Bulletin,* vol. 24 (1998), pp. 1167–76.
Salamon, Julie. "A Stark Explanation for Mankind from an Unlikely Rebel." *New York Times,* September 24, 2001.
Sanderson, S. *Religion, Politics, and Morality: An Approach to Religion and Political Belief Systems and Their Relation Through Kohlberg's Cognitive-Developmental Theory of Moral Judgment.* Lincoln: University of Nebraska Dissertation Abstracts International 346259B, 1974.
Savolainen, Peter et al. "Genetic Evidence for an East Asian Origin of Domestic Dogs." *Science,* vol. 298 (November 22, 2002), pp. 1610–12.
Schaeffer, Francis A. *How Then Should We Live?: The Rise and Decline of Western Thought and Culture.* New York: Fleming H. Revell Co., 1976.
Schlessinger, Laura. Editorial. *The Calgary Sun.* September 9, 1997, p. 22.
———. *How Could You Do That?!: The Abdication of Character, Courage, and Conscience.* New York: HarperCollins, 1996.
Segal, Nancy L. *Entwined Lives: Twins and What They Tell Us About Human Behavior.* New York: Dutton, 1999.
Segerstråle, Ullica. *Defenders of the Truth: The Battle for Science in the Sociobiology Debate and Beyond.* New York: Oxford University Press, 2000.
Selig, S., and G. Teller. "The Moral Development of Children in Three Different School Settings." *Religious Education,* vol. 70 (1975), pp. 406–15.
Semendeferi, Katerina et al. "Prefrontal Cortex in Humans and Apes: A Comparative Study of Area 10." *American Journal of Physical Anthropology,* vol. 114 (2001), pp. 224–41.
———. "Limbic Frontal Cortex in Hominoids: A Comparative Study of Area 13." *American Journal of Physical Anthropology,* vol. 106 (1998), pp. 129–55.
Shermer, Michael. "Demon-Haunted Brain." *Scientific American* (March 2003), p. 32.
———. *How We Believe: Science, Skepticism, and the Search for God.* New York: W. H. Freeman, 1999; 2nd ed. New York: Henry Holt/Owl Books, 2003.
———. "The Captain Kirk Principle." *Scientific American* (December 2002), p. 32.
———. *In Darwin's Shadow: The Life and Science of Alfred Russel Wallace.* New York: Oxford University Press, 2002.
———. *Why People Believe Weird Things: Pseudoscience, Superstitions, and Other Confusions of Our Time.* New York: W. H. Freeman, 1997; 2nd ed., New York: Henry Holt/Owl Books, 2002.
———. "The Pundit of Primate Politics: An Interview with Frans de Waal." *Skeptic,* vol. 8, no. 2 (2000), pp. 29–35.
———. "The Annotated Gardner: An Interview with Martin Gardner—Founder of the Modern Skeptical Movement." *Skeptic,* vol. 4, no. 1 (1997), pp. 56–60.
———. *The Borderlands of Science.* New York: Oxford University Press, 1997.
———. "The Crooked Timber of History." *Complexity,* vol. 2, no. 6 (1997), pp. 23–29.
———. "Exorcising LaPlace's Demon: Chaos and Antichaos, History and Metahistory." *History and Theory,* vol. 34, no. 1 (1995), pp. 59–83.

———. "The Chaos of History: On a Chaotic Model That Represents the Role of Contingency and Necessity in Historical Sequences." *Nonlinear Science Today*, vol. 2, no. 4 (1993), pp. 1–13.

Shermer, Michael, and A. Grobman. *Denying History: Who Says the Holocaust Never Happened and Why Do They Say It?* Berkeley: University of California Press, 1999.

Shermer, Michael, and Frank J. Sulloway. "The Grand Old Man of Evolution. An Interview with Evolutionary Biologist Ernst Mayr." *Skeptic*, vol. 8, no. 1 (2000), pp. 76–83.

Shibutani, Tamotsu. *Improvised News: A Sociological Study of Rumor*. Indianapolis: Bobbs-Merrill, 1966.

Simon, Rita J., and David E. Aaronson. *The Insanity Defense: A Critical Assessment of Law and Policy in the Post-Hinckley Era*. New York: Praeger Publishers, 1988.

Smith, R. E., G. Wheeler, and E. Diener. "Faith Without Works: Jesus People, Resistance to Temptation, and Altruism." *Journal of Applied Social Psychology*, vol. 5 (1975), pp. 320–30.

Snelson, J. Stuart. *Win-Win Theory for Win-Win Success: Science, Theory, and Strategy of Win-Win Success for Family and Business, Community and Nation*. Forthcoming.

Sloan, R., E. Bagiella, and T. Powell. *The Lancet*, vol. 353 (February 20, 1999), pp. 664–67.

Smith, Adam. *An Inquiry into the Nature and Causes of the Wealth of Nations* (1776). New York: Modern Library, Random House, 1965.

Smith, John Maynard. *Did Darwin Get it Right?* New York: Chapman and Hall, 1992.

Snyder, Louis L., ed. *Hitler's Third Reich: A Documentary History*. Chicago: Nelson Hall, 1981.

Sober, Elliott, and David Sloan Wilson. *Unto Others: The Evolution and Psychology of Unselfish Behavior*. Cambridge, Mass.: Harvard University Press, 1998.

Solomon, Robert C., ed. *Existentialism*. New York: Modern Library, Random House, 1974.

Solzhenitsyn, Aleksandr. *The Gulag Archipelago, 1918–1956: An Experiment in Literary Investigations*. 3 vols. New York: Harper & Row, 1974–78.

Sorabji, Richard. *Animal Minds and Human Morals: The Origin of the Western Debate*. Ithaca, N.Y.: Cornell University Press, 1993.

Spacks, Patricia Meyer. *Gossip*. Chicago: University of Chicago Press, 1986.

Spencer, Herbert. *The Principles of Ethics*. Vol 1. Indianapolis: Liberty Classics, 1893.

———. *The Principles of Ethics*. Vol. 2. New York: D. Appleton, 1891.

———. *Social Statics; or The Conditions Essential to Human Happiness Specified, and the First of Them Developed*. London: John Chapman, 1851.

Stammer, L. "Anglican Leader Visits L.A." *Los Angeles Times*, May 25, 1996, pp. B1–3.

Stanford, Craig B. *The Hunting Apes: Meat Eating and the Origins of Human Behavior*. Princeton, N.J.: Princeton University Press, 1999.

Stark, Rodney, and William Sims Bainbridge. *Religion, Deviance, and Social Control*. New York: Routledge, 1997.

———. *A Theory of Religion*. New Brunswick, N.J.: Rutgers University Press, 1987.

Steadman, Henry J. *Before and After Hinckley: Evaluating Insanity Defense Reform*. New York: Guilford Press, 1993.

Sulloway, F. J. *Born to Rebel: Birth Order, Family Dynamics, and Creative Lives*. New York: Pantheon, 1996.

Symons, Donald. *The Evolution of Human Sexuality*. New York: Oxford University Press, 1979.

Szalanski, Andrea. "Columbine Report to Vindicate Nonbelievers." *Secular Humanist Bulletin*, vol. 16, no 2, 2000.

Tamarin, Georges R. "The Influence of Ethnic and Religious Prejudice on Moral Judgment." *New Outlook*, vol. 9, no. 1 (1966), pp. 49–58.

———. *The Israeli Dilemma: Essays on a Warfare State*. Rotterdam, Netherlands: Rotterdam University Press, 1973.

Tavris, Carol. "All Bad or All Good? Neither." *Los Angeles Times*, September 20, 1998, p. M5.

———. *The Mismeasure of Woman*. New York: Simon and Schuster, 1992.

Tavris, Carol, and C. Wade. *Psychology in Perspective.* 2nd ed. New York: Longman/Addison Wesley, 1995.

Taylor, Michael. *The Possibility of Cooperation.* Cambridge: Cambridge University Press, 1987.

Taylor, Shelley E. *The Tending Instinct: How Nurturing Is Essential to Who We Are and How We Live.* New York: Times Books, 2002.

Thompson, Paul. *Issues in Evolutionary Ethics.* Albany, N.Y.: SUNY Press, 1995.

Tierney, Patrick. *Darkness in El Dorado: How Scientists and Journalists Devastated the Amazon.* New York: W. W. Norton, 2000.

Tipler, Frank J. *The Physics of Immortality.* New York: Doubleday, 1994.

Trivers, Robert L. *Social Evolution.* Menlo Park, Calif.: Benjamin/Cummings, 1985.

———. "The Evolution of Reciprocal Altruism." *Quarterly Review of Biology,* vol. 46 (1971), pp. 35–57.

Trut, Lyudmila N. "Early Canid Domestication: The Farm-Fox Experiment." *American Scientist* (March/April 1999).

———. "Domestication of the Fox: Roots and Effects." *Scientifur,* vol. 19 (1995), pp. 11–18.

Van Lange, P. A. M., T. W. Taris, and R. Vonk. "Dilemmas of Academic Practice: Perceptions of Superiority Among Social Psychologists." *European Journal of Social Psychology,* vol. 27 (1997), pp. 675–85.

Vercors [Jean Bruller]. *You Shall Know Them.* Translated by R. Barisse. New York: Pocket Books, 1955.

Von Neumann, John, and Oskar Morgenstern. *Theory of Games and Economic Behavior.* Princeton, N.J.: Princeton University Press, 1980.

Vos Savant, Marilyn. "Ask Marilyn." *Parade,* September 9, 1990, February 17, 1991, and July 7, 1991.

Wade-Benzoni, Kimberly A. et al. "Cognitions and Behavior in Asymmetric Social Dilemmas: A Comparison of Two Cultures." *Journal of Applied Psychology,* vol. 87, no. 1 (2002), pp. 87–95.

Wallace, Alfred R. "Evolution and Character." *Fortnightly Review,* vol. 83 (1908), pp. 1–24.

———. "The Limits of Natural Selection as Applied to Man." In *Contributions to the Theory of Natural Selection.* London: Macmillan, 1870, pp. 391–92.

———. "The Origin of Human Races and the Antiquity of Man Deduced from the Theory of 'Natural Selection.'" *J. ASL,* vol. 2 (1864), p. 173.

Webster, William L. et al., *Appellants v. Reproductive Health Services et al., Appellees,* 1988.

Weisberg, Jacob. "What Do You Mean by 'Violence'?" *Slate*.com (May 15, 1999).

West, Charles K. *The Social and Psychological Distortion of Information.* Chicago: Nelson-Hall, 1981.

White, J. A., and S. Plous. "Self-Enhancement and Social Responsibility: On Caring More, but Doing Less, Than Others." *Journal of Applied Social Psychology,* vol. 25 (1995), pp. 1297–1318.

Whitfield, Stephen E., and Gene Roddenberry. *The Making of "Star Trek."* New York: Ballantine Books, 1968.

Wilkinson, Gerald S. "Reciprocal Food Sharing in the Vampire Bat." *Nature,* vol. 308 (1984), pp. 181–84.

Wilson, David Sloan. *Darwin's Cathedral: Evolution, Religion, and the Nature of Society.* Chicago: University of Chicago Press, 2002.

———. "Nonzero and Nonsense: Group Selection, Nonzerosumness, and the Human Gaia Hypothesis." *Skeptic,* vol. 8, no. 1 (2000), p. 85.

Wilson, E. O. "The Biological Basis of Morality." *Atlantic Monthly,* vol. 281, no. 4 (1998), pp. 53–70.

———. *Consilience: The Unity of Knowledge.* New York: Knopf, 1998.

———. *Biophilia.* Cambridge, Mass.: Harvard University Press, 1984.

———. *On Human Nature.* Cambridge, Mass.: Harvard University Press, 1978.

———. *Sociobiology: The New Synthesis*. Cambridge, Mass.: Harvard University Press, 1975.

Wilson, James Q. *The Moral Sense*. New York: Free Press, 1993.

Wiscombe, J. "'I Don't Do Therapy.' Dr. Laura Schlessinger, the Country's Top Female Radio Personality, Calls Herself a Prophet." *Los Angeles Times Magazine*, January 18, 1998, p. 11.

Wise, Steven M. *Rattling the Cage: Toward Legal Rights for Animals*. Boston: Perseus, 2000.

———. *Science and the Case for Animal Rights*. Boston: Perseus Books, 2002.

Wolf, Naomi. "Our Bodies, Our Souls." *The New Republic* (October 16, 1995).

Woodward, Kenneth L. "Today the Sheep . . . Tomorrow the Shepherd?" *Newsweek* (March 10, 1997).

Wrangham, Richard. "Is Military Incompetence Adaptive?" *Evolution and Human Behavior*, vol. 20 (1999), pp. 3–17.

Wrangham, Richard, and Dale Peterson. *Demonic Males: Apes and the Origins of Human Violence*. New York: Houghton Mifflin, 1996.

Wrangham, Richard, and David Pilbeam. "Apes as Time Machines." In *African Apes*, vol. 1: *All Apes Great and Small*, edited by Birute M. F. Galdikas et al., pp. 5–18. New York: Kluwer Academic/Plenum Publishers, 2002.

Wright, Robert. *The Moral Animal: The New Science of Evolutionary Psychology*. New York: Pantheon Books, 1994.

Wulff, David M. *Psychology of Religion: Classic and Contemporary Views*. New York: Wiley, 1991.

Zak, Paul J. "Trust." *Journal of Financial Transformation* (CAPCO Institute), vol. 7 (2002), pp. 18–24.

Zimbardo, Philip. "The Human Choice: Individuation, Reason, and Order Versus Deindividuation, Impulse, and Chaos." In *Nebraska Symposium on Motivation, 1969*, edited by W. J. Arnold and D. Levine. Lincoln: University of Nebraska Press, 1970.

———. *The Psychology of Attitude Change and Social Influence*. New York: McGraw-Hill, 1991.

Zoba, Wendy Murray. "Church, State, and Columbine." Excerpt from *Day of Reckoning* (Brazos Press). *Christianity Today*, vol. 45, no. 5 (April 2, 2001), p. 54.

Zuckerman, M. "Physiological Measures of Sexual Arousal in the Human." In *Technical Reports on the Commission on Obscenity and Pornography*, vol. 1. Washington, D.C.: U.S. Government Printing Office, 1971.

ILLUSTRATION CREDITS

Fig. 1. Food Sharing in Chimpanzees. Photograph by Frans de Waal. From Frans de Waal, *Good Natured* (Cambridge, Mass.: Harvard University Press, 1996). Reprinted with permission of author.

Fig. 2. The Evolutionary Origins of Morality 100,000 Years Ago. From Stephen Jay Gould, ed., *The Book of Life* (New York: W. W. Norton, 1993), pp. 218–19. Courtesy of W. W. Norton.

Fig. 3. A Group of New Guinea Hunter-Gatherers Prepares for Battle. From John Alcock, *Animal Behavior: An Evolutionary Approach* (Sunderland, Mass.: Sinaaer Publishers, 2001), p. xx. Courtesy of Film Study Center.

Fig. 4. The Relationship Between Group Size, Brain Size, and Evolutionary History. From Robin Dunbar, "Brains on Two Legs," in Frans de Waal, ed., *Tree of Origin* (Cambridge, Mass.: Harvard University Press, 2001), p. 180. Reprinted with permission of author.

Fig. 5. Gossip as the Transmission of Data Via a Verbal Mode. *Crock* by Bill Rechin and Don Wilder. Courtesy of Bill Rechin and Don Wilder.

Fig. 6: The Bio-Cultural Evolutionary Pyramid. Rendered by Pat Linse.

Fig. 7: The Expanding Circle of Inclusiveness. Rendered by Pat Linse, adapted from Richard D. Alexander, *Darwinism and Human Affairs* (Seattle: University of Washington Press, 1987), figure 4.

Fig. 8. A Shocking Experiment on Obedience to Authority. From Stanley Milgram, *Obedience* (film), 1965. Courtesy of Penn State Media Sales.

Fig. 9. Hitler, Himmler, and the Holocaust as the Embodiment of Pure Evil. (*top*) Photograph by Estelle Bechhoefer. Courtesy of United States Holocaust Memorial Museum. (*bottom*) Courtesy of Yad Vashem, Jerusalem, Israel.

Fig. 10. Napoleon Chagnon: The Man Who Called the Yąnomamö "Fierce." Courtesy of Napoleon Chagnon.

Fig. 11. The Fierce Side of Human Nature. (*top*) From Kenneth Good, *Into the Heart* (New York: HarperCollins, 1991), facing p. 140. Courtesy of Kenneth Good. (*bottom*) From Napoleon Chagnon, *Yąnomamö* (New York: HBJ, 1992), p. 188. Courtesy of Napoleon Chagnon.

Fig. 12. The Erotic Side of Human Nature. From Good, *Into the Heart,* facing p. 140. Courtesy of Kenneth Good.

Fig. 13. Splendor in Fierceness. From Chagnon, *Yąnomamö,* pp. 9, 197. Courtesy of Napoleon Chagnon.

Fig. 14. Political Organization and Frequency of Warfare. Derived from Table 2.1 in the appendix of Lawrence H. Keeley, *War Before Civilization* (New York: Oxford University Press, 1996).

Fig. 15. Death Rate from Warfare. Data from Table 6.2 in the appendix of Keeley, *War Before Civilization,* pp. 185ff.

Fig. 16. Pueblo Culture War Sites. Data from Steven LeBlanc, *Prehistoric Warfare in the American Southwest* (Salt Lake City: University of Utah Press, 1999) pp. 142, 170, 178, and 272, respectively.

Fig. 17. Models for Moa Extinction. Data from R. N. Holdaway and C. Jacomb, "Rapid Extinction of the Moas (Aves dinomithiformes): Model, Test and Implications." *Science,* vol. 287 (2000), pp. 2250–54.

Fig. 18. The Fuzzy Deterioration of John Hinckley. Courtesy of Associated Press.

Fig. 19. Genetic Range of Potential Model. Rendered by Pat Linse.

Fig. 20. The Massacre at Columbine High School. Courtesy of Associated Press.

Fig. 21. Absolute v. Relative Morality. © 2002 Wiley Miller. Distributed by Universal Press Syndicate. Reprinted with permission.

Fig. 22. Jeremy Bentham's Hedonic Calculus. Rendered by Pat Linse.

Fig. 23. Drawing the Line. Courtesy of GE Medical Systems.

Fig. 24. Cloning God. Courtesy of Hilary Price.

Fig. 25. Equal Rights for All Hominids? Illustration by Michael Rothman. Reproduced by permission.

Fig. 26. The Fuzzy Logic of Animal Rights. From Steven M. Wise, *Science and the Case for Animal Rights* (Boston: Perseus, 2000), p. 241. Courtesy of Steven M. Wise.

Fig. 27. Comparative Brain Size. From John M. Allman, *Evolving Brains* (New York: W. H. Freeman, 1999), p. 197. Reprinted with permission of author.

Fig. 28. The Paedomorphic Primate. From Allman, *Evolving Brains,* p. 197. Reprinted with permission of author.

Fig. 29. The Domesticated Primates. From Adolph H. Schultz, *The Life of Primates* (London: Weidenfeld and Nicholson, 1969). Reprinted by permission of Orion Publishing Group, Inc.

Fig. 30. Making War, Not Love. (*top*) Photograph by Robert Gardiner. Courtesy of the Peabody Museum. (*bottom*) Photograph by SFC Jack H. Yamaguchi. From Alcock, *Animal Behavior,* p. 494. Reprinted with permission.

Fig. 31. Making Love, Not War. Photograph by Frans de Waal. From de Waal, *Good Natured,* p. x. Reprinted with permission of author.

Fig. 32. The Myth of Early America as a Christian Nation. From Edwin S. Gaustad, Philip L. Barlow, and Richard Dishno, *New Historical Atlas of Religion in America* (New York: Oxford University Press, 2001), figure 4.16. Courtesy of Oxford University Press.

Fig. 33. Moral Modules in the Brain. Rendered by Pat Linse, adapted from Jorge Moll et al., "The Neural Correlates of Moral Sensitivity," *The Journal of Neuroscience,* 2002, p. 2733, figure 3.

ACKNOWLEDGMENTS

I begin by thanking my agents Katinka Matson and John Brockman, not only for their personal support of my work but for what they have done to help shape the genre of science writing into a "third culture" on par with other cultural traditions. In today's society if you want to be a literate, well-read person, you need to read science books. That is good news for science, scientists, and science writers. We can all thank John and Katinka for what they have done to help bring about this salubrious situation. And thanks as well to my editors, initially John Michel at W. H. Freeman, then David Sobel and Paul Golob at Henry Holt/Times books who inherited the project, and finally Robin Dennis who saw it through to completion and helped shape it into a much more readable treatise. I also acknowledge Carol Rutan for constructive copyediting of the manuscript, Lisa Fyfe for the powerful cover design, Victoria Hartman for interior design, and Chris O'Connell for overall production.

A number of people were gracious enough to read parts of this book during its development and/or provide critical feedback on some of the ideas developed within. David Sloan Wilson and Steven Pinker provided valuable information on the group selection debate in evolutionary theory; the late Stephen Jay Gould encouraged a conciliatory attitude toward religion and cautioned me about taking evolutionary theory too far in attempting to explain the origin of morality; Jared Diamond confirmed for me many aspects of the human condition in a state of nature based on stories he related about his experiences with the indigenous peoples of New Guinea; Daniel Dennett for clarifying the free will–determinism debate; Robbin Gehrke and Rob Moses, over countless dinner conversations on morality, religion, politics, and economics, helped me hone my thinking on moral issues; most importantly, in endless dialogue over just about everything, my colleague and confidant Frank "the Tank" Sulloway has taught me more about human psychology than any book could, and his knowledge and experience in research design is reflected in the results of the morality survey, presented in the epilogue to this book.

Special thanks go to *Skeptic* magazine art director Pat Linse for her important contributions in preparing the illustrations, graphs, and charts for this and my other works, as well as for her unmitigated and deeply appreciated friendship and support. The skeptical movement in general owes a debt of gratitude to Pat for her behind-the-scenes work that has irrevocably shaped modern skepticism into a viable social movement.

I also wish to recognize the staff of the Skeptics Society and *Skeptic* magazine. First and foremost is our office manager Tanja Sterrmann, for getting the job done so efficiently, and especially for her unadulterated good cheer that creates a pleasant working atmosphere for all; thanks too go to Matt Cooper for his work on the morality survey, for his critical feedback on moral issues, and for his valuable help in the Skeptics Society. Thanks also go to senior editor Frank Miele; senior scientists David Naiditch, Bernard Leikind, Liam McDaid, and Thomas McDonough; our Web meister Nick Gerlich; contributing editors Tim Callahan, Randy Cassingham, Clayton Drees, Steve Harris, Tom McIver, Brian Siano, and Harry Ziel; editorial assistants Gene Friedman, Sara Meric, and the late Betty McCollister, one of the most humane humanists in the history of that movement; photographer David Patton and videographer Brad Davies for their visual record of the Skeptics' Caltech Science Lecture Series; and database manager Jerry Friedman. I would also like to recognize *Skeptic* magazine's board members: Richard Abanes, David Alexander, the late Steve Allen, Arthur Benjamin, Roger Bingham, Napoleon Chagnon, K. C. Cole, Jared Diamond, Clayton J. Drees, Mark Edward, George Fischbeck, Greg Forbes, the late Stephen Jay Gould, John Gribbin, Steve Harris, William Jarvis, Lawrence Krauss, Gerald Larue, William McComas, John Mosley, Richard Olson, Donald Prothero, James Randi, Vincent Sarich, Eugenie Scott, Nancy Segal, Elie Shneour, Jay Stuart Snelson, Julia Sweeney, Frank Sulloway, Carol Tavris, and Stuart Vyse.

Thanks as well for the institutional support for the Skeptics Society at the California Institute of Technology go to David Baltimore, Susan Davis, Chris Harcourt, and Kip Thorne; Larry Mantle, Ilsa Setziol, Jackie Oclaray, Julia Posie, and Linda Othenin-Girard at KPCC 89.3 FM radio in Pasadena have been good friends and valuable supporters for promoting science and critical thinking on the air; thanks to Linda Urban at Vroman's bookstore in Pasadena for her support; Robert Zeps and John Moores have been especially supportive of both the Skeptics Society as well as the skeptical movement in America, and Bruce Mazet has been a good friend to the skeptics who has influenced the movement in myriad unacknowledged ways. Finally, special thanks go to those who help at every level of our organization: Stephen Asma, Jaime Botero, Jason Bowes, Jean Paul Buquet, Adam Caldwell, Bonnie Callahan, Tim Callahan, Cliff Caplan, Randy Cassingham, Shoshana Cohen, John Coulter, Brad Davies, Janet Dreyer, Bob Friedhoffer, Michael Gilmore, Tyson Gilmore, Andrew Harter, Terry Kirker, Diane Knudtson, Joe Lee.

I would also like to acknowledge John Rennie and Mariette DiChristina at *Scientific American* for providing skepticism a monthly voice that reaches so many people, as well as for keeping me on my toes about accuracy, style, and succinctness. I look forward each month to writing my column more than just about anything else I do in my working life.

Finally, I thank my daughter Devin who is now old enough to begin thinking about moral issues and helped me see clearly and through a child's eyes how the world should be; and to my life partner Kim . . . for everything.

INDEX

ABOUT THE AUTHOR

DR. MICHAEL SHERMER is the founding publisher of *Skeptic* magazine, the director of the Skeptics Society, a contributing editor and monthly columnist for *Scientific American*, the host of the Skeptics Lecture Series at Caltech, and the cohost and producer of the thirteen-hour Fox Family television series *Exploring the Unknown*. He is the author of *In Darwin's Shadow,* about the life and science of the codiscoverer of natural selection, Alfred Russel Wallace; *The Borderlands of Science,* about the fuzzy land between science and pseudoscience; and *Denying History,* on Holocaust denial and other forms of historical distortion. The first two books in his trilogy on belief are *Why People Believe Weird Things* and *How We Believe: Science, Skepticism, and the Search for God.*

Dr. Shermer received his B.A. in psychology from Pepperdine University, his M.A. in experimental psychology from California State University, Fullerton, and his Ph.D. in the history of science from Claremont Graduate University. He lives in southern California.